THE PRESENT STATE OF PHYSICS

*A symposium presented on December 30, 1949
at the New York meeting of the
American Association for the
Advancement of Science*

Arranged by
FREDERICK S. BRACKETT

Essay Index Reprint Series

BOOKS FOR LIBRARIES PRESS
FREEPORT, NEW YORK

Copyright © 1954 by
The American Association for the Advancement of Science
Reprinted 1970 by arrangement

STANDARD BOOK NUMBER:
8369-1542-9

LIBRARY OF CONGRESS CATALOG CARD NUMBER:
75-99617

PRINTED IN THE UNITED STATES OF AMERICA

PREFACE

During the last two decades the problem of elementary particles, their nature, origin, and interaction with matter has been the center of interest for both theoretical and experimental physicists. The three papers in the first part deal with these problems.

The paper on the magnetic moment of the electron determined from atomic beam resonance measurements of the hyperfine structure of hydrogen gives the first experimental evidence that this moment is slightly larger than the value predicted from the Dirac theory. Refinement in newer measurements carried out by Kusch and collaborators since 1949 yielded a value in excellent agreement with the newer theoretical developments.

Until recently most of our knowledge about elementary particles, other than the electron, proton, and neutron, came from cosmic ray investigations. The next paper discusses techniques used and results obtained at balloon altitudes. At these altitudes, as Bradt and Peters, F. Oppenheimer, Ney, and others showed, one observes also heavy nuclei (*not* elementary particles) which provide a clue for the origin of the cosmic radiation.

The third paper by Street discusses the μ-meson, its mass, and decay process, the discovery of the π-meson and its interaction with nuclei and its decay. There is a brief paragraph about the primaries and their reactions at very high altitude.

This field has developed so rapidly that the interpretation is changing continuously, and so far we have no real theoretical understanding of the multiplicity of particles and processes. There is a vast amount of observational material, but so far it is difficult to unravel and clarify this complex information. The theory of elementary particles is a theory of the future.

In contrast to this field of physics with its complexities and the perplexing aspect of the theory, the fields of solid state physics, chemical physics, and biophysics discussed in the second, third, and fourth parts, respectively, rest on well understood foundations. As Eyring put it in his discussion of chemical kinetics of biological systems, "the advance in nuclear theory may be expected to have no more effect on the quantum theory of the molecules, than quantum mechanics itself had on Newtonian mechanics for systems with large quantum numbers."

Physicists are convinced that the wave mechanical theory of molecules and the wave mechanical treatment of periodic structures in crystals are essentially correct and that the problem is to find experimental evidence, corresponding to the theoretical predictions.

In the part on the physics of the solid state the clarification of the bulk properties of elementary semiconductors as discussed in the paper "The New Electronics" is a good example.

That in this field many new phenomena will be found that are both theoretically and technically of interest is shown in the next paper in the discussion of the injection mechanism by J. Bardeen, its discoverer, and a brief outline of the fundamentals of transistor physics and in the discussion by A. von Hippel of ferroelectrics, the electrical analog to ferromagnetics discovered only as late as 1917.

The two papers in chemical physics by Debye and Eyring lead into the problems of biophysics, where one is dealing with large molecules. Debye's discussions of structure of high polymers as investigated by light scattering show how much information about such complicated structures can be deduced from the classical observations of such quantities as turbidity, symmetry, or asymmetry of light scattering or interference effects in very large molecules. Lumry and Eyring's discussion is as much biophysics as chemical physics. It aims to show that the laws of thermodynamics, together with simple molecular models suggested by physical analogies lead to the understanding of biological processes, if one is willing to assume that the same few building blocks, that which may be arranged folded or unfolded (see Debye's discussion of elastomers) exist in all proteins. Surface reactivity, electrical phenomena, such as polarizability and effective dielectric constant, and geometrical organization are used to describe enzymatic catalysis, neural phenomena and photosynthesis.

In the part on biophysics the success of the application of the reaction rate theory to biological systems is demonstrated in Brink's article on the physical and chemical properties of axon related to conduction of nerve impulses and Johnson's discussion of bioluminescence.

Biophysics, as Johnson points out, "is still in its infancy and only one of the many aspects of its currently developing potentialities could be enlarged upon in this discussion. It is growing rapidly, and one may look forward with confidence that biophysics will bring results equally spectacular [as nuclear physics] and of no less consequence to mankind."

KARL LARK-HOROVITZ

CONTENTS

ELEMENTARY PARTICLES

The Magnetic Moment of the Electron	P. KUSCH	3
Cosmic Ray Experiments at High Altitude	EDWARD P. NEY	11
Developments in Cosmic Radiation, 1945-1950	J. C. STREET	28

PHYSICS OF THE SOLID STATE

The New Electronics	KARL LARK-HOROVITZ	57
Flow of Electrons and Holes in Semiconductors	J. BARDEEN	128
Barium Titanate Ferroelectrics	A. VON HIPPEL	150

CHEMICAL PHYSICS

The Structure of Polymers	PETER J. W. DEBYE	169
Implications of the Chemical Kinetics of Some Biological Systems	RUFUS LUMRY AND HENRY EYRING	189

BIOPHYSICS

Some Physical and Chemical Properties of Axons Related to Conduction of Nerve Impulses	FRANK BRINK, JR.	213
Bioluminescence and the Theory of Reaction Rate Control in Living Systems	FRANK H. JOHNSON	238
Index		263

CONTRIBUTORS

J. BARDEEN, Department of Physics, University of Illinois, Urbana, Illinois

FRANK BRINK, JR., Department of Biophysics, The Johns Hopkins University, Baltimore, Maryland

PETER J. W. DEBYE, Cornell University, Ithaca, New York

HENRY EYRING, Department of Chemistry, University of Utah, Salt Lake City, Utah

A. VON HIPPEL, Laboratory for Insulation Research, Massachusetts Institute of Technology, Cambridge, Massachusetts

FRANK H. JOHNSON, Department of Biology, Princeton University, Princeton, New Jersey

P. KUSCH, Department of Physics, Columbia University, New York

KARL LARK-HOROVITZ, Department of Physics, Purdue University, Lafayette, Indiana

RUFUS LUMRY, Department of Chemistry, University of Utah, Salt Lake City, Utah

EDWARD P. NEY, Department of Physics, College of Science, Literature, and the Arts, University of Minnesota, Minneapolis, Minnesota

J. C. STREET, Department of Physics, Harvard University, Cambridge, Massachusetts

ELEMENTARY PARTICLES

THE MAGNETIC MOMENT OF THE ELECTRON

P. KUSCH

Columbia University, New York

INTRODUCTION

The classical electron is a particle to which are ascribed only the properties of mass (m) and charge (e). Classical theory, of course, states that such a charged particle gives rise to a magnetic moment when it moves in an orbit, but such a moment is the consequence of the motion of the charge and is not an intrinsic property of the electron. Indeed, with the advent of the Bohr theory of the atom, which describes the stable orbits that electrons may assume within an atom, it was possible to find the magnetic moments which certain atoms in various states may have. These magnetic moments may interact with an applied magnetic field and give rise to the normal Zeeman effect, i.e., a splitting of spectral lines in an applied magnetic field. In addition to the normal Zeeman effect, there was observed an anomalous Zeeman effect which could not be explained simply in terms of the magnetic moment produced by the orbital motion of electrons.

Let us consider, briefly, the quantities which are involved. The motion of the electron is limited to those orbits in which the orbital angular momentum of the electron is an integral multiple of $h/2\pi$, where h is Planck's constant. The associated orbital magnetic moment is then the same integral multiple of $eh/4\pi mc$. This last quantity is designated as the Bohr magneton, μ_0, and is the magnetic moment produced by an electron in an orbit of unit angular momentum in terms of $h/2\pi$. In all considerations of angular momenta and magnetic moments we are concerned not with the absolute values of these quantities but only with their values in terms of $h/2\pi$ and the Bohr magneton, μ_0, respectively.

The observation of increasingly fine detail in the spectra of atomic systems led to the empirical postulate that the electron cannot be considered as a simple point charge, but that the electron itself possesses a spin, which may be interpreted as a rotation about some internal axis. The postulated angular momentum was $\frac{1}{2}(h/2\pi)$ and the spin of the electron is, therefore, $\frac{1}{2}$. With this spin, there must be some associated magnetic moment which may be interpreted as the result of a charge distribution rotating about an axis of rotation. All experimental evidence led to the conclusion that the spin magnetic moment of the electron is equal to the Bohr magneton, the orbital magnetic moment of an electron in the state of lowest, non-zero angular momentum.

To introduce some terminology, we define the g value as the ratio of the magnetic moment measured in units of the Bohr magneton to the angular momentum, measured in units of $h/2\pi$. This definition leads to the conclusion that the orbital g value of the electron, g_L, is equal to 1 and that the spin g value of the electron, g_S, is equal to 2.

With the advent of the relativistic quantum mechanics of Dirac, it was shown that the spin of the electron and the associated spin magnetic moment are not new properties to be ascribed to the electron in addition to the classical properties of mass and charge. The non-classical properties of the electron may be deduced from the mass and charge of the electron in consequence of the relativistic invariance of the Dirac theory of the electron. The spin of the electron is predicted to be ½, and the spin magnetic moment is predicted to be exactly one Bohr magneton, in agreement with the values postulated from experimental evidence.

The simple Dirac theory assumes the electron to be a particle which may interact with applied electric and magnetic fields produced by adjacent particles. However, it entirely ignores the interaction of the electron with the quantized radiation field. Prior to the discovery of the Lamb shift in certain of the atomic energy states of hydrogen and the discovery of the anomalous magnetic moment of the electron, there was no experimental data whose interpretation required the inclusion of this interaction. What is more, until recently, formidable theoretical difficulties prevented the detailed analysis of effects arising from the inclusion of the interaction of the electron with the quantized radiation field.

It is the purpose of this paper to present some of the evidence which led to the suggestion that the simple Dirac theory does not adequately describe the electron. Experimentally it is observed that the spin magnetic moment of the electron is not equal to one Bohr magneton, but differs from this normal magnetic moment by some small amount which we call the anomalous magnetic moment.

Experimental Procedure

Certain properties of the atom and of its components (nuclei and electrons), as well as the nature of the interactions between the component particles of the atom, may be determined by an analysis of the characteristic atomic spectrum of an atom. The traditional procedures in spectroscopy observe lines in the visible, the ultraviolet, and infrared regions of the spectrum. These lines occur as the result of the transition of the atom between widely separated energy levels, whose large energy separation is determined by the gross details of the electronic configuration. Each of such energy levels is, however, split up into a number of sublevels (hyperfine splitting) whose relative spacing and number is determined by the properties of the atomic nucleus. The relevant properties are the spin, the magnetic moment, and the electric quadrupole moment of the nucleus. In general the spin may have any half integral value and the magnetic moment some value much smaller than that of the electron and not simply related to it.

In the observation of optical spectra the effect of this splitting of atomic energy levels is to give rise to the hyperfine structure of spectral lines. Unfortunately the differences of the frequencies between the hyperfine structure components of a line are several orders of magnitude less than the line frequencies themselves and in most cases the hyperfine structure cannot be observed with high precision or else cannot be observed at all because of the great natural width of the optical line.

In the experiments to be discussed in this paper we employ the atomic beam method for the observation of the hyperfine structure of atoms. The lines which are observed arise from transitions between the various sublevels in any one atomic energy state. The observations may be made at zero magnetic field, or, equally well, at some finite magnetic field when the Zeeman effect of hyperfine structure is observed.

The frequencies at which lines are observed may be very low, say, from zero to 3000 megacycles per second as compared to the frequencies of optical lines which are of the order of 600,000,000 megacycles per second. The limitation in the frequency range is determined only by the limitation in the availability of suitable electronic frequency generators. It is not possible to observe the low-frequency radiation that is characteristic of the energy difference between two hyperfine structure levels of an atom since the lifetime of each of the sublevels in the ground state of the atom is so great that the intensity of radiation is prohibitively low. By subjecting the atom, however, to an oscillating magnetic field of the appropriate frequency, transitions may be stimulated between two states; the process will go in both directions, i.e., as many atoms will absorb radiation of the specified frequency as will emit this radiation and hence there is no net absorption or radiation of energy. The fact that spontaneous emission does not occur is of enormous value in observing details of spectra, since the long lifetime of the states permits the observation of lines of a width determined only by instrumental factors. In optical spectra the line width is determined by the lifetime of the upper state, usually of the order of 10^{-8} sec. In the present case the width of the line is determined by the time during which the process of emission or absorption occurs. This time is determined by the instrumental arrangements, more specifically by the time an atom in free flight spends in the region in which the oscillating field exists. The real limitation in the possible reduction of the width of a line is determined only by the details of instruments. In practice line widths of the order of 2 kc/sec have been obtained. This, in the notation of optical spectroscopy corresponds to a width of 10^{-8} cm^{-1}.

Since no net absorption or emission of energy occurs as a result of the transitions, special techniques are required for the observation of a transition and for the measurement of the frequency at which such a transition occurs. For this purpose we employ the method of atomic beams. In essence the method serves to detect a small change in the properties of the atom when it undergoes a transition, rather than the radiation emitted in the transition.

Consider a very narrow, ribbon-shaped beam of atoms which leaves an oven (or other source), passes through appropriate slits to define the beam, and then falls on an appropriate detector. The actual techniques are extremely complex; the interested reader will find discussions of technique in the literature. An atom in each of the possible states in which it may exist is characterized by some definite magnetic moment, contributed by the orbital motion of the electron, by the spin of the electron and by the nuclear spin. It is then possible to arrange in the path of the beam two magnetic fields which are characterized by a gradient (some rate of variation of the magnetic field with position) such that the first of the two fields deflects atoms of a given moment in one direction and the second of the two fields deflects the particles in the opposite direction. The net deflection at the detector can be made zero provided that the magnetic moment of the particular atom has not changed during the motion of the atom through the apparatus. If in a region between the two deflecting fields a radiofrequency field of suitable frequency and amplitude is arranged, transitions will occur between two states of different moment, the moment of a particular particle will in general be altered, and the particle will no longer strike the detector. As many particles will make transitions from state P to state Q as will make transitions from Q to P. The net transfer of energy is zero, but each particle will have a new moment after the transition.

It is only necessary, then, to observe the intensity of the beam at the detector as a function of the applied frequency. Variations in beam intensity at once indicate the presence of a line. Since the observation of the Zeeman effect of hyperfine structure is extremely fruitful, an additional experimental parameter is the magnitude of the applied magnetic field in which transitions occur.

Experimental Observations

Atomic hydrogen, in its lowest state, that is, the state into which all atoms fall, is in a $^2S_{1/2}$ state. By this notation the spectroscopist means that there is no net orbital angular momentum of the electron and that the total spin angular momentum is equal to $\frac{1}{2}$. If the nucleus of the hydrogen atom, the proton, did not itself possess a magnetic moment the state would be single. However, the proton does have a spin of $\frac{1}{2}$. The spin of the electron and that of the proton may be either parallel or antiparallel. No intermediate possibilities exist. The effect of these two possibilities is to split up the ground state of hydrogen into two levels which are separated from each other by some finite energy interval. The frequency of radiation which must be absorbed or emitted to effect the transition is called $\Delta \nu$.

Nafe and Nelson (1) have measured this hyperfine structure separation, $\Delta \nu$, of hydrogen by the application of methods which have been described in the previous section. The quantity is about 1420 megacycles per second and has been measured with an accuracy of about one part in two hundred thousand. Previous measurements of Millman and Kusch had determined the value of the nuclear magnetic moment of the proton in terms of the spin magnetic moment of the electron. Now it is possible, in the case of an atom as simple as the hydrogen atom, to calculate the $\Delta\nu$ from the observed magnetic moment of the hydrogen nucleus by the application of the appropriate quantum mechanical analysis. When this was done, the discrepancy between the calculated and observed values of $\Delta\nu$ was found to be about $\frac{1}{4}\%$. This discrepancy was beyond the experimental uncertainty of any of the observational data which entered into the comparison and also beyond the uncertainty in the values of any of the physical constants which enter into the calculation. The existence of the discrepancy led Breit (2) to suggest that the spin magnetic moment and the orbital magnetic moments of the electron are not identical.

Kusch and Foley (3) have performed a long series of experiments which not only indicates the existence of the anomalous spin magnetic moment of the electron by a procedure which is more direct than that indicated previously, in the sense that it requires less intervention of theory, but also determines the magnitude of the spin magnetic moment to very high precision.

In general the total angular momentum, J, as well as the magnetic moment, contributed by the electrons of an atom in any state is the vector sum of a contribution arising from the orbital motion and the spin. The moment of the electronic configuration, μ_J, may then be expressed as a linear combination of the orbital and spin moments of the electron, where the coefficients of the component moments are in general calculable from the quantum numbers which describe the atomic energy state. To give an example, the g_J value (the ratio of the magnetic moment in units of the Bohr magneton to the spin) is equal to g_S for a $^2S_{1/2}$ state and is equal to $\frac{4}{3}g_L - \frac{1}{3}g_S$ for a $^2P_{1/2}$ state. In the simple case, when the nuclear spin is zero, the moment of the electronic configuration is the moment of the atom.

Suppose that such an atom is now placed in a magnetic field. The angular momentum of the atom (J) will be so oriented in the field that the projection of the momentum along the field (m_J) will take on values J, $J-1$, $J-2$, $-(J-1)$, $-J$. The component of the magnetic moment along the field will be $m_J g_J \mu_0$. The state, which at zero field is specified by a single J, is specified at a non-zero field by the quantum numbers J and m_J. The sublevels at a non-zero field depart from their zero field energy at different rates, and transitions are allowed between the sublevels, provided the change in m_J is ± 1. The frequency of the line is $g_J \mu_0 H/h$, where H is the magnetic field. Evidently if both the magnetic field and the values of the fundamental constants which enter into μ_0, the Bohr magneton, are known, it is possible to determine g_J. If g_J is determined for each of two different electronic configurations, where g_J is in each case a known linear combination of g_L and g_S, it is possible to determine the ratio g_S/g_L. In actual practice it is very difficult to measure a magnetic field in terms of absolute standards. However, by making observations of lines which arise in each of two atomic energy states of different J at the same magnetic field, it is possible to find g_{J1}/g_{J2} and from this the ratio g_S/g_L.

The previous considerations apply to an atom which has a nucleus of spin zero. The atoms that are available to us for study, however, contain a nucleus of some finite spin and magnetic moment. The general methods to be employed in determining the ratio g_S/g_L are the same as indicated above for the case of an atom with a nucleus of zero spin but the details are very much more complex. Essentially it is necessary to consider the coupling between the various angular momenta within the atom. At zero field, of course, the total angular momentum, F, is made up of contributions from the orbital motion of the electrons, from the electron spin and from the nuclear spin. Correspondingly, the total g value of the atom, g_F, is a linear combination of the g values of the nucleus, the orbital electronic g value and the electron spin g value. In general, there will be several F values associated with the lowest electronic state of the atom, each with its own characteristic g value, depending on the relative orientation of the electronic and nuclear angular momentum. It is no longer true that the line frequencies will be equal to $g_F \mu_0 H/h$, except at the lowest fields. This occurs because at low fields the electron spin angular momentum is coupled to the orbital angular momentum to give a total electronic angular momentum which is in turn coupled to the nuclear angular momentum to give a total atomic angular momentum. The quantum number which characterizes this angular momentum is a real quantum number and must have quantized components along the direction of the magnetic field. At high fields, however, the electronic and the nuclear angular momentum are separately quantized with respect to the field, because the interaction of the two angular momenta, with their associated magnetic moments, is much weaker than the interaction of each angular momentum with the applied field. Actually, of course, neither of these two limiting cases occurs, and an analysis of the rather complicated "intermediate field case" is required.

The calculation of the ratio g_S/g_L can be made by the straightforward application of the appropriate quantum mechanical relationships. In order that this can be done, it is necessary to make a prior determination of the nuclear spin and the nuclear magnetic moment, as well as of the zero field splitting of the hyperfine structure components of the ground state of the atom. Finally, it is necessary to make the assumption that the states in question are pure states, unperturbed by an admixture of other states. The last assumption can be checked by the consistency of the results of various experiments.

As an example of a check of the validity of the last assumption, consider the case of the ground states of the various alkali atoms. The spectroscopic character of these states is identical ($^2S_{1/2}$) and the electronic g value of these states is simply equal to the spin g value of the electron. If the states are perturbed by an admixture of other states, the perturbation would be different for the different alkali atoms. An intercomparison of the g_J values of the alkali atoms has been made by Kusch and Taub (4), and it is found that the g_J values of the atoms of lithium, sodium, and potassium are identical to within the error of measurement, about one part in fifty thousand. It may, therefore, be inferred that the g_J value of each of these atoms is exactly equal to g_S within the error of measurement.

Kusch and Foley have made an intercomparison of the g_J values of the $^2S_{1/2}$ state of sodium, the $^2P_{1/2}$ states of indium and gallium, and the $^2P_{3/2}$ state of gallium. A determination of the ratio of the g_J values of any two of the states will yield a value of the ratio g_S/g_L. A tabulation of the results follows:

State	g_J	g_J Nominal Value
$^2S_{1/2}$	g_S	2
$^2P_{1/2}$	$\tfrac{4}{3}g_L - \tfrac{1}{3}g_S$	$\tfrac{2}{3}$
$^2P_{3/2}$	$\tfrac{2}{3}g_L + \tfrac{1}{3}g_S$	$\tfrac{4}{3}$

Comparison	Nominal	Observed	g_S/g_L
$g_J(^2P_{3/2}\text{ Ga})/g_J(^2P_{1/2}\text{ Ga})$	2	$2(1.00172 \pm 0.00006)$	$2(1.00114 \pm 0.00004)$
$g_J(^2S_{1/2}\text{ Na})/g_J(^2P_{1/2}\text{ Ga})$	3	$3(1.00242 \pm 0.00006)$	$2(1.00121 \pm 0.00003)$
$g_J(^2S_{1/2}\text{ Na})/g_J(^2P_{1/2}\text{ In})$	3	$3(1.00243 \pm 0.00010)$	$2(1.00121 \pm 0.00005)$

Whereas the value of the ratio g_S/g_L as determined from any pair of atoms does not constitute evidence for the statement that the ratio does not have its nominal value of 2, because of the possibility of perturbations, the agreement of the three independently determined ratios is overwhelming evidence of the fact that the ratio does depart from the value of 2 by the indicated amount. The agreement among the results is so good that a rather striking auxiliary result is the remarkable spectroscopic purity of the states in question. The three independent results agree to within the error of measurement. It is, of course, entirely possible that part of the small discrepancies between the separate values arises from perturbations of the indicated atomic energy states.

Additional experiments have verified the conclusions discussed above. Taub and Kusch (5) have measured the ratio of the nuclear magnetic moment of the proton to the electronic magnetic moments of both cesium and indium. The ratios are, of course, different because even the nominal values of the electronic magnetic moments differ. However, the result may be used to find g_S/g_L and for this quantity the value $2(1.00119)$ is obtained. Thomas, Driscoll, and Hipple (6), at the National Bureau of Standards, have made an accurate determination of the magnetic moment of the proton in absolute units. If this result is combined with the result above, in which the proton moment is measured in terms of the spin magnetic moment of the electron, it is possible to find a value of $g_S/g_L = 2(1.00084)$, using the accepted value for the Bohr magneton. Essentially the procedure is to measure the spin g value of the electron in a measured magnetic field, where the proton moment serves only as a comparison and transfer mechanism.

Alternatively, if the value of g_S/g_L determined by Kusch and Foley is assumed, it is possible to find a value of e/mc of considerably greater accuracy than any previous value.

Conclusions

It would be possible on phenomenological grounds to insert the observed anomalous magnetic moment of the electron into the Dirac theory as an *intrinsic* property of the electron. The theory would then describe observed data much more adequately than does the simple theory.

A much more significant approach than this one is to look for a more fundamental interpretation of the new data. This approach modifies the relativistic quantum theory to include terms which arise from the interaction of the electron with the quantized radiation field. The quantization of the field is manifested by the existence of fluctuating field strengths in space. The interaction of the electron with this field can itself alter the apparent magnitude of the spin magnetic moment. In the presence of the quantized electromagnetic field an electron will continuously emit and absorb quanta of energy. The emitted quanta of energy may, before reabsorption, produce electron-positron pairs, which may then annihilate each other to produce quanta of radiation. The process may be extremely complicated and many steps may occur before the energy finally returns to the electron. These processes have been subjected to rigorous analysis by Schwinger (7) and others, and it is indeed found that the interaction terms arising from the processes give rise to a correction in the spin magnetic moment of the electron just equal to the observed anomalous magnetic moment of the electron. In detail, it is found that the anomalous magnetic moment, $\delta\mu = (\alpha/2\pi)\mu_0$ where α is a dimensionless combination of fundamental constants. ($\alpha \cong 1/137$). This makes the ratio $g_S/g_L = 2(1 + \alpha/2\pi) = 2(1.00116)$, in excellent agreement with the experimental result. The series of experiments to be discussed constitutes one of the principal lines of evidence for the validity of the new approach in quantum mechanics.

BIBLIOGRAPHY

Articles of general interest in this field are as follows:
Stern, A. W., *Physics Today* **2**, No. 5, 21 (1949).
Weisskopf, V. F., *Revs. Mod. Phys.* **21**, 305 (1949).
Kellogg, J. M. B., and Millman, S., *Revs. Mod. Phys.* **18**, 323 (1946).

REFERENCES

1. Nafe, J. E., and Nelson, E. B., *Phys. Rev.* **73**, 718 (1948).
2. Breit, G., *Phys. Rev.* **72**, 984 (1947).
3. Kusch, P., and Foley, H. M., *Phys. Rev.* **74**, 250 (1948).
4. Kusch, P., and Taub, H., *Phys. Rev.* **75**, 1477 (1949).
5. Taub, H., and Kusch, P., *Phys. Rev.* **75**, 1481 (1949).
6. Thomas, H. A., Driscoll, R. L., and Hipple, J. A., *Phys. Rev.* **75**, 902 (1949).
7. Schwinger, J., *Phys. Rev.* **73**, 415 (1948).

COSMIC RAY EXPERIMENTS AT HIGH ALTITUDE

Edward P. Ney

University of Minnesota, Minneapolis, Minnesota

This was to be a discussion of heavy nuclei in cosmic rays, but since it is a seminar on elementary particles and since elementary particles include practically everything except heavy nuclei, the discussion will treat of the types of particles and processes one observes in cloud chambers and photographic emulsions at balloon altitudes.

During 1950 and 1951, mainly through the efforts of the Office of Naval Research and the General Mills Company, constant altitude balloons have been developed which make experiments possible for prolonged periods of time at altitudes of 90,000 to 95,000 ft. The residual pressure at this altitude is about 1.2 cm of mercury. In the vertical direction the absorbing thickness of air is 15 g/cm². That this is really in the region of primary cosmic rays can be appreciated by comparing the 15 g/cm² air path with the mean free path for absorption of the N component of 118 g/cm² as measured by Tinlot on Mount Evans. The mean free path for a proton in air corresponding to the geometric cross section of the average air nucleus is 65 g/cm². At the balloon altitudes one is less than one-fourth of a mean free path for nuclear absorption of protons from the top of the atmosphere.

Figure 1 is a photograph of a balloon launching.

The experiments at Minnesota have been carried out with photographic emulsions and with cloud chambers with and without counter control. The principal contributing group has consisted of Phyllis Freier, E. J. Lofgren, F. Oppenheimer, J. T. Tate, Charles Critchfield, and myself. In any discussion about heavy nuclei the outstanding work of Bradt and Peters must be included.

Figure 2 is a photograph of a counter-controlled cloud chamber for high altitude work. The chamber has a frontal area of 8 by 8 in. and is filled with ¼-in. absorber plates. The entire equipment weighs 85 lb with a battery supply sufficient for ten hours operation. The power requirement is about 10 watts.

Several cloud chamber pictures show the qualitative difference between mountain altitudes and balloon altitudes. At mountain altitudes (3000 meters) the typical star observed in the chamber has low-energy and short-range prongs. Figure 3 shows a star at 30,000 meters. It can be seen that most of the star particles are energetic enough to penetrate several ¼-in. lead plates. The star production rate at 30,000 meters is about 200 times greater than that at 300 meters for those stars which can be observed with ¼-in. lead plates. Because of the higher average energies of the high-

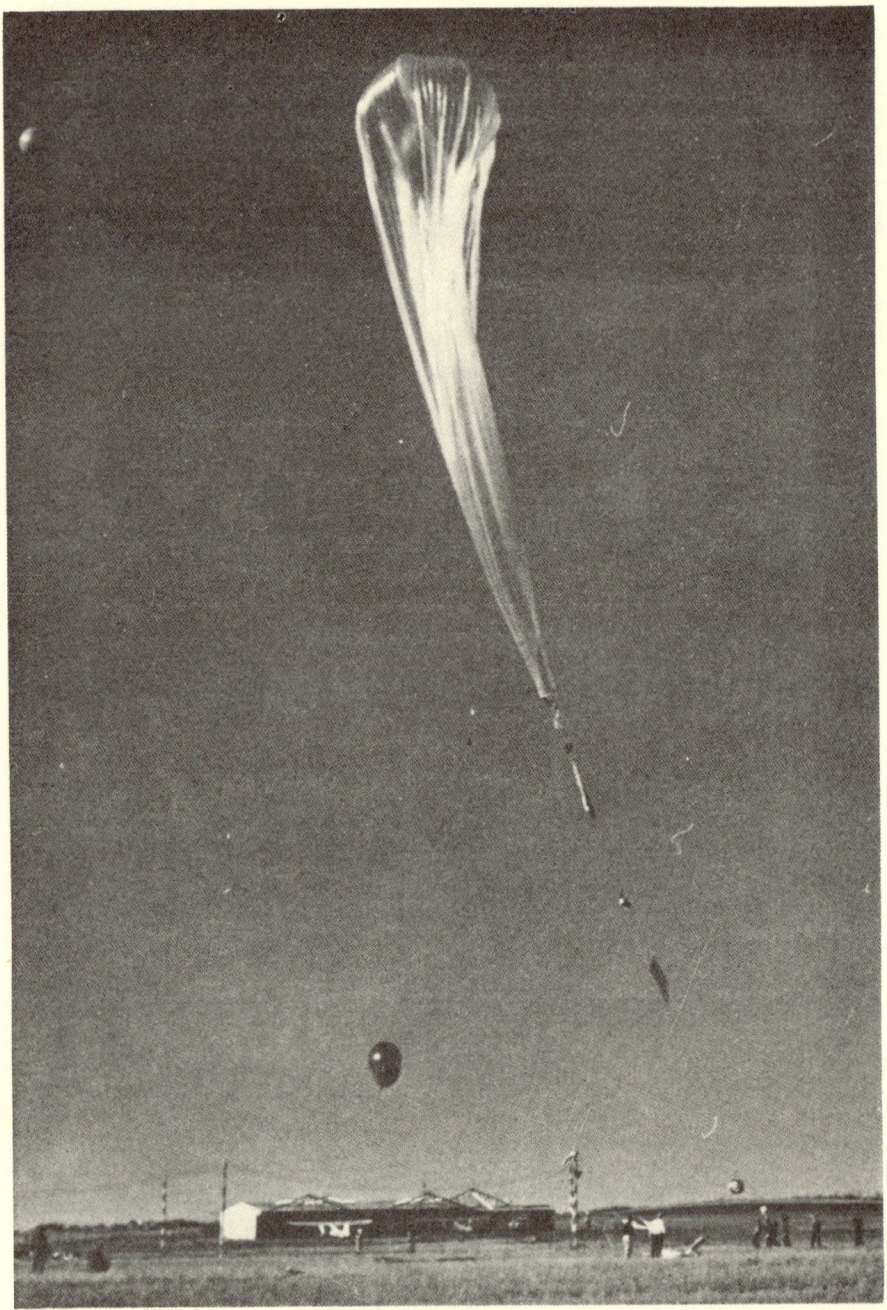

FIG. 1. Balloon at the instant of takeoff. The gas bubble at the top represents 2% of the expanded volume of the balloon. The instrument load (large sphere at the bottom of the picture) has just been released. (Courtesy of C. B. Moore, General Mills Aeronautical Research Laboratory.)

altitude stars more prongs can probably be seen emerging from the lead than at lower altitudes. The average particle energy in the 3000-meter stars is about 10 Mev for a proton prong and for 30,000-meter stars, it is not uncommon for a proton of 200 Mev to come out. A large fraction of the stars at mountain altitudes have been shown by Powell and Hazen to be neutron produced, whereas a much larger proportion of stars at 30,000 meters are made by charged primaries. The neutrons at lower altitudes are probably produced as star secondaries and by proton exchange reactions in the atmosphere.

FIG. 2. Counter-controlled cloud chamber for high-altitude work. The chamber with lead absorber plates is in the background. In the foreground is the diaphragm pump which compresses the cloud chamber and drives the camera through a gear train. The equipment requires 10 watts of electric power.

The same qualtitative difference in stars is observed in photographic emulsions. Here, however, one can get better quantitative figures. If one considers only large stars (>10 heavy prongs) and compares the production rate at 30,000 meters with the production rate for similar stars measured by Bernadini at lower altitudes, the mean free path for absorption of the star-producing radiation is about 150 g/cm^2 or corresponds to about one-half the geometric cross section. The fact that smaller stars seem to show the same absorption between 3000 and 30,000 meters is misleading since many such stars are secondary produced and the present indication seems to be that there is a maximum for small stars in the atmosphere.

At 3000 meters one is impressed by the number of energetic electron showers usually associated with Auger air showers; at 30,000 meters one is equally impressed by the lack of events of this sort. Whereas primary events are greatly increased at the top of the atmosphere, secondary events are like-

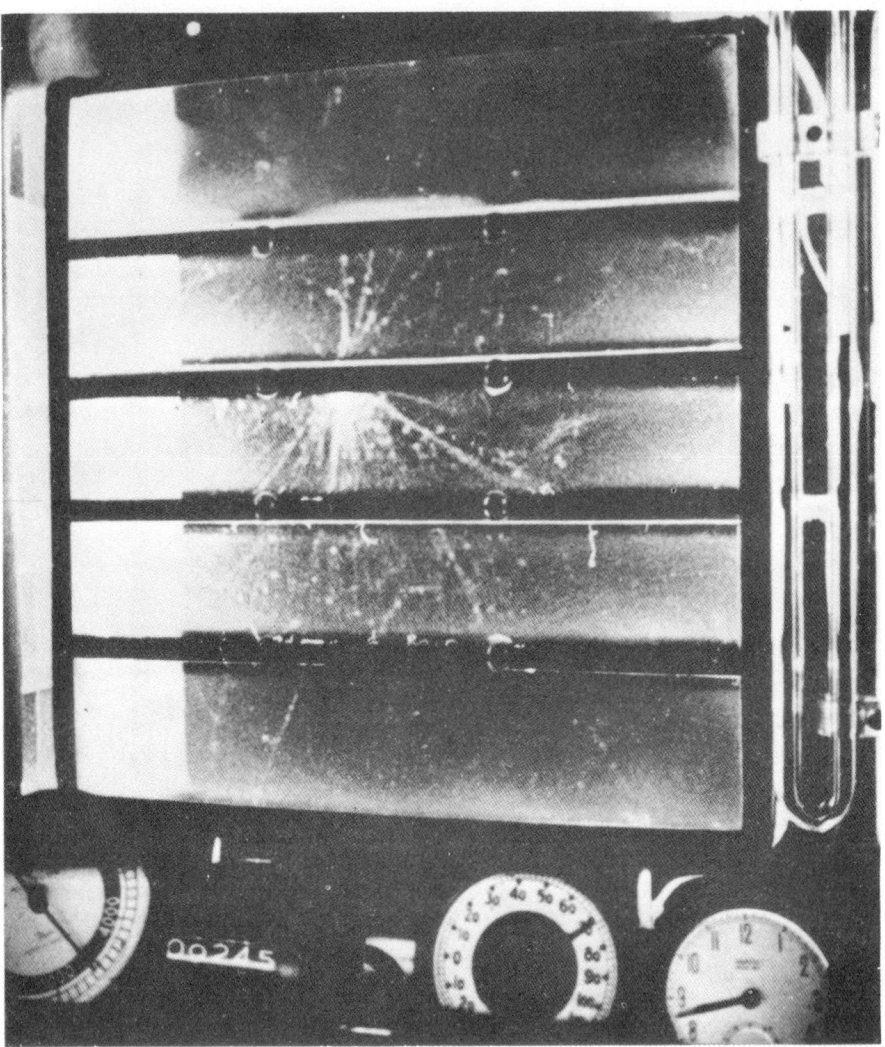

Fig. 3. Typical high-altitude star. Observe the relatively large number of penetrating prongs.

wise decreased. Figure 4 shows a large electron shower at 3000 meters which is a typical air shower event. Figure 5 shows a similar event at 30,000 meters. The penetrating particles in extensive showers at low altitudes have been shown by Fretter, Cocconi, and others to comprise only about 2% of the particles because of the large multiplication of the electrons. At 30,000 meters of the order of half the particles in the mixed showers are penetrating, indicating that the precursors of the electrons are probably present but not the electrons or γ rays themselves. It is generally postulated that the

Fig. 4. Large air shower observed at 3000 meters (top of Mount Evans). The conditions of operation of the chamber are essentially the same as during a balloon flight.

electrons come from the decay of neutral mesons although this has not been demonstrated. Figure 6 shows an example which may be this kind of event.

A pair of electron showers originate in the glass roof of the cloud chamber along with at least one penetrating particle. The energies and angle between the showers are consistent with the decay of a neutral meson into two γ rays.

In cases where electrons originate with penetrating particles, the angle between the electron showers is often too large to be explained by pair formation from a γ ray. If the electrons are formed as a pair from a γ ray, the angle between the showers should be of the order of mc^2/E where E is the

Fig. 5. Air shower type of event at balloon altitude. Six particles come into the chamber. Two of these can be seen to be electrons by their multiplication. Three particles penetrate the lead plates without multiplication.

energy of the γ ray and m is the mass of the electron. If the showers are formed by the decay of a neutral meson to γ rays, the angle should be of the order of $\mu c^2/E$ where μ is the meson mass. Rossi, Gregory, Tinlot have also recently observed cases of this kind at 3000 meters, in locally produced penetrating showers.

For the last several years there has been some controversy about whether mesons are made multiply, in single nucleon-nucleon encounters, or whether

FIG. 6. A pair of electron showers with large angle between them originate in the roof of the cloud chamber. A possible explanation is that mesons were made in the glass roof of the cloud chamber (one penetrating particle is observed near the core of the large shower) and that a neutral meson decayed to γ rays which initiated the showers.

they dribble out one at a time in a series of nucleon-nucleon interactions. Multiple production of the first sort has been advocated by Heisenberg and plural production by Heitler. The hard showers, which probably consist largely of mesons, were first observed by Fussel and later by a number of other experimenters. Quite extensive experiments have been done on the composition and nature of these showers produced in lead by Rochester in England and Greisen and his collaborators, Brown and Mackay and Walker,

FIG. 7. Approximately thirty foreward-direction particles are produced in an interaction in the second from the top lead plate.

and in photographic emulsions by Salant and Hornbostle. We have observed in the cloud chamber at 30,000 meters locally produced penetrating showers in lead and in carbon. Figure 7 shows an example of a lead shower in which about thirty particles are involved. Figure 8 shows a shower produced in *carbon* in which thirteen particles are produced by a singly charged primary. At least nine of these particles penetrate 7 g/cm² of lead. Since no particles multiply and hence are not electrons, one can say that because of charge conservation at least two and probably more than six of the

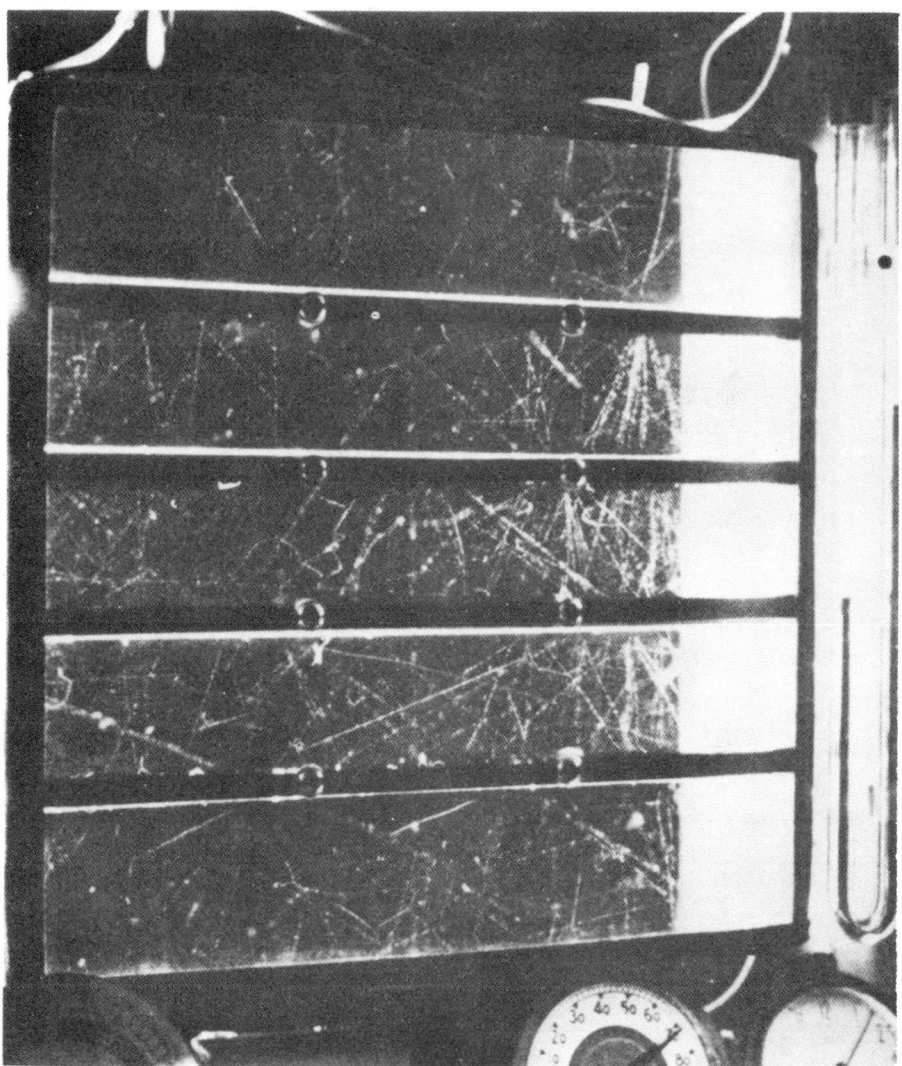

FIG. 8. A singly charged particle interacts in the upper right of the carbon plate and produces thirteen particles in a narrow shower.

particles are mesons. This argues for a plural-multiple type of process in which several mesons are made in each of a series of interactions. The average multiplicity of particles in such showers in lead was 10.1 and in carbon was 6. The larger multiplicity in lead probably comes from the larger number of multiple processes in the same nucleus. The average energy of primaries making the meson showers was estimated from the shower angles to be about 5 Bev.

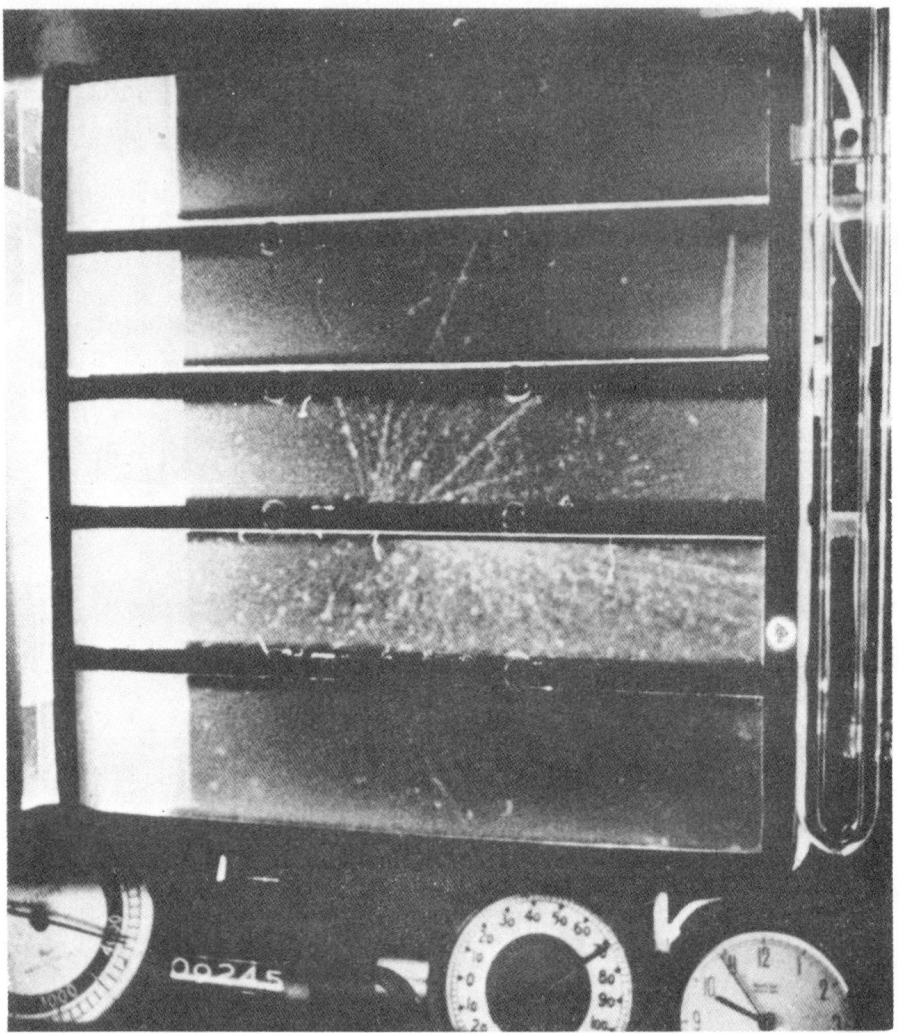

FIG. 9. A large "minimum ionization spray" produced in a ¼-in. lead plate. The spray seems to have star particles associated with it. All the tracks in the picture are counter age, and it is probable that they all are genetically related. This picture was taken at geomagnetic latitude 30° where the background of the low-energy particles is quite small.

A short time ago we reported the frequent occurrence of a kind of process which we called minimum ionization sprays at 30,000 meters.

This interaction as yet not understood at all occurs with the production of from several to fifty or sixty minimum ionization particles in a lead plate.

From their ranges they can be shown to be in the mass range from electrons to twenty mass mesons and are probably electrons. The radiation which produces them is strongly absorbed in lead, more sprays being made in the top of the chamber than in the bottom plates. The mean free path for absorption of the spray-producing radiation is less than the mean free path for absorption of the star-producing radiation. This, together with the fact that only one has been observed at lower altitudes (by Ralph Schut), indicates that they are due to some kind of primary interaction. In most cases there does not appear to be an initiating ionizing particle for the sprays. Figure 9 shows an example of such a spray which contains about fifty particles, none of which penetrates ¼ in. of lead. As in this case almost all the sprays show a predominantly downward transfer of momentum. Only a few per cent of the sprays have associated visible star tracks.*

Perhaps the most spectacular event that is observed in the pictures and plates at 30,000 meters is the occurrence of a heavy primary. These heavy particles come in from outer space with abundances at least as great as their abundance in terrestrial or stellar matter. The heavier nuclei are rapidly absorbed in the upper atmosphere and only the α particles can penetrate to lower levels. The nuclei when observed are completely stripped of electrons and of course lose energy by ionization Z^2 times the rate for a proton at the same speed. Their cross section for nuclear interaction is also larger than that for a proton. Table 1 shows the abundances of these particles at 30,000 meters and 55° latitude.

TABLE 1   Cosmic Radiation

Element	Atomic Number	Cosmic Abundance[a]	Relative Abundance at 55° Latitude	Flux of Heavy Nuclei
Hydrogen	1	1000	1000	0.1 part/cm² ster. sec
Helium	2	150–300	250	0.025
Lithium, beryllium, boron	3,4,5	Small	?	
Carbon, nitrogen, oxygen	6,7,8	0.4 –2.5	12	1.2×10^{-3}
10	10	0.04–0.5	2.9	2.9×10^{-4}

[a] From Harrison Brown *Revs. Mod. Phys.* **21**, 625 (1940).

The hydrogen helium abundances were determined in the cloud chamber; the other fluxes were measured in the photographic plates. The numbers are those measured at Minnesota, but they are in quite good agreement with the values measured by the Rochester group and the hydrogen helium ratio obtained by Pomerantz and Hereford with inefficient counters. It will be evident later that for $Z \geqq 10$ these abundances may be somewhat too small.

---

*It has been shown by Rau and Critchfield, Ney, and Oleksa that the sprays are produced by γ rays from neutral meson decays.

The primary nature of the heavy nuclei was established by the following facts. First, no nuclei have been observed to originate in the stack of photographic plates or to travel upward. (About 300 have been followed in plates.) Second, they are always energetic enough to have penetrated the earth's magnetic field. Third, there are nuclei present which are not present in the upper atmosphere (for example, iron). Finally, their energies are too high to have been imparted by proton interactions with air nuclei.

If has also been shown that the more energetic particles enter the earth's atmosphere completely stripped. There have been no particles observed at 30° geomagnetic latitude which do not have energy enough to overcome the earth's field cutoff for completely stripped nuclei. The corresponding critical energies for partially stripped nuclei are much lower. The observation of a latitude effect between 30° and 55° shows that lower energy particles are present, but they are probably also completely stripped.

TABLE 2

		Atomic Number			
	1	2	6	10	26
1. Nucleus	Proton	Helium	Carbon	$Z = 10$	Iron
2. Relative rate of energy loss at $v \cong c$, Mev cm^2/g	2	8	76	200	1352
3. Rate of energy loss at $v \cong c$ per nucleon, Mev cm^2/nucleon gram	2	2	6	10	26
4. Ionization range [a] at earth's field cutoff energy at 55°, g/cm^2	370	70	23	14	7
5. Ionization range [b] at earth's field cutoff energy at 30°, g/cm^2	4500	1700	570	340	130
6. Per cent energy lost by ionization in first mean free path for nuclear collision at energy of (4) [c]	12–22	70–100	100	100	100
7. Per cent energy lost by ionization in first mean free path for nuclear collision at energy of (5) [c]	1.5–3	2.4–4.8	5–10	5–10	11–22
8. Bradt & Peters mfp (air) geom. cross section [d]	66	41	35	27	17
9. Freier mfp (air) one-half geom. cross section [d]	132	82	70	54	34

[a] 1 Bev for protons, 0.35 Bev/nucleon for heavy nuclei: same magnetic rigidity.
[b] 7 Bev for protons, 3.5 Bev/nucleon for heavy nuclei: same magnetic rigidity.
[c] The two figures in 6 and 7 refer to mean free paths of geomagnetic and one-half geometric cross section.
[d] Overlap = range of nuclear forces.

The fast absorption by nuclear processes and by ionization of the heavy nuclei is shown in Table 2. The reason that the heavy nuclei, with the possible exception of helium, have not been observed at mountain altitudes is evident from this table. The relative rates of energy loss by ionization are given in column 2 for particles at the same speed. A convenient unit to use in comparing heavy nuclei with protons is their energy per nucleon. The rate of energy loss per nucleon is given in item 3. The heavy nuclei have been studied at 30° and 55° geomagnetic latitude. Item 4 gives the ionization range for particles just able to enter the earth's field in the vertical direction at 55°, and item 5 the same quantity for 30°. The abundances for $Z \geq 10$, which were shown in a previous slide, are somewhat underestimated at 55° because the ionization range at the magnetic field cutoff energy is just equal to the air absorber above the balloon. This is also borne out by the fact that the angular distribution for $Z \geq 10$ corrected for nuclear absorption drops off quite rapidly near the zenith.

At 55° almost one-half of the particles which disappear in the stack of plates stop by ionization. This can be seen from item 6 which shows the fractional loss of energy by ionization in the first mean free path for nuclear collision. At energies of the order of 0.5 Bev per nucleon most of the energy loss is by ionization. The reverse is true for higher energies of the order of 3 Bev per nucleon, when the particles make a collision on the average before they are stopped by ionization. Item 7 shows that only a small fraction of the energy is lost by ionization per mean free path for collision. At 30° the angular distribution should reflect only the loss of particles by nuclear collisions and Bradt and Peters have shown that if one corrects the observed angular distribution for nuclear absorption the result is an isotopic distribution.

The mean free path for nuclear interaction of heavy nuclei has been measured by Bradt and Peters and by Phyllis Freier. The particles involved in Bradt and Peter's measurement had energies above 3.5 Bev per nucleon and in Freier's experiment above 0.35 Bev per nucleon. Bradt and Peters find that the mean free path corresponds to the geometric cross section and Freier finds one-half the geometric cross section. It is quite probable that the difference is due to the difference in energy. Figure 10 shows the interactions observed by Freier in emulsion. In addition to these seventeen interactions in emulsion, she has used seventy-one collisions in glass to obtain a mean free path of 70 g/cm² of glass for carbon, nitrogen, and oxygen nuclei and 50 g/cm² of glass for $Z \geq 10$.

The energy spectrum of the heavy nuclei can be obtained from their latitude effect and in principle from their angular distribution at 55° geomagnetic latitude where their ionization range compares with the thickness of absorber at oblique angles. The ratio of the intensities at 30° and 55° for particles with $Z \geq 10$ has been found to be 3.0. This is in agreement with the ratio of 3.5 for large stars measured by Salant and Hornbostle and by

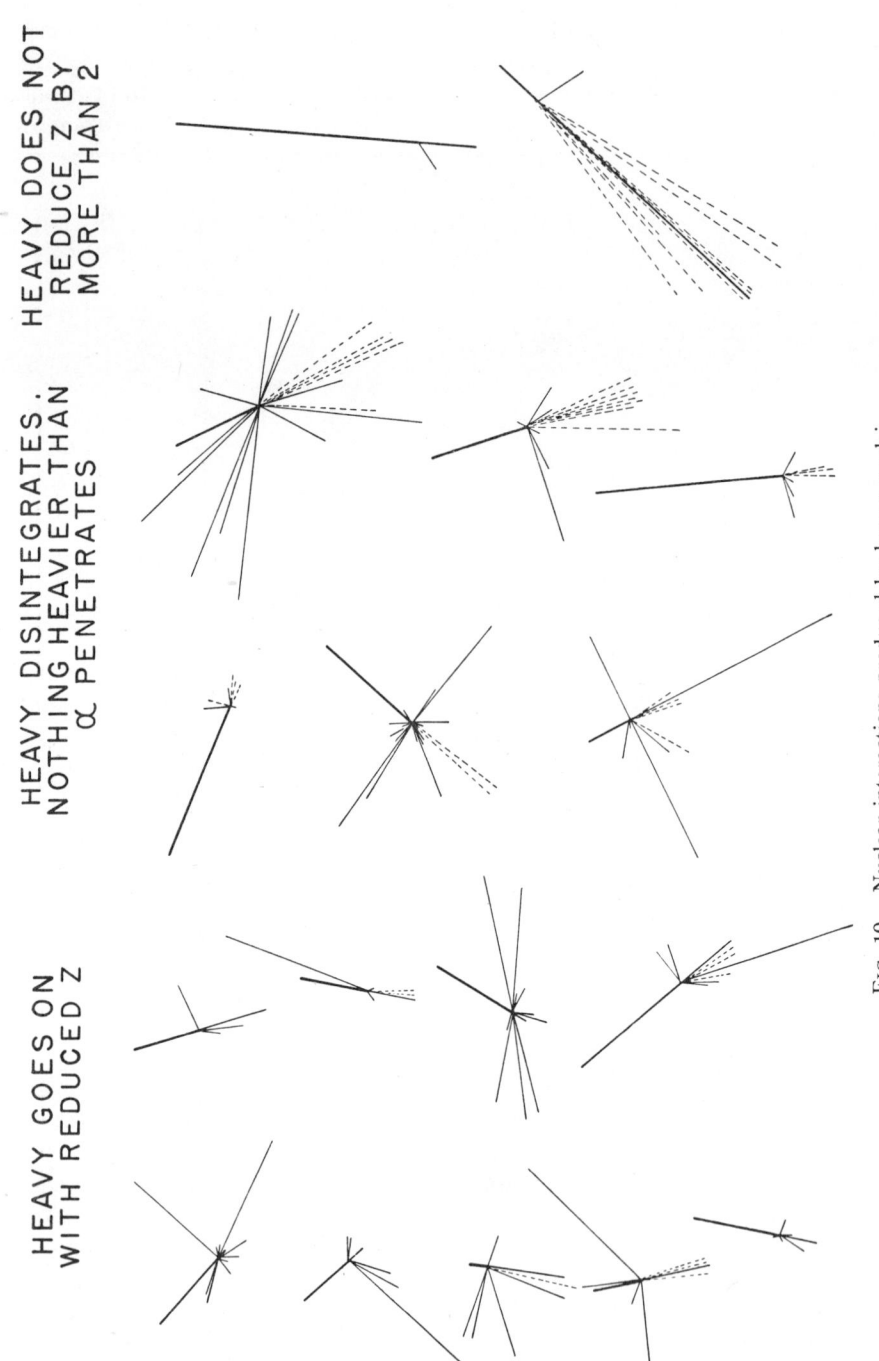

Fig. 10. Nuclear interactions produced by heavy nuclei.

Freier. The large stars are probably a good measure of the proton latitude effect. The factor of 3 in the integral particle intensity between 0.35 and 3.5 Bev per nucleon implies a flat energy spectrum. If the differential particle spectrum is expressed as $NdE = KdE/E^\gamma$, γ turns out to be 1.5. For energies above 5 Bev for protons the experiments of Millikan and others have in-

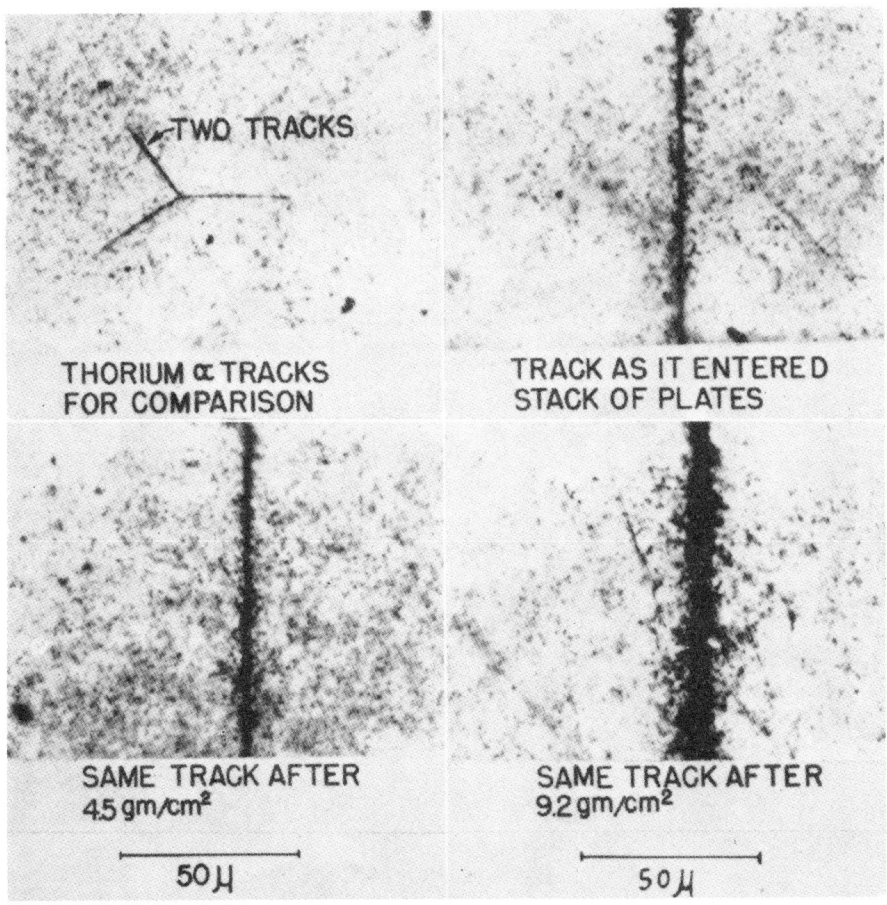

Fig. 11. The heaviest nucleus observed. The atomic number is estimated as 40 by δ ray counting.

dicated an exponent of about 2.7. The angular distribution at 55° indicates a γ of 1.7 for heavy nuclei between 0.5 and 1.0 Bev per nucleon in agreement with the latitude effect.

Figures 11, 12, 13 show examples of heavy nuclei. Figure 11 shows the heaviest nucleus so far observed in photographic emulsion. Its Z is estimated as 40. Figure 12 shows a nuclear interaction in the cloud chamber by a nucleus of the lithium beryllium boron group. Figure 13 shows a nucleus of

$Z \geqq 10$ passing through three lead plates and one carbon plate without interacting. The atomic number of 10 is estimated from the δ ray count.

The heavy nuclei have given some clues about the origin of cosmic radiation. The fact that they exist at all argues against acceleration mechanisms

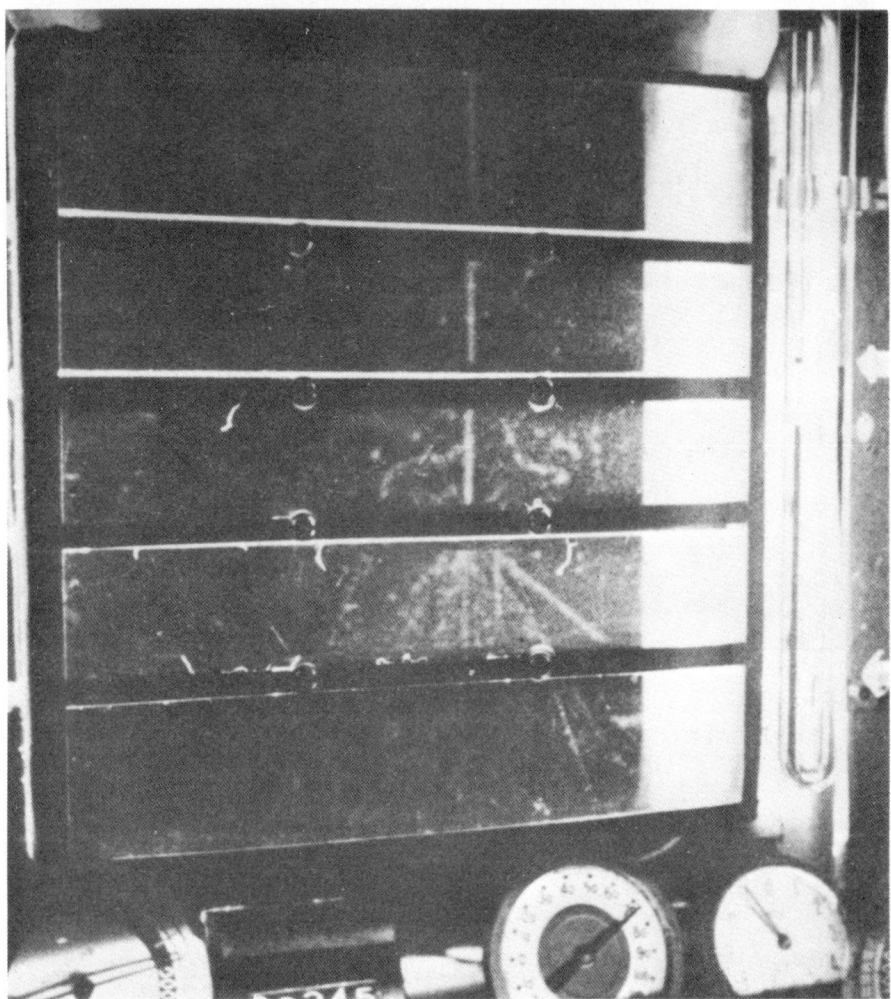

FIG. 12. A heavy particle (probably boron) interacts in a lead plate giving rise to a star with forward and backward prongs.

which work only for protons, such as that recently suggested by Fermi, unless we are willing to ask for two acceleration mechanisms. Since heavy nuclei of the same energy per nucleon exist in approximately normal cosmic abundances, one might expect some kind of electric or electromagnetic acceleration. For example, a huge van de Graaff accelerator with all elements present in its source would give essentially the observed result, as would acceleration near the sun by, for example, the cygnatron suggested by Swann and Vallarta. Teller, Alfven, and Richtmeyer have postulated that cosmic rays accelerated near the sun are held in the planetary system by magnetic fields. The condition that the heavy nuclei impose on this theory is that the nuclei do not travel long compared to their collision time with

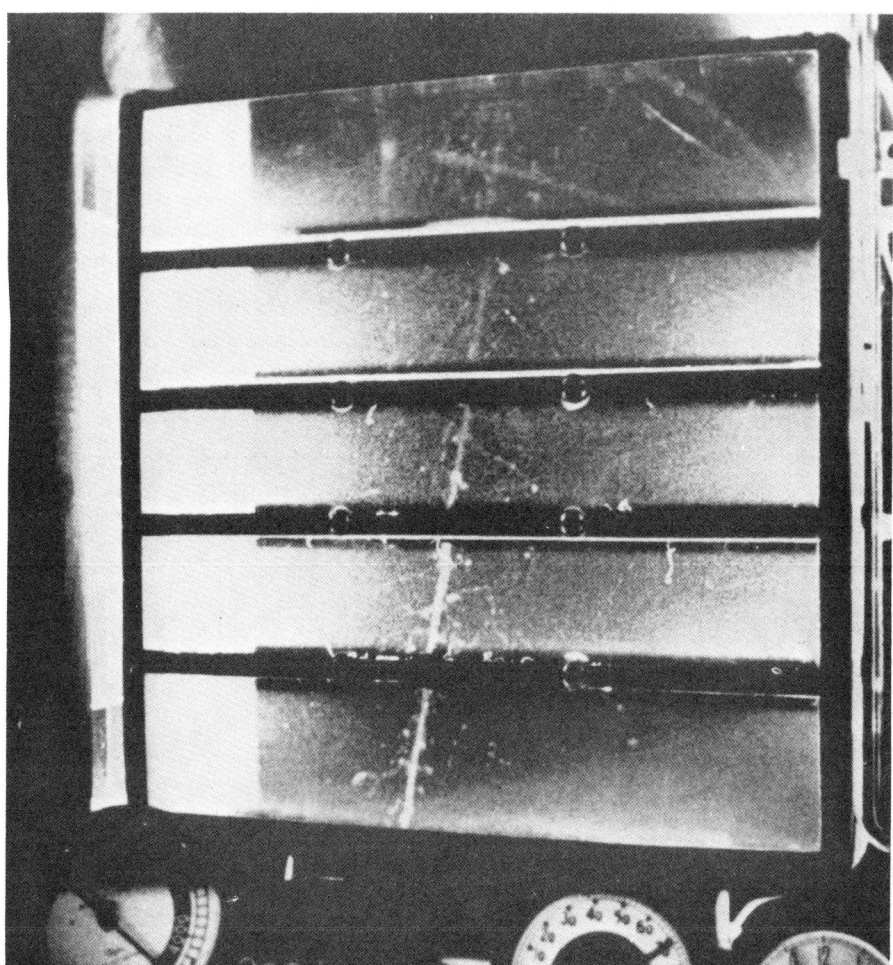

FIG. 13. A nucleus at $Z \geqq 10$ penetrating 2 g/cm² of carbon and 21 g/cm² of lead without interacting.

protons in space. Otherwise the heavier nuclei would not appear in normal abundance. A day-night effect for heavy nuclei would support the solar origin, whereas the absence of such an effect would not disprove it. An experiment on the diurnal effect has been carried out at Minnesota by Phyllis Freier, and she will report the results as soon as they are complete. It would be quite satisfying if the sun, in addition to giving us everything else that is good, also gave us cosmic rays.

DEVELOPMENTS IN COSMIC RADIATION, 1945-1950

J. C. STREET

Harvard University, Cambridge, Massachusetts

Since World War II the number of workers in the cosmic ray field has increased at least tenfold; techniques of observation and measurement have been remarkably improved and extended. For these reasons the advance in our knowledge of the field proceeds at such a pace that it is difficult for even the specialist in this work to keep abreast of recent developments. In this paper I will attempt to survey the key experiments of the past few years with, however, no pretension of completeness or of quantitative thoroughness. I owe acknowledgment for my material to many workers in various laboratories, and I will try to cite sources as far as it is practical.

SITUATION IN 1945

At the end of World War II we were familiar with two general components of the cosmic radiation as measured at sea level and at moderate altitudes:

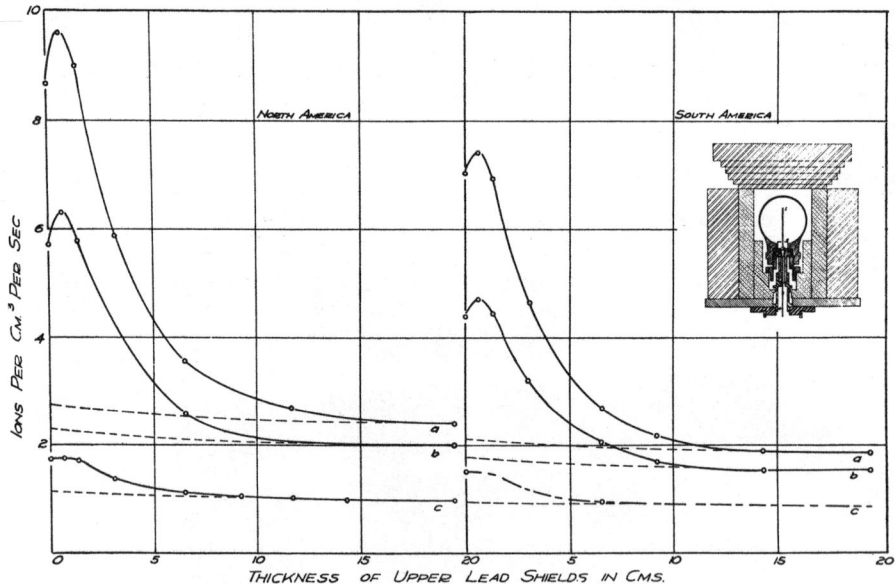

FIG. 1. The ionization produced by cosmic rays as a function of the thickness of lead shields above the ion chamber. The data is shown for three altitudes with barometric pressures (a) 45, (b) 51, and (c) 76 cm Hg and for two geomagnetic latitudes, North America 49° N and South America 1° S. The observations were made with the arrangement shown in the insert. The plotted values represent ionization arising from only that part of the radiation coming in through the upper lead shields (1). The absorption of the soft component in lead and its rapid increase with altitude are evident.

(1) a soft component mainly composed of electrons and photons; (2) a hard component consisting largely of mesons. Figure 1 gives an illustration of the experimental separation of hard and soft components with an ionization chamber. The particles of the soft component multiply rapidly in passage through a heavy absorber giving rise to cascade showers and are therefore quickly absorbed in lead shields. There is a rapid change of intensity of this component even in the atmosphere. (See Fig. 1.)

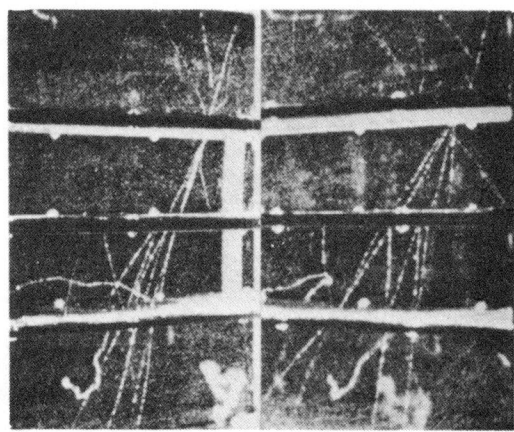

FIG. 2. An electron produces a shower in several stages. The plates in the chamber are lead, approximately 6, 6, and 1 mm from top to bottom (2).

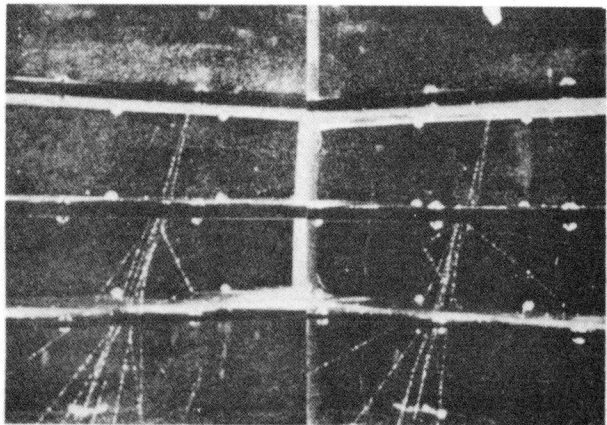

FIG. 3. A photon produces a small shower in several stages (2).

Figures 2–6 are cloud chamber photographs illustrating the multiplication of particles of the soft component in lead plates. These are examples of local electronic showers. The interactions of these particles, electrons and photons, radiation by the electrons, pair creation, and Compton collisions by the photons, has been adequately described by quantum electrodynamics as developed in the 1930's (4). Noteworthy, however, were observations of giant air showers of electrons and photons containing millions of energetic particles extending over areas many meters on a side. In some cases, it is estimated that the energy required to produce the events runs to about 10^{17} electron volts. These are known as Auger showers after the original observer (5). The appearance of a cloud chamber which expands on selection of a large air shower is shown in Fig. 7.

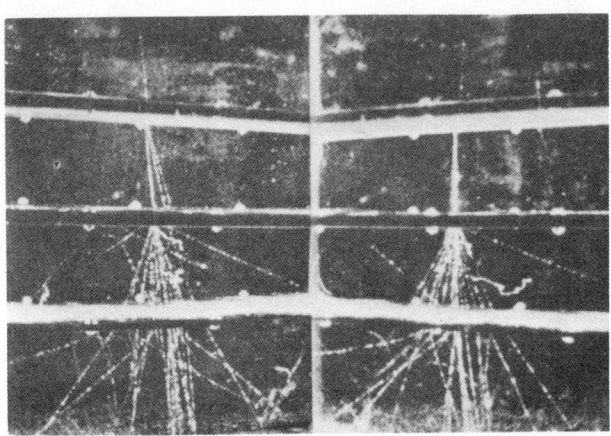

FIG. 4. A more energetic electron produces a shower (2).

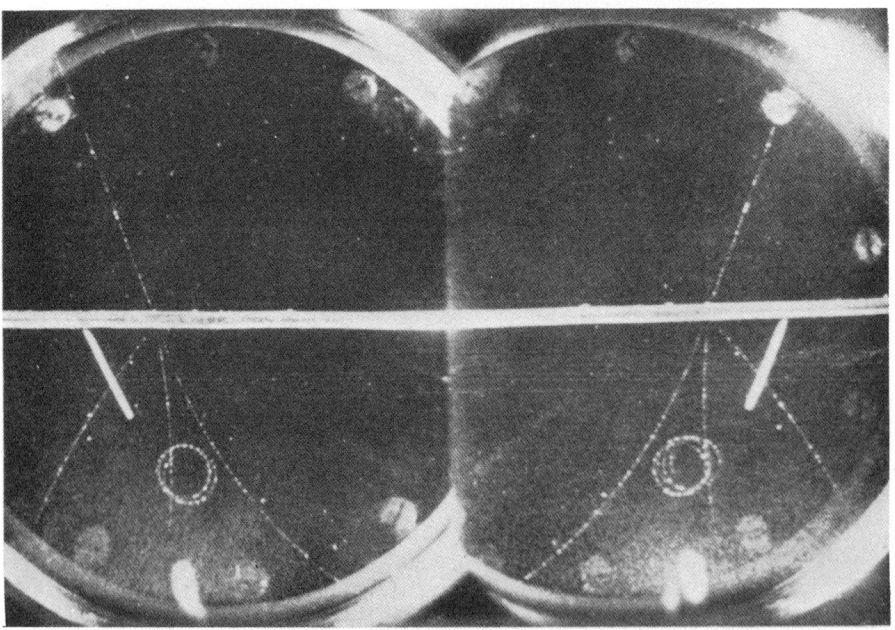

FIG. 5. Appearance of a shower in a magnetic field (3).

It had been shown that the hard component as observed at moderate altitudes consisted largely of mesons (6, 7). These particles are very penetrating as is seen from the slow increase of this component with altitude and

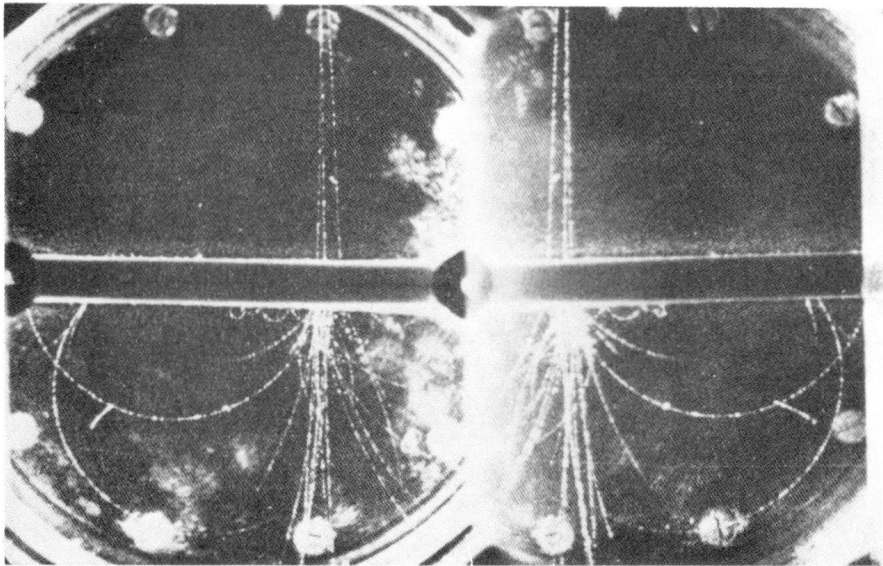

FIG. 6. Another case of a shower in a magnetic field (3).

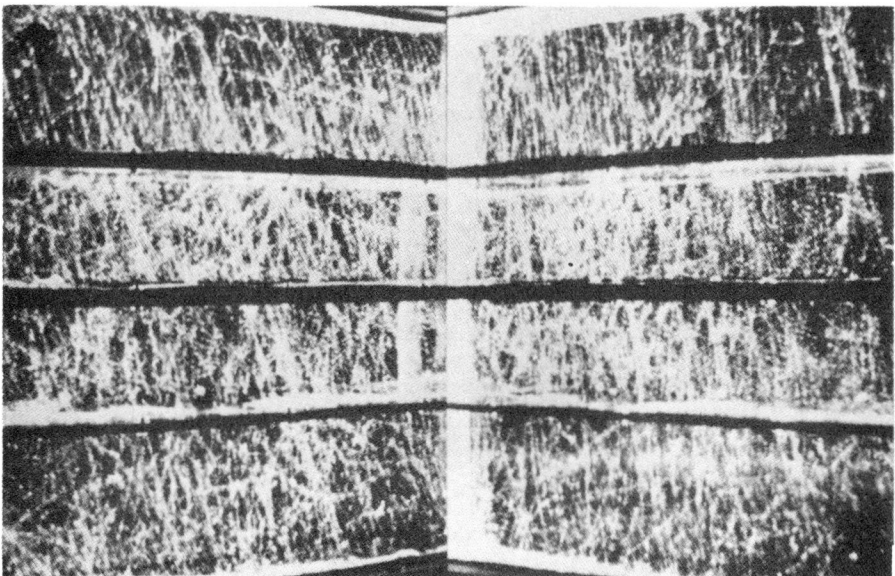

FIG. 7. General appearance of an air shower in a cloud chamber. The chamber shows only a small part of the shower. Many cases have been observed where parts of an air shower have greater track density than this.

the very slight absorption in heavy shields. Essentially their loss of energy is by electron collisions, ionization. Experiments during the late 1930's and early 1940's had shown that the mesons are unstable particles which decay, giving off electrons, with a mean life at rest of about 2 microseconds (8). In addition it was known that at sea level and mountain elevations a few protons, 1 to 10%, are mixed with the hard component, and it was suspected that these would be a more important fraction at higher elevations.

Experimentally, energetic electrons can be separated from mesons and protons since the electrons have high probability of producing showers on penetration of lead plates (or other heavy material). Below a momentum of

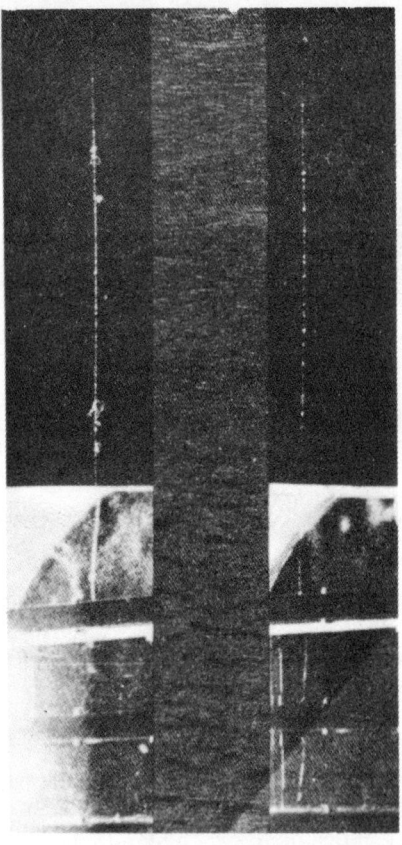

FIG. 8. The tracks shown are proton on the left, and meson on the right. They were observed in two cloud chambers, one above the other and with about 4 cm of lead between. The upper chamber was in a magnetic field and indicates that the momentum of the two particles was about the same (600 Mev/c). The lower chamber contains lead plates and shows the proton track becoming very dense and stopping in the top plate while the meson passes through the three 1-cm plates at minimum ionization. The difference of ionization between the two tracks is quite clear even in the upper chamber (9).

about 700 Mev/c, observed by magnetic deflection in a cloud chamber, protons can be easily distinguished from mesons because the ionization along the track of the proton will be perceptibly greater than that of the meson for the same magnetic curvature. Figure 8 illustrates this distinction clearly. It was known from observations of the sort illustrated in Fig. 8 that the mesons are singly charged, some positive and some negative, and that their rest mass is approximately 200 times that of the electron. It was not clear, however, that a unique mass could be assumed.

Because both soft and hard components exhibited a latitude effect it was believed that the primaries to both groups were charged particles subject to deflection by the earth's magnetic field. The azimuthal asymmetry, East minimum, of the hard component indicated positive primaries. Preliminary

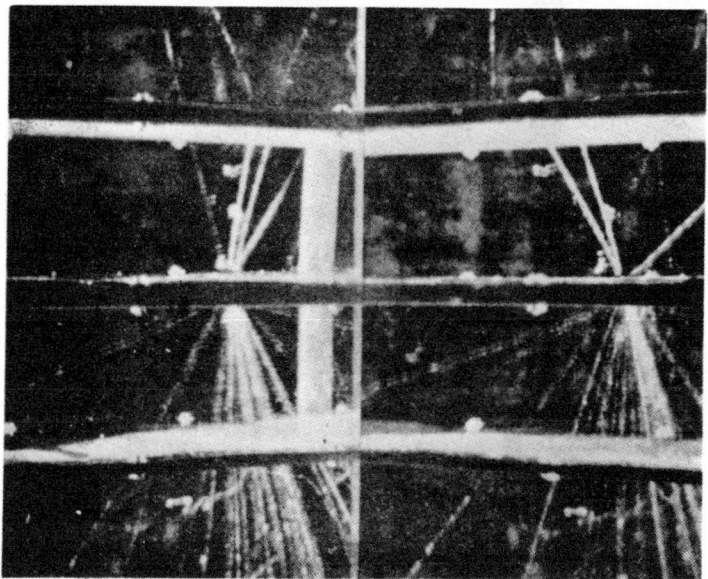

Fig. 9. Cloud chamber example of a cosmic ray nuclear explosion observed at sea level (2).

experiments showed little if any asymmetry of the shower component, and one wondered if we had to deal with a different set of primaries. (See for example T. H. Johnson, 1939.) (10)

A few rather sketchy experiments with cloud chambers and photographic emulsions gave evidence of violent nuclear reactions. Because of their general appearance these were called nuclear stars or just stars for short. These were rarely observed at sea level, but the experiments showed that their frequency increased rapidly with altitude, something like a factor of 20 to 14,000 ft. Illustrations of such events are given in Figs. 9 and 10.

This is certainly not a complete picture but does give the highlights of the situation which existed when cosmic ray research began to be pushed again under the favorable circumstances of the postwar period.

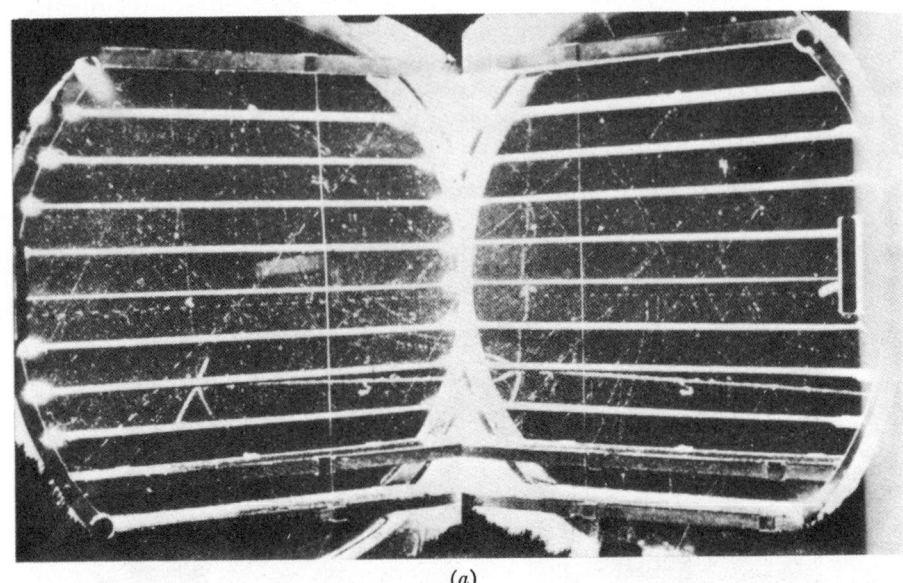

(a)

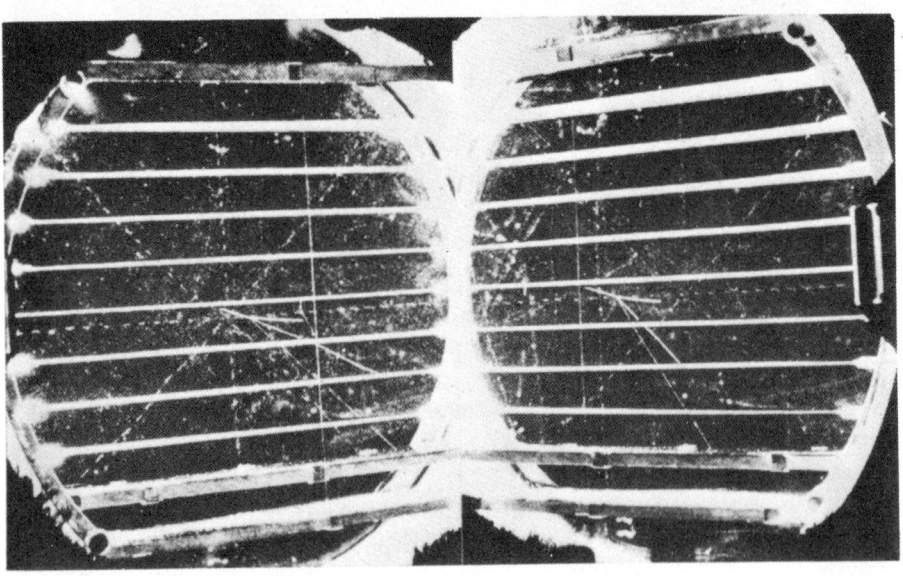

(b)

FIG. 10. Showing typical nuclear stars as seen in a cloud chamber at 11,000 feet. (a) Star probably caused by an ionizing particle (proton?) (b) Star probably caused by a neutron. (Unpublished photographs, Cool, Fowler, and Street.)

For the survey of recent work I will consider briefly the following topics; no attempt is made to document the discussion completely.

1. Studies of the μ meson: Mass. Details of the decay process. Interaction with the nucleus. Intensity in the atmosphere.

2. The improved photoplate technique and the discovery of the π meson: Decay and nuclear interactions of the π meson.

3. Nuclear interactions and the origin of the π mesons: Fragments from nuclear reactions.

4. Experiments relating to the primary radiation: Geomagnetic effects. Nature and intensity of primary particles.

5. Reactions at high altitude: Origin of the secondary radiations.

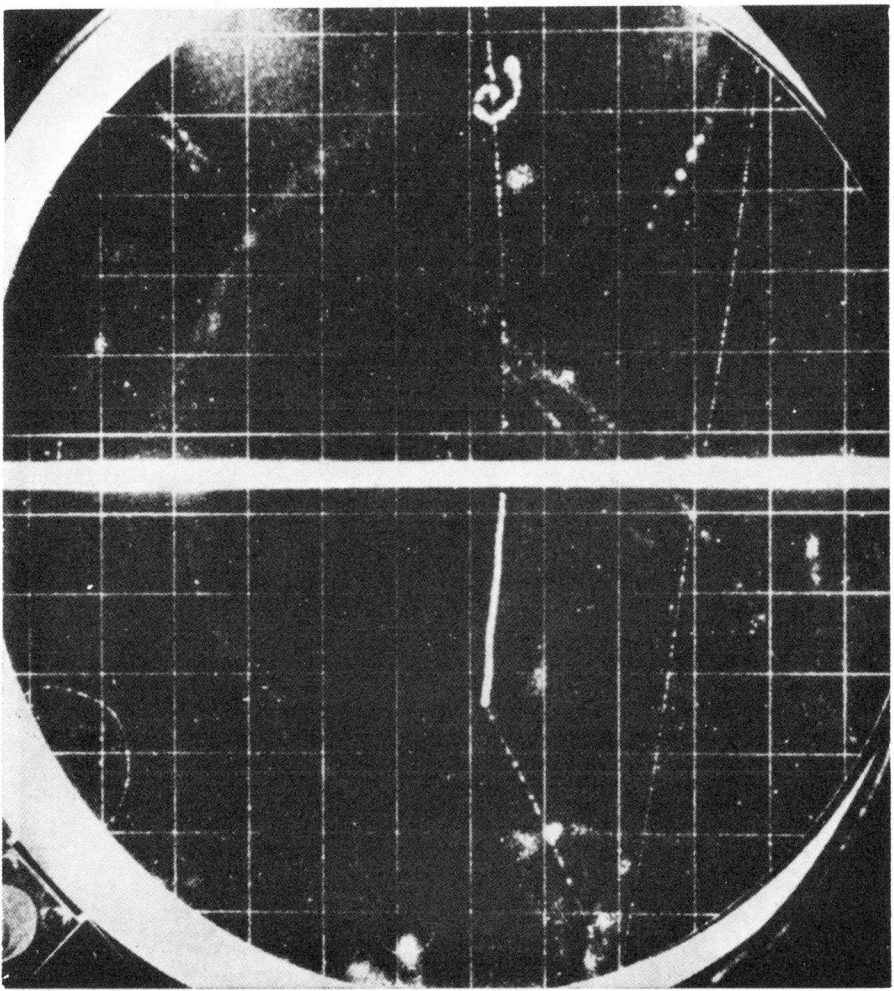

FIG. 11. Cloud chamber photograph obtained by R. W. Thomson showing the decay of a μ^+ meson in the gas of the cloud chamber. He gives the energy of the decay electron as rather uncertain, 20−50 Mev (13).

THE μ MESONS

The mass of the μ meson (the one mentioned before) has been studied by Brode and his colleagues (11) at Berkeley. They used the method introduced by Stevenson and myself in 1937. (See Fig. 8.) Two cloud chambers are used, one above the other, both triggered by counters above and below the top chamber. This chamber, which is in a magnetic field, gives a mea-

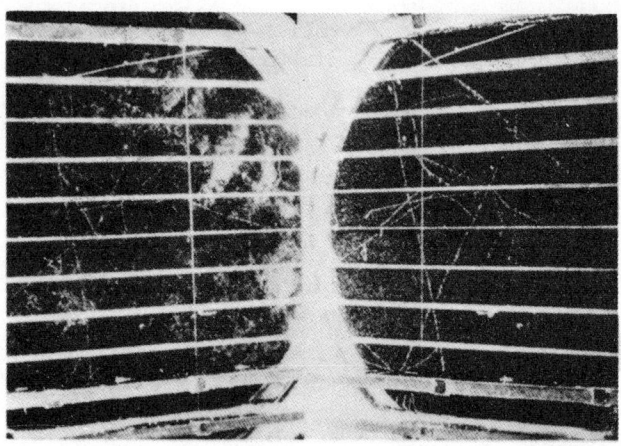

FIG. 12. A meson stops in a thin aluminum foil just below a lead plate at the top of the chamber and the emergent decay electron goes down through seven thin foils (0.02 cm each) and stops in a 9-mm lead plate. The decay electron energy estimated from scattering is 15 Mev (14).

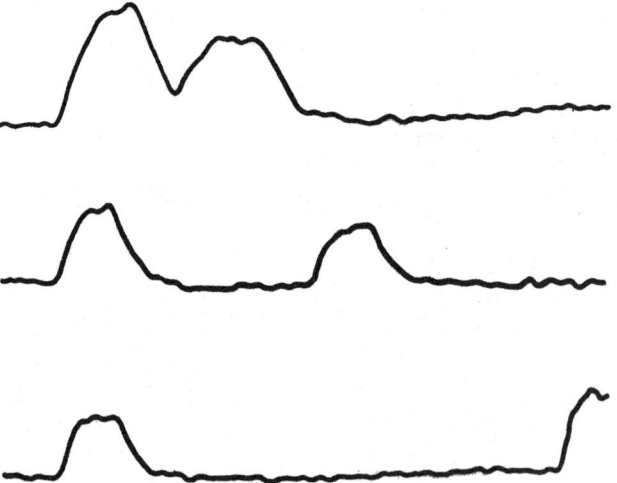

FIG. 13. The decay of μ mesons observed with a silver chloride crystal counter. The first pulse arises from a meson stopping in the crystal while the second pulse is from the decay electron which produces ionization in the same detector at a delayed time. The time length of the sweep is 6 microseconds (15).

sure of the momentum of the particle. The lower chamber contains several metal plates spaced an inch or so apart and therefore indicates the range of those particles which stop. On the assumption that energy is lost by ionization alone, it is possible to deduce the mass of a particle from these observations. The method of calculation depends on theoretical energy loss functions which have been checked experimentally to fair accuracy. Results for penetrating particles (hard component) at sea level give a consistent unique mass 216 electron masses (216 m_e) equivalent to 110 Mev rest energy.

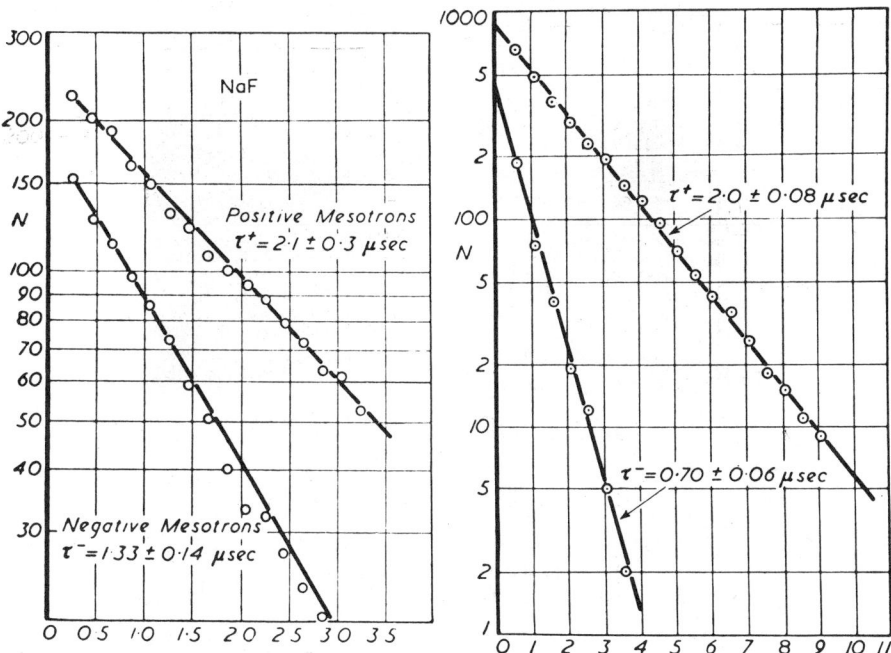

Fig. 14. Illustration of the effect of nuclear capture on the observed mean life of negative μ mesons. *Left*, disintegration curves in sodium fluoride (21) and *right*, in aluminum (20). The abscissa scales are in microseconds.

Perhaps a better value of the mass is 210 m_e obtained in experiments with photoplates using artificially produced mesons (12).

The decay of μ mesons has been studied by counting methods which select particles stopping in a block of material and simultaneously record the emergence of the decay electrons at measured times after the mesons stop. Cloud chambers have also been used to study the details of the process, in particular the energy spectrum of the decay electrons. Figures 11, 12, and 13 illustrate typical observations in a graphic way.

The first measurement of the mean life for decay by direct counting methods was made by Rasetti (16). The most careful measurements, those of Rossi and Nereson (17), give a value of 2.15 microseconds. Conversi,

Pancini, and Piccioni were the first to show that, while positive mesons give decay positrons when stopped in any material, the negative mesons give decay electrons only when stopped in materials of low atomic number (18). It had already been noted by Tomonaga and Araki (19) that, if mesons had the nuclear interaction expected on the basis of meson theory of nuclear forces, negative mesons should invariably be captured when stopped in either light or heavy materials. The results of Conversi *et al.* have been checked in detail by Valley (20) and by Ticho and Schein (21). For stopping materials of atomic number around 10, there is roughly matching competition between the decay and capture processes. As a result of the combined effect, the observed mean life is shortened. (See Fig. 14.)

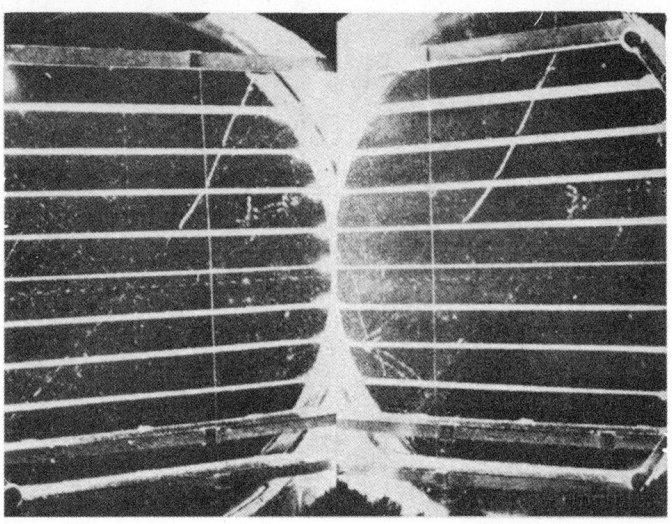

FIG. 15. A cloud chamber photograph shows a μ meson stopping in a thin aluminum foil with no visible star or other reaction. Many such cases have been seen.

What happens to the rest energy of the μ meson when captured by a nucleus? The experiments of Hinks and Pontecorvo, of Sard, and of Piccioni (22) have shown that no energetic photons are emitted in the process. No visible nuclear explosions are observed associated with such capture, which has been studied in detail by Chang and others with the cloud chamber and by Voorhies using the silver chloride conduction crystal (23). (See Fig. 15.)

Thus it appears that we need to assume that most of the available energy (100 Mev) is given to a light, unobservable, neutral particle; in this predicament it is usual to call on the neutrino. With this assumption we have for the capture process in the nucleus

$$\mu + P \longrightarrow N + \nu + \text{Kinetic energy} \quad (110 \text{ Mev})$$

To conserve momentum in the detail process most of the energy would be expected to go to the neutrino. A low excitation of the nucleus would be expected which would possibly give rise to emission of a neutron but not a proton. Sard *et al.* report positive results in a search for such neutrons using lead as stopping material (24). All the evidence combines to show that μ mesons are not nuclear mesons. They provide no clue to nuclear forces and bear no resemblance to the particles postulated by Yukawa to explain such forces.

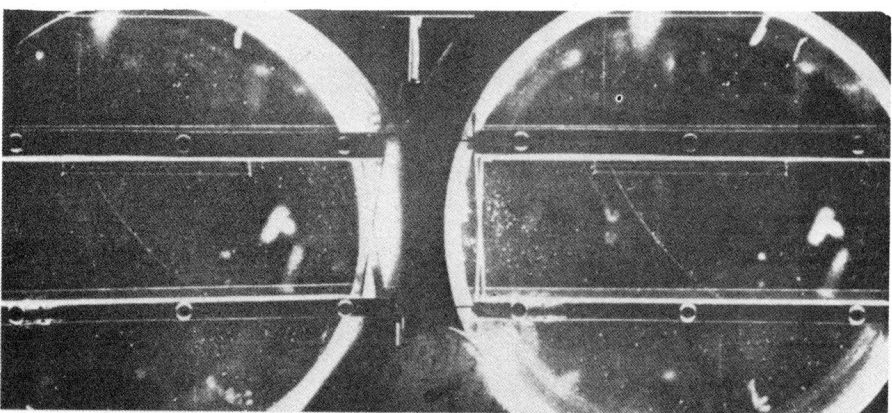

FIG. 16. Decay of μ^- meson with electron energy 42 Mev (25).

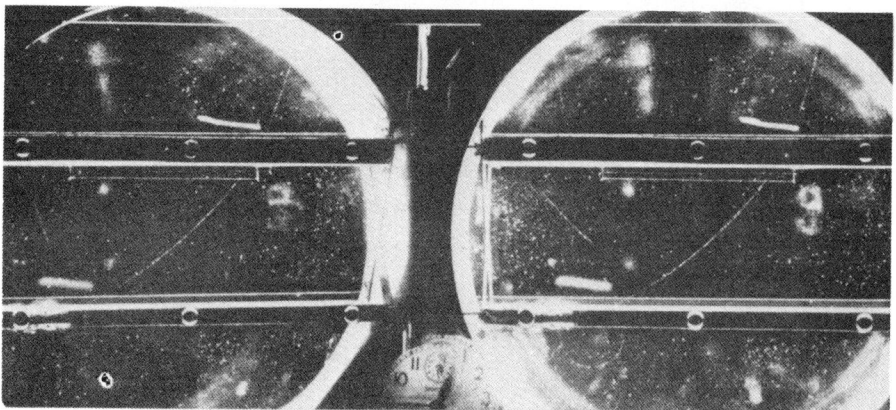

FIG. 17. Decay of μ^+ meson with electron energy 9 Mev (25).

The decay process is also interesting and surprising. Leighton, Anderson, and Seriff have measured the energies of many decay electrons and have found, contrary to expectation, a spectrum rather than a unique value (25). The energies range from a few Mev to an upper limit of about half the rest energy of the meson $\sim$ 55 Mev. (See Figs. 16 and 17.)

Since there is no unique energy it must be assumed that there are at least three fragments of the decay reaction. Again we assume the unobservable particles are neutrinos. They must be essentially massless to fit with the upper limit of the energy spectrum.

$$\mu \longrightarrow e + \nu + \nu + \text{Kinetic energy} \quad (110 \text{ Mev})$$

If this reaction is correct the spin of the μ meson is odd, most probably one-half. This would check with the spin expected from the capture reaction already discussed.

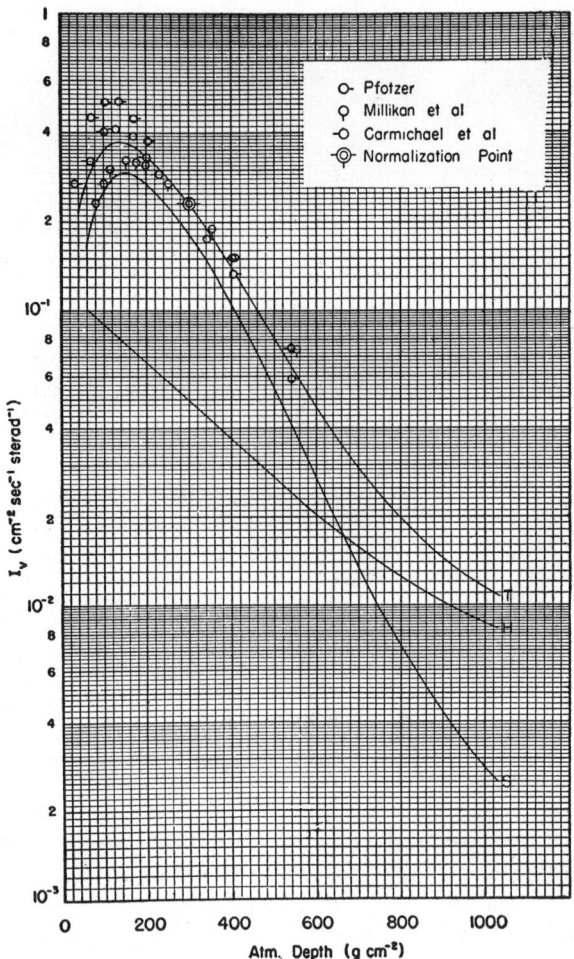

FIG. 18. Cosmic ray vertical intensity observed with counter telescopes at various depths in the atmosphere. Hard component H, soft component S, total radiation T for latitudes greater than 45° (26).

Extensive studies have been made of the intensity and energy distribution of μ mesons from high in the atmosphere to a considerable depth under water and underground. Detailed discussion of these is not possible in a short report, but I will show three figures from a summary prepared by Rossi in 1948 (26).

Photoplate Technique and Discovery of the π Meson

The direct recording photoplate technique, known since 1911, has been remarkably improved in recent years both by reduction of false background and increase of sensitivity for lightly ionizing tracks. In the hands of C. F. Powell and the group at Bristol, England, and more recently by many

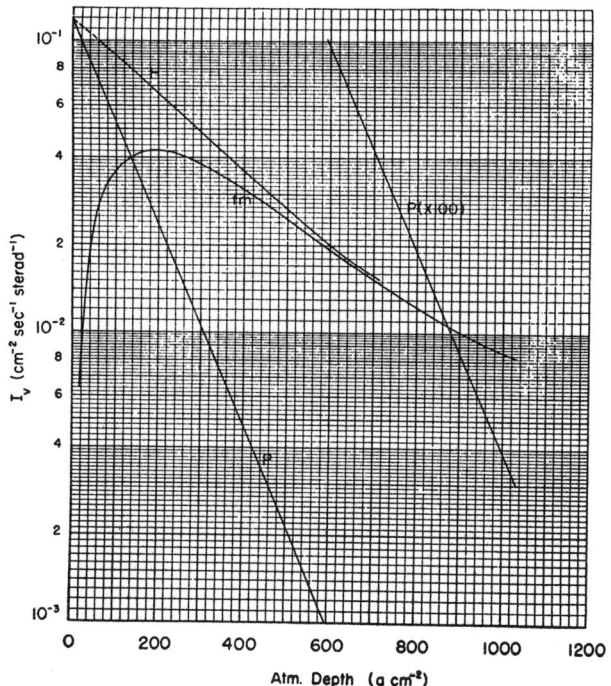

Fig. 19. Analysis of the hard component. Protons P, fast mesons f_m with range greater than 167 g cm^{-2} lead (26).

others, this technique has yielded very important results. Microscopic study of plates exposed at mountain altitudes led in 1947 to identification of a new type of meson now generally termed the π meson (27).

The studies undertaken to investigate cosmic ray stars and μ mesons were clear in showing tracks of particles of intermediate mass which were at first considered to be examples of the sort of mesons we have already considered. In many examples the mass of the particles could be roughly determined if the tracks ended in the emulsion. This is accomplished by consideration of density of developed grains, range, and scattering. That some of the intermediate mass particles are quite different is most clearly seen by noting what happens at the ends of certain of the meson tracks. Two types of events are observed: (1) A meson stops and from the end point a second meson of rather low energy and smaller mass emerges and stops after traveling a rather definite range (600 microns of emulsion); or (2) a meson stops and gives rise to a small star. Analysis shows that in the first process the secondary meson is identical with the μ meson and indeed Powell has recently published examples which show (*a*) the meson stopping, (*b*) the secondary meson, and (*c*) the decay electron from the μ meson (28).

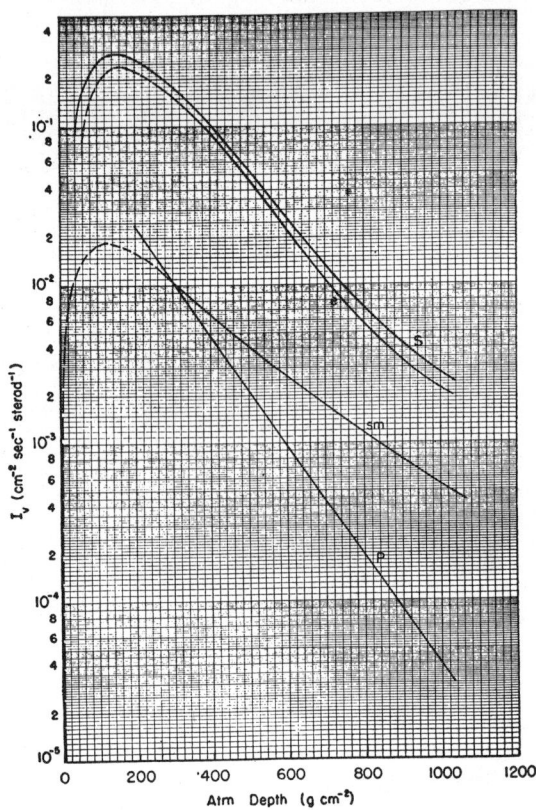

FIG. 20. Analysis of the soft component. Slow mesons *sm*, electrons positive and negative *e*, and slow nucleons *P* (26).

The fact that the μ mesons from this process have a definite range, that is, a discrete energy, suggests a two-fragment decay process for the π decay. For the past few years these new mesons have been generated artificially by means of the large Berkeley cyclotron. At Berkeley, Lattes and Gardiner have been able to determine the π meson mass from magnetic deflection and range to fair precision and give for its value 286 m_e about 1.32 times the mass of the μ meson. This provides enough information to analyze numerically the π to μ decay.

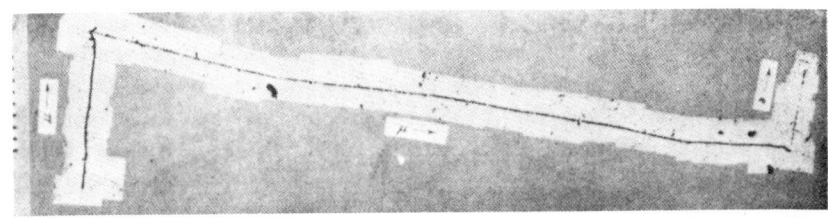

FIG. 21. Examples of the successive decay: $\pi^+ \to \mu^+ \to e^+$. Kodak NT 4 emulsion (28).

The kinetic energy of the μ^+ particle is found from the range (600 microns) and known mass and is according to Powell 4.25 Mev. Its momentum is 31 Mev/c. Thus the ν particle has a total energy of 32 Mev and a momentum

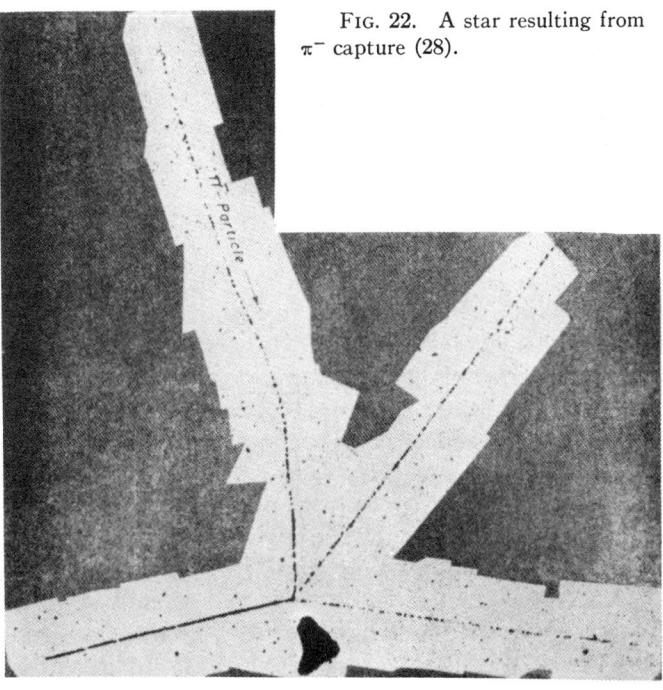

FIG. 22. A star resulting from π^- capture (28).

of 31 Mev/c. These are consistent (within experimental error) with the assumption that the particle is massless, presumably a neutrino.

In the case of the process in which the stopped meson gives rise to a small star the natural assumption is that we have to deal with a π^- meson captured by a nucleus and resulting in rather strong nuclear excitation. This to be contrasted with the capture of the μ^- meson with very low excitation. (See Fig. 22.)

The process which seems probable is a proton to neutron transformation.

$\pi^- + P +$ Other particles $\longrightarrow N +$ Other particles $+$ 146 Mev in nucleus

The stars from this process have been investigated by Perkins and by the

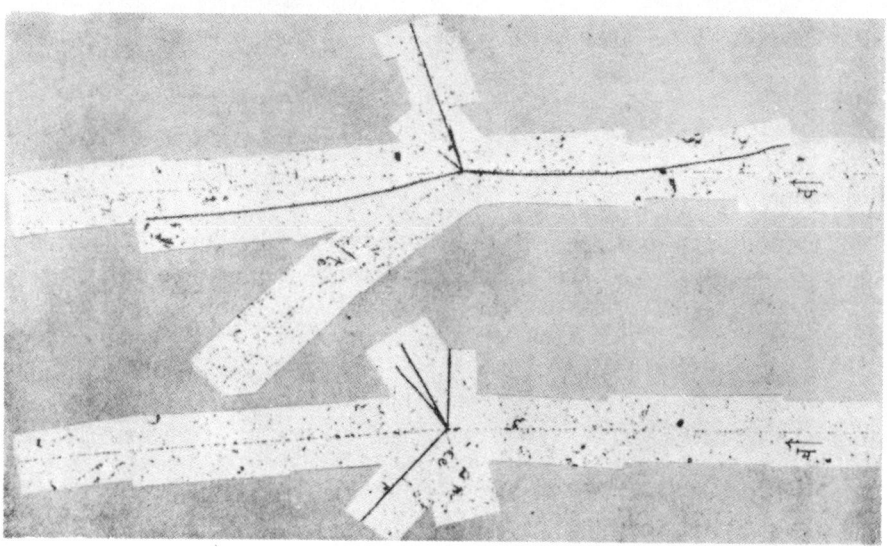

FIG. 23. Stars which are probably caused by an energetic proton making a "glancing" collision with a nucleus (Bristol Group, see reference 30).

group at Bristol and it appears that the average nuclear excitation is about 40 Mev in light nuclei, 80 Mev in heavy. It is concluded that about half the available energy is carried off by one or two neutrons, a result in agreement with direct studies of nuclear excitation by neutron bombardment.

From the above discussion we conclude that π mesons have integral spin and react strongly with nuclei. Experiments at Berkeley on the decay of artificial π mesons indicate that the mean life for decay is about 10^{-8} sec. Even with this shorter mean life, the mesons are invariably captured in both light and heavy nuclei. It will be of considerable interest to examine the stopping of mesons in hydrogen for in this case the capture process which has been postulated cannot take place so simply—there is no third body to balance momentum.

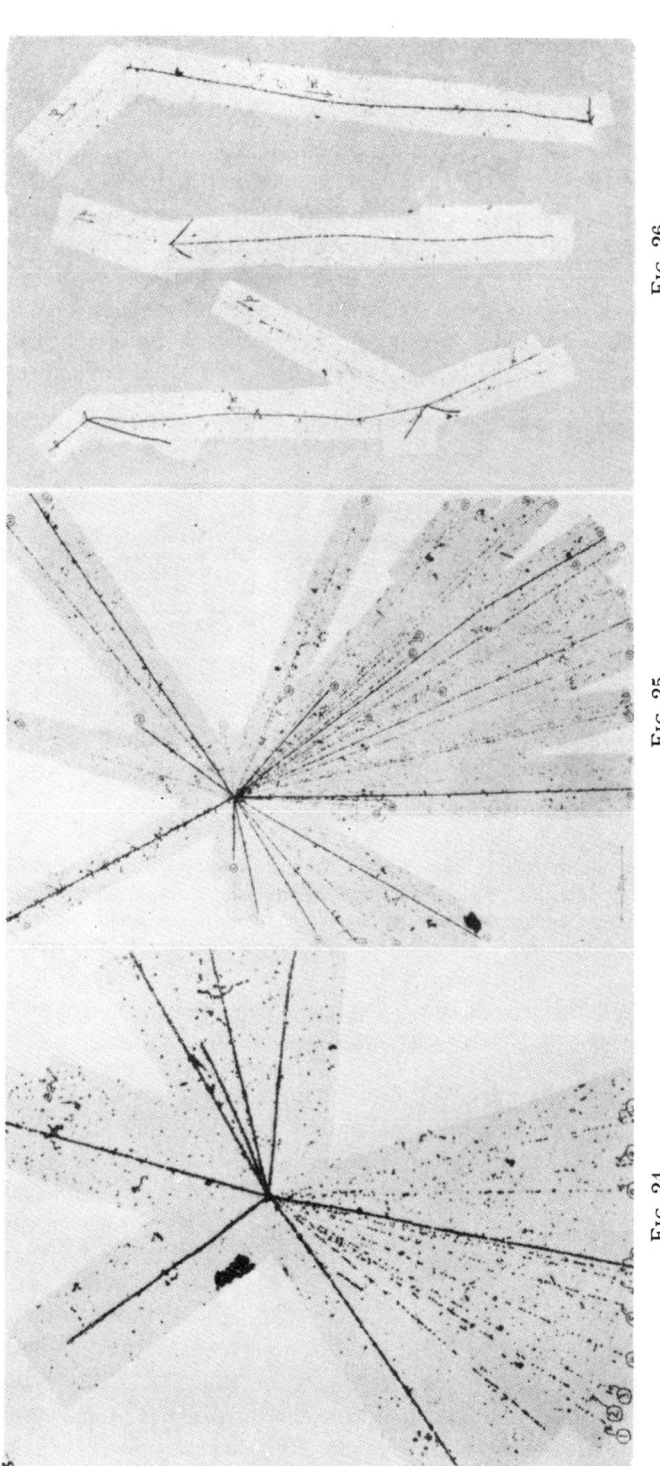

FIG. 24. A star with many energetic particles. There is no thin track above and the assumption is made that a neutron caused the star (30).

FIG. 25. A very striking star is probably caused by an energetic chlorine nucleus. The high charge of the chlorine projectile exhibits itself through the many delta rays along the track. This star was observed in a plate exposed at very high altitude in a balloon (31).

FIG. 26. Stars with slow π^- meson secondaries (30).

FRAGMENTS FROM NUCLEAR REACTIONS

Photoplates, cloud chambers, and counting techniques have been used to study nuclear reactions and their products at various altitudes. Such reactions evidenced by stars increase rapidly with altitude increasing by a factor of 2 or 3 for every 100 g/cm² of the atmospheric layer. Many of the stars are of amazing complexity showing the emission of dozens of particles. There are about fifteen stars per cubic centimeter of emulsion per day at 11,000 ft and about 2000 per cubic centimeter per day near the top of the atmosphere. It is known that the emitted particles include π mesons, pro-

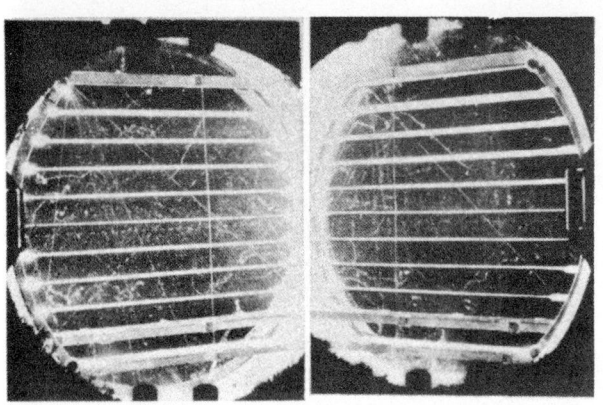 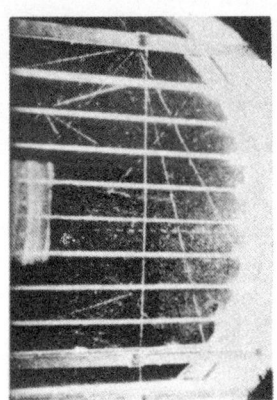

FIG. 27   FIG. 28

FIG. 27. A nuclear explosion in the upper lead plate at the left edge has both thin (fast) and dense (slow) secondaries. Some additional information can be obtained by studying the behavior of the secondaries in passing through the foils and lower lead plates. (Cool, Fowler, and Street, unpublished photographs.) See Fig. 12 for foil and plate thickness.

FIG. 28. A double star in the cloud chamber. The lower star takes place in the second aluminum foil. The two stars are probably connected by a thin track.

tons, deuterons, tritons, alpha particles, heavier nuclear fragments, and no doubt neutrons. The identification of a particular particle is difficult, if not impossible, except for those which stop, or unless special circumstances allow application of special techniques such as scattering or delta ray count or magnetic deflection. Detailed consideration of the stars is beyond the scope of this report, but I will show certain interesting cases. Stars with a few dense tracks are most frequent. (See Fig. 10b.) These are probably produced by moderate energy neutrons. Many stars have thin tracks above (Fig. 10a), probably indicating a high energy incident proton. Fairly frequently these protons appear to go on after collision (Fig. 23). Stars with many relativistic secondaries, perhaps a mixture of mesons, protons, and neutrons (not visible) are also observed (Figs. 24, 25). Slow π^- mesons which stop in the emulsion are observed as secondaries from some stars (Fig. 26).

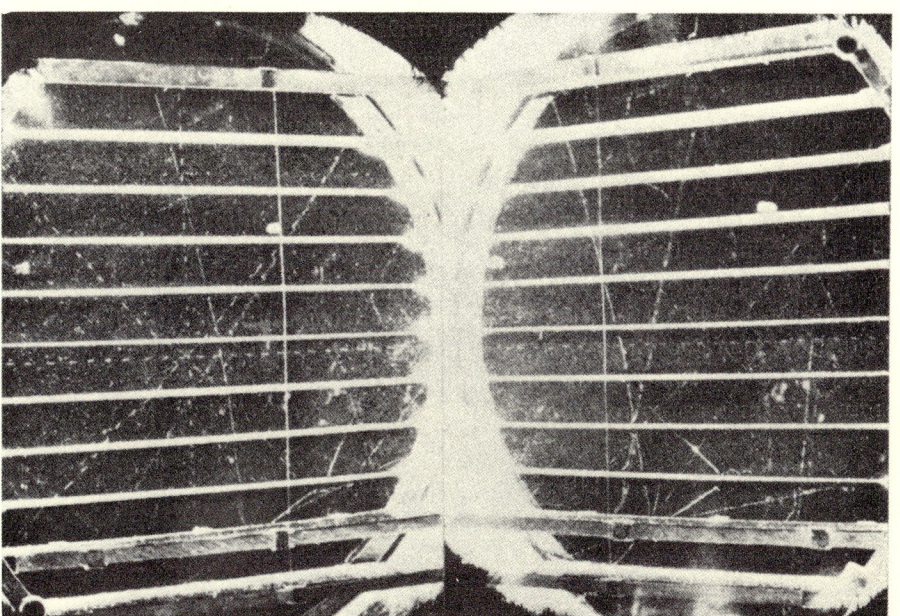

Fig. 29. A star is produced in the second lead plate, probably by the lightly ionizing particle coming in from the upper right. Two dense tracks go up. Analysis of the scattering and track densities indicates that both are probably tracks of fairly slow mesons.

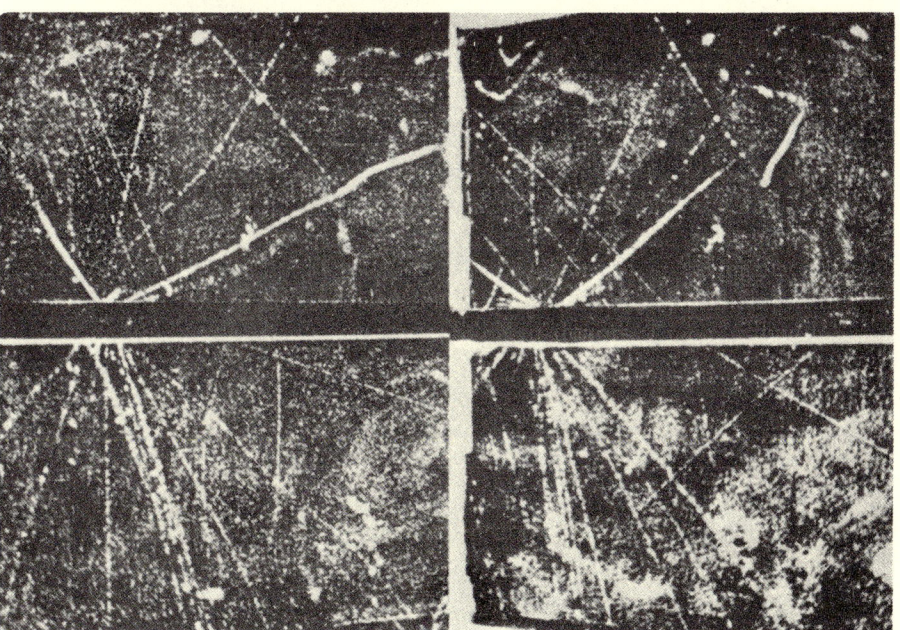

Fig. 30. An old photograph taken by Fussel in 1937 not understood at the time (2). A nuclear interaction takes place in the lead plate. It appears likely that the heavily ionizing particle going up to the right is a π^+ meson which passes out of the lighted region at the front and stops in the glass window. There it decays, sending the μ^+ decay particle back into the chamber down and to the left. The μ^+ particle stops in the gas and decays; the decay electron track is clearly visible going up and to the left.

A few double stars have been observed in photoplates in which an energetic secondary from one star produces another star. These are fairly common in a cloud chamber with many thick metal plates, especially when using counter selection of "penetrating" showers. With such thick plates many of the particles from the stars are absorbed and not observed. Figures

FIG. 31. A nuclear explosion and associated cascade shower below. (Unpublished photograph, E. C. Fowler, Street, W. B. Fowler, and Sard). Many such examples have been observed in cloud chambers with lead plates. One of the first observations was that of Fussel (2).

27, 28, and 29 are interesting examples of stars observed in a cloud chamber at 11,000 ft. Figure 30 is a beautiful example of a complete sequence: a star with many shower particles has a π^+ meson secondary which is observed to decay through successive stages.

Negative π mesons are often observed to emerge from stars in photoplates, but positive π's slow enough to stop are not seen as would be expected since they are repelled by the nuclear coulomb field and have small chance of stopping in the emulsion. Both positive and negative π's have been produced by high energy α and proton bombardment in the Berkeley cyclotron and by x rays from the 300-Mev synchrotron.

No doubt high energy mesons are present among the shower particles of the stars and in certain cases have been partially identified by scattering. Because these mesons exhibit strong nuclear reaction it is no doubt possible

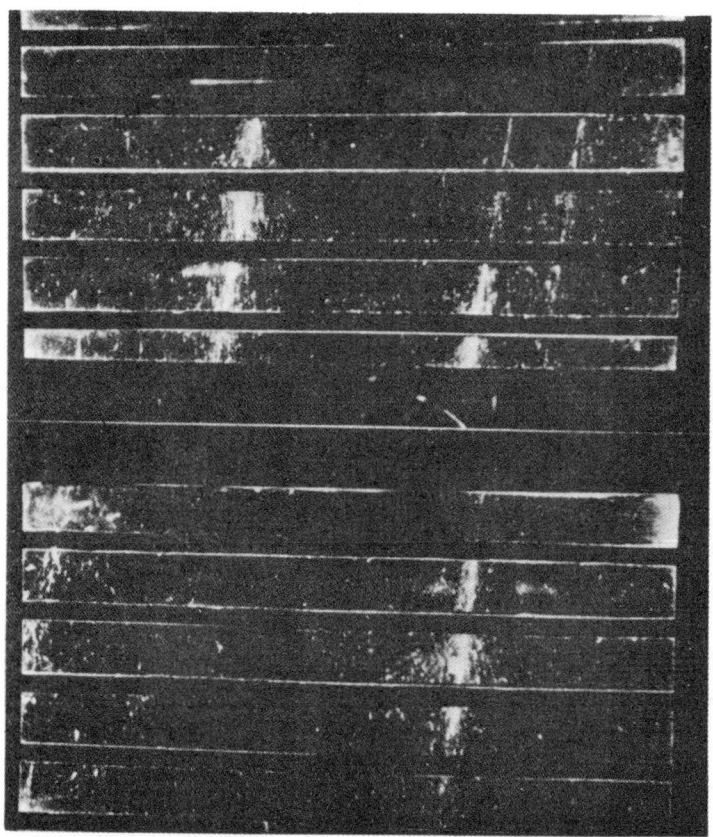

FIG. 32. Two cloud chambers each containing ½-in. lead plates are separated by an additional 6 in. of lead. Expansions were selected to detect large air showers. It would be expected that the electronic component would be completely absorbed but the illustration shows new cascades starting out again in the lower chamber. (Unpublished photograph, courtesy of K. Greisen.)

for them to produce stars themselves. However, as a result of their short life, they cannot be cosmic ray primaries nor, for the same reason, can they contribute very much to the stars produced in the atmosphere. These must be due mainly to energetic protons and neutrons. It is to be noted that the decay in flight of energetic π mesons created in stars provides a reasonable explanation of the origin of the fast μ mesons which are the main cosmic ray component at sea level.

We need to discuss one more product of the nuclear collisions before we are prepared to give a sketch of the entire series of events originating with the incidence of the primaries on the top of the atmosphere. Examination of cosmic ray nuclear reactions at high energy shows that energetic photons as well as the particles we have discussed are frequently emitted. These in turn give rise to electronic cascade showers. Figure 31 is an illustration of such a phenomenon. The production of such photons has also been observed in the Berkeley cyclotron (32). There is little quantitative data on

FIG. 33. Similar to Fig. 32. A very powerful cascade shower starts from the lower part of the lead between the chambers (Greisen).

this process but tentatively we take it to be the explanation of the origin of the soft component. There are theoretical reasons for believing the photons come from the decay of neutral mesons created in the nuclear reaction, but since the expected mean life is of the order of 10^{-18} sec, there is little chance to observe the effect directly.

Thus we now believe that the hard and soft components arise from the same class of original primaries through the intermediary of nuclear reactions of great complexity. The majority of such events take place in the high atmosphere, but it is still quite feasible to observe phenomena of this type at mountain elevations. Auxiliary experiments support this view. In particular new and more extensive studies of the East-West asymmetry arising from the earth's magnetic field show that it is quite reasonable to assume the same primary radiation for hard and soft components (33). The observations seem to have come into line on this point in indicating positive primaries for the soft as well as the hard component. In addition, Rossi and

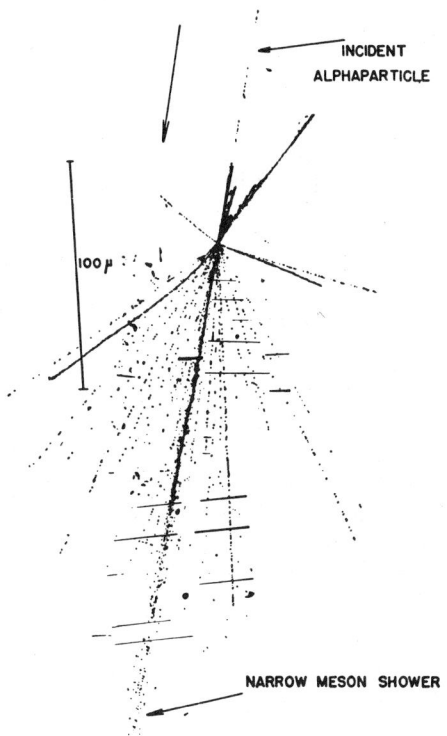

FIG. 34. An extremely energetic primary cosmic ray alpha particle makes a collision with a heavy nucleus (Ag or Br) of the emulsion. The collision gives rise to a very narrow (2.5°) penetrating core of some 23 relativistic singly charged particles. The core is surrounded by a more diffuse shower of 33 relativistic particles and in addition 18 slower-moving fragments carrying at least 23 units of charge. The narrow core is observed in another emulsion after passage through 2 cm of glass and it is found that electron pairs are created, indicating the presence of very energetic gamma rays. It is supposed that most of the charged particles in the narrow core are π mesons while probably some of the more diffuse tracks are caused by fast secondary protons. It is estimated from the angular spread of the narrow cone that the alpha-particle energy is about 10^{13} electron volts. (Kodak NTB 3 emulsion flown at 100,000 ft (39).

Hulsizer have been able to show from observations at high elevations, using balloons, that the number of photons and electrons in the primary radiation is negligible (34).

Another check lies in the closer examination of the giant air showers. It has often been remarked that these showers almost exclusively consist of electrons and photons. If, now, they arise from some original nuclear collision or collisions, they should contain high and low energy nucleons, nuclear fragments, and mesons. The Cornell group Cocconi, Greisen, et al. have recently shown conclusively that these showers do contain neutrons, mesons, and particles of sufficient energy and penetration to start powerful cascades over again, even under several inches of lead, a thickness sufficient to stop the electronic component (35). This restarting fits well with the hypothesis of very energetic nucleons repeating over again collisions of the same type that started the original shower. Cloud chamber examples are shown in Figs. 32 and 33.

Survey of Processes in the Cosmic Radiation

We may now review by giving the sequence of events from incidence of primaries to the final absorption of μ mesons deep in the earth.

1. The primary rays are protons, alpha particles, and heavier nuclear fragments ranging in energy from the cutoff due to the earth's magnetic field on the low side (2 to 3 Bev at this latitude for protons; 10 to 15 Bev at the equator) up to energies at least 10^{17} electron volts as evidenced by the giant showers. Roughly the average particle energy is about 10 Bev at this latitude since the number of very high energy rays is not large. The intensity of primary particles at the top of the atmosphere is about 0.1 /cm²/sec/ steradian (36). From a study of the latitude effect and analysis of the air showers it is estimated that the energy distribution of the primaries varies as $E^{-1.8}$ at moderate energies and as $E^{-2.8}$ at high energies (E, the energy) (37). Recent work of Bradt and Peters and of Oppenheimer, Ney, et al. shows that the number of alpha particles in the primary radiation is about 20% of the entire group (38). These together with heavier primaries bring in about as many nucleons as the proton component.

2. These primary particles make nuclear collisions of various degrees of complexity producing secondary nucleons of various energies and creating positive, negative, and neutral mesons. In turn the secondary nucleons give rise to other stars, the positive and negative π mesons decay to μ mesons which are the hard component at lower elevations, the neutral mesons immediately decay into two photons each, and these build up into the cascade showers of the soft component. Kaplon, Bradt, and Peters reported a graphic illustration of the start of the sort of sequence described above (39). Their example and discussion of it are given in Fig. 34.

REFERENCES

1. Young, R. T., and Street, J. C. *Phys. Rev.* **52**, 552 (1937).
2. Street, J. C. *J. Franklin Inst.* **227**, 765 (1939).
3. Neddermeyer, S. H., and Anderson, C. D. *Revs. Mod. Phys.* **11**, 191 (1939).
4. Heitler, W., Quantum Theory of Radiation. Oxford University Press, 1944.
5. Auger, P., Ehrenfest, P., Maze, R., Daudin, Robley, Freon, A., *Revs. Mod. Phys.* **11**, 288 (1939). See also Janossy, L. *Cosmic Rays.* Oxford University Press, 1948. pp. 318-343.

6. Street, J. C. *J. Franklin Inst.* **227**, 782-787 (1939).
7. Neddermeyer, S. H., and Anderson, C. D. *Revs. Mod. Phys.* **11**, 195-196 (1939).
8. Rossi, B. *Revs. Mod. Phys.* **11**, 296 (1939). See also Janossy, L. *Cosmic Rays*. Oxford University Press, 1948. pp. 20-22.
9. Street, J. C., and Stevenson, E. C. *Phys. Rev.* **51**, 1005 (1937).
10. Johnson, T. H. *Revs. Mod. Phys.* **10**, 228 (1938); **11**, 208 (1939).
11. Retallack, J. G., and Brode, R. B. *Phys. Rev.* **75**, 1716 (1949).
12. Smith, F. M., et al. Paper D2, Stanford Meeting, *Am. Phys. Soc.* Dec. 29, 1949.
13. Rossi, B. *Cosmic Radiation* (Colston Papers). Interscience Publishers, New York, 1949. p. 47.
14. Fowler, E. C., Cool, R. L., and Street, J. C. *Phys. Rev.* **74**, 101 (1948).
15. Voorhies, H. G., and Street, J. C. *Phys. Rev.* **76**, 1100 (1949).
16. Rasetti, F. *Phys. Rev.* **60**, 198 (1941).
17. Rossi, B., and Nereson, N. *Phys. Rev.* **62**, 417 (1942).
18. Conversi, M., Pancini, E., and Piccioni, O. *Phys. Rev.* **71**, 209 (1947).
19. Tomonaga, S., and Araki, G. *Phys. Rev.* **58**, 90 (1940).
20. Valley, G. E. *Phys. Rev.* **72**, 772 (1947).
21. Ticho, H. K., and Schein, M. *Phys. Rev.* **73**, 81 (1948).
22. Piccioni, O. *Phys. Rev.* **74**, 1754 (1948).
23. A brief discussion of earlier papers on this subject is given in reference 15.
24. Sard, R. D., Ittner, W. B., and Conforto, A. M. *Phys. Rev.* **74**, 97 (1948).
25. Leighton, R. B., Anderson, C. D., and Seriff, A. J. *Phys. Rev.* **75**, 1432 (1949).
26. Rossi, B. *Revs. Mod. Phys.* **20**, 537 (1948).
27. Lattes, C. M. G., Muirhead, H., Occhialini, G. P. S., and Powell, C. F. *Nature* **159**, 694 (1947).
28. Powell, C. F. *Cosmic Radiation* (Colston Papers) Interscience Publishers, New York, 1949. p. 86.
29. Perkins, D. H. *Phil. Mag.* **40**, 601 (1949).
30. Brown, R. H., Camerini, U., Fowler, P. H., Heitler, H., King, D. T., and Powell, C. F. *Phil. Mag.* **40**, 862 (1949).
31. Camerini, U., Coor, T., Davies, J. H., Fowler, P. H., Lock, W. O., Muirhead, H., and Tobin, N. *Phil. Mag.* **40**, 1073 (1949).
32. York, H. F., Moyer, B. J., and Bjorklund, R. *Phys. Rev.* **76**, 187 (1949).
33. Winckler, J. R., Stroud, W. G. *Phys. Rev.* **76**, 1012 (1949).
34. Hulsizer, R. I., and Rossi, B. *Phys. Rev.* **73**, 1402 (1948).
35. Cocconi, G., Tongiorgi, V. C., and Greisen, K. *Phys. Rev.* **76**, 1020 (1949).
36. See reference 26, page 574, and references cited there.
37. Dwight, K., Sabin, R., Stix, T., and Winckler, J. R. Paper G9, New York Meeting, Am. Phys. Soc., Feb. 2-4, 1950. See also reference 38.
38. Private communication, but see: Pomerantz, M. A. *Phys. Rev.* **77**, 836 (1950) and references cited there.
39. Kaplon, M. F., Peters, B., and Bradt, H. L. *Phys. Rev.* **76**, 1735 (1949); Marshak, R. E. *Phys. Rev.* **76**, 1736 (1949).

PHYSICS OF THE SOLID STATE

THE NEW ELECTRONICS

KARL LARK-HOROVITZ

Purdue University, Lafayette, Indiana

> ... we may hope hereafter to discover by experiment the law which probably holds together all the above effects with those of the *evolution* and the *disappearance* of heat by the current and the striking and beautiful results of thermo-electricity, in one common bond.* M. FARADAY.

In recent years a number of electronic devices, such as current control mechanisms with negative or variable temperature coefficients of resistance (50), plate and point contact rectifiers (7, 10, 17) (crystal diodes), transistors (25, 46) (crystal triodes), photoconductors (9, 11, 20), photovoltaic cells (7, 16), and numerous applications of phosphors (22, 23), have shown that there is a new type of electronics in the making. This new electronics, made possible by the controlled production, release, and direction of electrons in solids and at boundaries between solids, promises to have great theoretical and practical importance. The main topic of this discussion will be the electronic conduction in nonmetals. In the first part is given a historical outline of the most important developments in typical semiconductor problems. In the second part the behavior of germanium semiconductors in bulk will be discussed.

Although this development started over 100 years ago, consideration of its various phases shows that only recently has it been possible to investigate these phenomena with a unifying idea, based on theoretical developments which originated in the quantum mechanical theory of solids.

DEVELOPMENTS IN SEMICONDUCTOR PROBLEMS

VARIABLE CONDUCTORS OF ELECTRICITY[1]

In 1833 for the first time, Faraday (125) discovered in silver sulfide a substance which had metallic conductivity at elevated temperatures, but a temperature dependence opposite to the one observed in metals: with decreasing temperature the resistance increased; at high temperatures the compound conducted as well as a metal, but at low temperatures approached an insulator in resistivity.

Hittorf (173) claimed that Ag_2S is a solid electrolyte, and since the conductivity of electrolytes increases with temperature, this particular com-

---

* *Experimental Researches in Electricity*, Series xii, § 1342, Feb. 8, 1838.
[1] The name given to semiconductors by Koenigsberger (15).

pound did not seem to be an exception. About fifty years later Streintz (321) reinvestigated metal oxides and sulfides and again concluded that Ag_2S has metallic conductivity (see also 156), as Hittorf had admitted for PbS, FeS, NiS, HgS, and Ag_2Se.

The first step in the investigation of the electrical properties of any nonmetallic substance must be the clarification of the type of conductivity, whether electrolytic or metallic. One usually concludes that electrolytic conductivity exists if Faraday's law of electrolysis is fulfilled and the transport number, the fraction of the total charge carried by any one ion, corresponds to the value expected for electrolytic conductivity. One also must compare the measured conductivity with the maximum conductivity[2] which can be expected for an electrolyte (167).

Since the conductivity of silver sulfide and cuprous sulfide is far larger than the maximum conductivity expected for any electrolyte (for instance, the conductivity of molten silver chloride is only 5 mho/cm while that of molten copper sulfide (Cu_2S) at its melting point is 1.3×10^3 mho/cm), it is unlikely that these high conductivity materials are electrolytes. If one interprets these high conductivities as electrolytic and calculates mobilities accordingly, mobility values are obtained which are by orders of magnitude larger than those observed in ordinary electrolytes. For this reason Baedeker (42) again classified Ag_2S as a metallic conductor.

A new phase of this discussion started when Tubandt and his collaborators (327) measured the transport number and the Faraday equivalent with an accuracy that seemed to establish silver sulfide definitely as a solid electrolyte. However, the following facts are then difficult to explain: (a) the high electrical conductivity is difficult to interpret as ionic; (b) it is possible to measure the Hall effect (204), which indicates electronic conductivity by its magnitude and sign; if the conductivity were ionic, the Hall effect would be too small to measure; (c) if one directly measures diffusion in silver sulfide and calculates the mobility μ by use of the Einstein relation[3] $D = \mu_i kT$, the value so obtained is about 1000 times smaller than the mobility calculated from the conductivity (328); (d) measured emf of a cell $Pt \mid S \parallel \alpha - Ag_2S \mid Ag$ is only 0.002–0.005 v instead of the 0.2 v calculated from the heat of formation and the emf calculated on the basis of ionic conduction (338).

---

[2] Maximum conductivity is obtained if the electrolyte is completely dissociated and the mobility has the theoretically maximum possible value. However, extraordinarily high ionic conductivity is found in crystal structures with defect lattices (Strock, 322) with the equivalent lattice points not completely filled or occupied by different particles (e.g., α-AgI or α-Ag_2HgI_4).

[3] D is the diffusion coefficient, μ_i the ionic mobility per unit force, k the Boltzmann constant, and T the absolute temperature. If the mobility is measured in cm/sec/volt/cm the equation $eD = \mu kT$ has to be used.

$$D \left[\frac{cm^2}{sec} \right] = \mu \left[\frac{cm^2}{volt\text{-}sec} \right] \cdot kT/e \text{ [volts]}$$

The clarification of this contradictory and confusing evidence is due to Wagner (338, 341), who concluded from theoretical investigations of the tarnishing reaction Ag | Ag$_2$S | S that the observed apparent fulfillment of Faraday's law is due to a secondary diffusion reaction and not to the action of the current (341). The fact that this compound exists in two modifications, α-Ag$_2$S, an electronic conductor stable at temperatures above 179°C, and β-Ag$_2$S, a mixed conductor with 20% electronic conductivity at the transformation point (327) but almost pure electrolytic conductivity at low temperatures, adds to the difficulty of interpretation.

This case has been discussed in detail because it shows the difficulties involved in the clarification of the conduction mechanism in semiconductors such as Ag$_2$S.

Conductivity

The electrical conductivity, σ, is equal to $ne\mu$ (109, 235, 313) where n is the number of carriers per cubic centimeter, e the electronic charge, and μ the mobility, which can be expressed as $\mu = (el/m^*v) \cdot F^4$. Because of the dependence on n and μ, both of which can be temperature dependent, interpretation of the conductivity is complicated. However, the behavior at the melting point can differentiate, at least qualitatively, between electronic and electrolytic conductors. Upon melting an increase in conductivity by orders of magnitude indicates electrolytic conductivity (167). In most electronic conductors the conductivity decreases upon melting.[5]

In the following discussion electronic conductivity will be considered. The conductivity depends on the form and purity of the material used. In many cases in which compressed powders, sintered materials or natural minerals are used (155, 335), the measured conductivity σ may be greatly influenced by transitional resistances, foreign layers, or barriers. Gudden (148) has particularly emphasized these points, and he and his collaborators (114, 165) have shown that reported deviations from Ohm's law and field dependence of conductivity are often due to such secondary causes (see also 107, 189). It is therefore necessary to use, whenever possible, materials in stoichiometric composition and in the form of single crystals. However, even without these precautions, it has been possible from resistance measurements alone to derive conclusions which were confirmed by later more careful investigations. Figure 1 represents measurements of the resistivity ρ of silicon (297, 314) as a function of temperature, indicating the com-

---

[4] Where l is the mean free path, m^* the effective mass, v the average thermal velocity of the carriers, and F a numerical factor determined by statistical considerations.

[5] Exceptions may be observed when the coordination number in the liquid is greater than the coordination number in the solid, as, for example, in germanium. [J. J. Keyes, Phys. Rev. **84**, 367 (1951); A. Epstein, Purdue Prog. Rept. Oct.–Dec. 1951.] In tellurium also the resistivity drops by more than a factor of 10 at the melting point. [A. Epstein, Purdue Prog. Rept., 1952.]

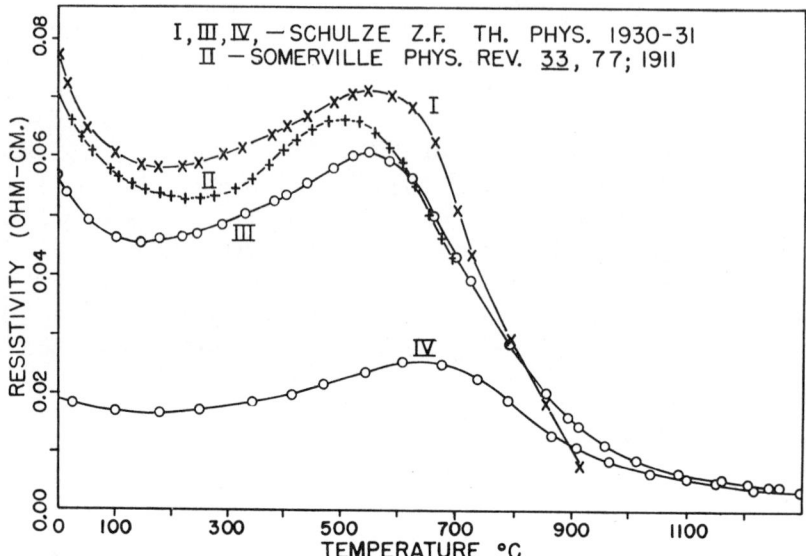

FIG. 1. Resistivities of various silicon samples plotted as functions of temperature. Note the large differences at both low and high temperatures.

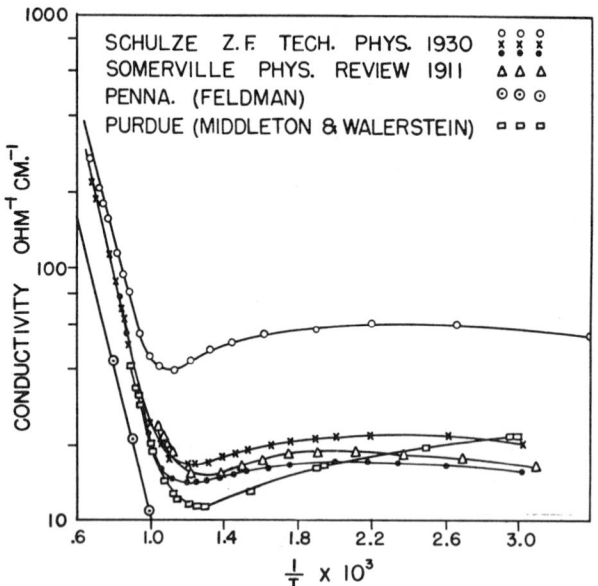

FIG. 2. The logarithms of conductivities of the same silicon samples (Fig. 1) plotted vs $1/T$. Observe the variety of slopes in the low-temperature region and the practically identical slope in the high-temperature region. Thus even the results obtained between 1911 and 1930 indicate intrinsic conductivity in silicon (17).

plicated behavior typical of semiconductors. The explanation given by Koenigsberger (15) still holds today: the number of carriers follows a temperature excitation law approximated by $n = n_0 e^{-E/kT}$ (n, the number of carriers per cubic centimeter; E, the dissociation energy; T, the absolute temperature; and k, Boltzmann's constant). The observed behavior of the resistivity curves is due to the combination of this temperature dependence of the number of carriers and the temperature dependence of mobility.

Unfortunately, Koenigsberger interpreted his results not on the basis of theory but by using a semi-empirical formula and therefore did not find the simple law governing the apparently complicated resistivity behavior. If one replots Fig. 1 as log $1/\rho$ (= log σ) vs $1/T$ (Fig. 2), it becomes apparent that there are two temperature regions to be distinguished (17).[6] In the high-temperature region the slope is identical for all silicon samples of widely different origin, but the low-temperature slopes vary from sample to sample, as does the position of the conductivity minimum.[7]

It is obvious that the high-temperature slope indicates a "dissociation energy" which is common to all samples (intrinsic) and depends only on the basic material, whereas the low-temperature slope is dependent on the type of preparation (impurity, crystal size, tempering, etc.).

The shape of the log σ vs $1/T$ curves can be explained in the following way. At low temperatures carriers are released with a small "dissociation energy" determined by the preparation of the material. When all the available carriers with this small dissociation energy are released, the conductivity reaches a maximum. With further increase in temperature the conductivity decreases because of a decrease in mean free path, until at still higher temperatures it begins to increase again as new carriers are produced by dissociation characteristic of the lattice itself.

One can write an empirical equation that includes these two types of conductivity:

$$\sigma = A_1 e^{-E_1/kT} + A_2 e^{-E_2/kT}, E_1 \gg E_2 \qquad (1)$$

where E_1 represents the high-temperature or intrinsic dissociation energy and E_2 the low-temperature dissociation energy. A_1 and E_1 should have the same values for all samples of a given material, whereas A_2 and E_2 vary from sample to sample.[8] ($A_{1,2}$ are functions of temperature.)

---

[6] Measurements from the Purdue group and the University of Pennsylvania have been added to those of Fig. 1.

[7] Far more accurate measurements on silicon have been obtained by the Bell Telephone Laboratories group (266).

[8] The alkali halides also show electrolytic conduction and it has been shown by Lehfeldt (229) that one has to distinguish between electrolytic conductivity due to the material itself and that due to the crystal imperfections and impurities. The conductivity curves thus obtained indicate a behavior similar to the conductivity curves which are obtained in electronic semiconductors, a high-temperature intrinsic part and a low-temperature part that is structure sensitive.

Streintz (321) observed that the negative temperature coefficient of resistivity is greater for high-resistivity materials. The connection between the conductivity and activation energy of impurity semiconductors was discussed in detail by Meyer and Neldel (242). For semiconductors with $\sigma_{20°C} < 10^{-2}$ ohm^{-1}cm^{-1} they find $\log A_2 = \alpha + \beta E_2$; the activation energy increases for low-conductivity semiconductors with increasing concentration of impurity centers. But for $\sigma_{20°C} > 10^{-2}$ ohm^{-1}cm^{-1} E_2 becomes very small and E_2 decreases with increasing number of impurity centers. Various attempts have been made to explain this behavior (85, 139), but it is generally conceded that the interaction of impurity centers necessary to understand this behavior needs further theoretical study (266; 273; 32, Mott, p. 1; Castellan and Seitz, p. 8; 33, Henisch, p. 41).

A more complete understanding of conductivity requires a knowledge of the number of carriers as a function of temperature. This is provided by measurement of the Hall effect.

Hall Effect

Under the action of a magnetic field charged carriers are deflected at right angles to the direction of the current and the field. Thus a transverse field (161) is developed such that the electric field just compensates for the action of the magnetic field and the carriers flow uniformly through the sample as if no field were present. By formulating the condition that the total transverse force (Lorentz force) on the carriers disappears, one derives, for one type of carrier only, that the Hall coefficient R, the transverse electric field developed per unit current density and unit magnetic field, is given by

$$R = \pm \frac{r}{ne} \qquad (2)[9]$$

In this formula $\pm$ indicates the presence of either negative carriers (electrons), or positive carriers (holes),[10] with the sign of R being that of the carrier. Conductors from here on will be referred to as N type or P type, depending upon whether the carriers are negative or positive, respectively.

The factor r depends on the condition under which the Hall effect is measured. It depends upon whether the mean free path of electrons in the material is dependent on energy. It is customary to treat a semiconductor as a classical electron gas, for which $r = 3\pi/8$, and thus $R = \pm(7.37 \times 10^{18})/n$ cm^3/coulomb. However, this holds only for a mean free path independent of energy. For other cases the factor r is different; for a Fermi gas the factor is 1 (metals, also degenerate semiconductors). In the case of

---

[9] n is the number of carriers per cubic centimeter and r a numerical factor discussed later under *Conductivity and Hall Effect*.

[10] Electron semiconductors are called N type, or excess, semiconductors; hole conductors are called P type, or defect, semiconductors (see 17, 25, 28).

scattering due to charged impurity centers, the mean free path is proportional to the square of the energy and r is equal to $315\pi/512 = 1.93$[11].

In the derivation one must assume that $(el/mv) \cdot H \ll 1$ or that the radius of curvature in the magnetic field H is much larger than the mean free path l; el/mv is of the order and dimensions of a mobility, and the condition may not be fulfilled at very low temperatures or in strong magnetic fields. One also must assume that there is no temperature gradient $(dT/dx = dT/dy = 0)$ to obtain the so-called isothermal Hall effect R_i (137). If one admits the existence of a temperature gradient, one obtains the so-called adiabatic Hall effect R_a (137, 313). Recent calculations[12] indicate that, in material with a high thermal conductivity, such as germanium, silicon, and even tellurium, the fractional difference $(R_a - R_i)/R_i$ is of the order of 0.1% or less and can therefore be neglected.[13] The reason that Koenigsberger came to the conclusion that this difference is large, by orders of magnitude, because the fact is overlooked that, in semiconductors, only a fraction of the thermal conductivity is of electronic origin.

Before the development of the wave mechanical treatment (164, 268) the fact that Hall coefficients of either sign exist caused considerable difficulty, and for this reason the Hall effect has not been used as frequently as it might have been. The first systematic investigation of Hall effect and conductivity in a semiconductor of controllable properties was carried out by Baedeker and Steinberg (43, 318), who investigated both conductivity and Hall effect in one and the same sample of cuprous iodide (CuI).

Since the conductivity σ equals $ne\mu$, it is clear that the product $R\sigma$ gives a direct measure of the mobility μ and also makes it possible to estimate the mean free path in the semiconductor. Mobility values, expressed in (centimeters per second)/(volts per centimeter), range from $\mu < 1$ to $\mu > 3000$[14] (113).

In the case of the silver sulfide mentioned above, the Hall coefficient for the alpha form existing above 179°C is equal to 0.6, whereas below the transition point, the beta form has an R value of 20.1 (204). This indicates that the electron density changes at the transition point by more than 30 times. Semiconductor Hall coefficients are much greater[15] than the ones observed in metals, for which the number of carriers is of the order of the

---

[11] Bardeen (266) and Schottky (private communication) arrived at the same conclusion independently.

[12] Johnson and F. Shipley, *Phys. Rev.* **85,** 724 (1952), F. Shipley Thesis, Purdue University, 1952.

[13] Older observations (210) are not substantiated by modern experiment.

[14] In some cases of abnormally small values of μ one is dealing with an apparent mobility due to trapping (172).

[15] In 1912 in discussing Baedeker's (44) "artificial metallic conductors," this suggested to Ebert the possibility of using the Hall effect in semiconductors to measure magnetic fields.

number of atoms per cubic centimeter. Since the conductivity depends on e^2, it is not possible to derive the sign of the carrier from conductivity measurements alone. In electronic conductors, Hall effect and thermoelectric behavior distinguish between the two signs of the carrier.

By observing both Hall effect and conductivity it is possible to distinguish between changes due to mean free path or mobility variation and changes due to the carrier density variation.

If carriers of both signs are present[16] (n_1 electrons per cubic centimeter and n_2 holes per cubic centimeter), the formula for the Hall coefficient can be derived in a fashion similar to that for one carrier; the result (99, 128, 223, 269) is

$$R = \frac{-r(\mu_1^2 n_1 - \mu_2^2 n_2)}{e(\mu_1 n_1 + \mu_2 n_2)^2} \qquad (3)$$

This formula shows that under the condition $\mu_1^2 n_1 = \mu_2^2 n_2$ the Hall effect is zero.[17] It also shows that the Hall coefficient changes sign if one of the two carriers has a higher mobility than the other one, and if the high mobility carrier is relatively scarce at low temperatures but becomes comparable in numbers with the other type of carrier at high temperatures (17, 296).

Systematic investigations of semiconductors by study of not only the Hall effect and conductivity, but also of thermoelectric power and optical properties, have been carried out by Koenigsberger and his collaborators (15). It was found that both the Hall effect and thermoelectric power are far larger for semiconductors than for metals and that both have the same sign and go through zero at approximately the same temperature if there is a sign reversal.

Most of the early investigations were handicapped, as already pointed out by Baedeker (42), primarily by the fact that materials used were of dubious purity or contained impurities which were not accurately known. For this reason the pioneer work of Baedeker on CuI was of great importance. By

---

[16] Setting $dn_1/dy = dn_2/dy = 0$, the condition of stationary flow gives for the Hall field F_H, in the Y direction:

$$n_1\mu_1 (F_H + \mu_1 F_x H_z) = n_2\mu_2 (-F_H + \mu_2 F_x H_z)$$

or

$$F_H = \frac{n_2\mu_2^2 - n_1\mu_1^2}{n_1\mu_1 + n_2\mu_2} F_x H_z$$

$$= \frac{n_2\mu_2^2 - n_1\mu_1^2}{n_1\mu_1 + n_2\mu_2} \frac{i_x H_z}{\sigma}$$

$$R = \frac{F_H}{i_x H_z} = \frac{-(n_1\mu_1^2 - n_2\mu_2^2)}{e(n_1\mu_1 + n_2\mu_2)^2}$$

This is the formula (3) except for the factor r, which depends on the statistics (99, 223)

[17] Thus a very small Hall coefficient at one particular temperature may not indicate a large number of carriers; an R vs T curve is necessary to establish that fact.

adding iodine in various amounts he was able to vary the conductivity over a wide range, and by measuring Hall effect, thermoelectric power, and conductivity as functions of iodine content, he established for the first time the fact that deviation from stoichiometric behavior can markedly influence the electrical behavior (44; see also 236 and 33, p. 147).

Koenigsberger already realized from resistivity, Hall effect, and thermoelectric measurements that the elements silicon, selenium, and tellurium, as well as a number of oxides and sulfides, have to be classified as semiconductors (15). The metals titanium and zirconium, listed by him as semiconductors, obviously showed anomalous behavior due to impurities or grain boundary effects. X-ray investigations showed that the electrical behavior of semiconductors is not due to phase transformations as Koenigsberger assumed for silicon and Haken (160) for tellurium (261).

Thermoelectric Effects

The emf developed in a circuit containing two heterogeneous metallic conductors when the junctions are held at different temperatures (299) is called the thermoelectric emf. Seebeck noted that significantly larger effects were produced when tellurium replaced one of the metals, and that even greater effects occurred with the use of some of the sulfides and oxides in mineral form. He pointed out (299, p. 306) that the elements can be arranged in a "thermoelectric series" with bismuth at the negative extreme and tellurium at the positive. He found that only a sample of galena (PbS) was numerically higher than bismuth in the thermoelectric series and that the place of Cu_2S is near tellurium. Seebeck also realized that the small number of effective thermoelectric minerals lie either at one end or the other of the thermoelectric series (one would say now that they are either N type or P type), and finally he stressed that, when using mineral samples, their homogeneity and purity should be considered. For a summary of thermoelectric phenomena observed in minerals see (29, 34). It is clear from the figures quoted in Wiedemann's (29) summary that the thermal emf in tellurium and selenium (with respect to silver) is of an order of magnitude higher than for the combination Bi-Ag and 100 times greater than for the combinations Ag-Cu, Ag-Pb. All these observations agree with modern results.

The thermoelectric behavior of a semiconductor was first systematically investigated by Baedeker (44), who found that there is correlation among the thermoelectric power, resistivity, and Hall coefficient of a semiconductor. In general he established qualitatively that the smaller the number of carriers in the semiconductor, the larger are the resistivity, the thermoelectric power, and the Hall effect. He gave a formula for the differential thermoelectric power Q, per degree temperature difference, of a thermocouple composed of semiconductors of the same basic material but of various conductivities, or carrier concentrations N_1 and N_2, as equal to

$$Q = \frac{R}{F} \ln\left(\frac{N_1}{N_2}\right) \qquad (4)$$

where R is the gas constant and F the Faraday constant (see also 317, 340). Baedeker's experiments on cuprous iodide indicated good agreement between theory and experiment.

The thermoelectric power, $Q = dE_t/dT$, is defined so that the signs of the Hall coefficient, R, and of Q are the same as the sign of the carriers in the impurity range (one carrier only). The thermoelectric power Q of semiconductors can be approximated by the formula usually quoted in texts:

$$Q = \frac{\mp E}{2eT} \quad (6, \text{ p. } 245) \qquad (5)$$

where E is the activation energy of the carriers $\mp$, and the choice of sign depends on the sign of the carriers. (It must be emphasized that this formula does not hold for low temperatures; also, while the order of magnitude of the observed values agrees with prediction from theory, the observed temperature dependence does not follow, in general, the expected one (24, 37, 81, 101, 174, 211).

If both kinds of carriers are present, the expression for thermoelectric power is much more complicated (see below, *Measurement and Theory of Thermoelectric Power in Germanium Semiconductors*) and just like the Hall coefficient, it passes through zero if the sign of the dominant carrier changes. Such observations have been made for cuprous oxide, lead sulfide, and other materials. In PbS (171), Tl$_2$S (172), and SnS (36) the change of the sign of thermoelectric power has been used to show transition from P to N type (315).

Contact Rectification

The unidirectional conductivity of metal-semiconductor contacts was first observed by Braun (75a) at contacts of metals with sulfides and oxides and by Schuster (298) with copper vs tarnished copper (probably copper oxide) contacts. Braun was able to show that this is neither an electrolytic process, nor a thermoelectric effect, and it was with great intuition that he described the process (75c) by saying that the more he studied these effects the more apt seemed the analogy to the behavior of gas discharges.

When, some sixty years later, Schottky (290) introduced his contemporaries to an understanding of the space charge layer in the barrier between metal and semiconductor, he again used the same analogy as Braun, that of a discharge in gases.

Rectifiers. The observations of Braun and Schuster found renewed interest and applications in the early days of radiotelegraphy, (68, 70, 76, 272, 324, 326), when it was found that the point contact rectifier could be used as a detector of radio waves.

In the early years of radiotelegraphy a great many semiconductor-metal combinations were tried as detectors. These investigations turned out to be rather inconclusive. The action of the detector was not completely understood and, with the mass production of radio tubes, interest in crystal detectors decreased. However, theoretical work continued (284) and Schottky (288) came to the conclusion that the lack of inertia in the action of the detector suggests an electronic process, i.e., a kind of "cold emission."

The interest in contact rectifiers revived with the discovery of the copper-cuprous oxide rectifier (7, 145, 146). This discovery stimulated the investigation of bulk cuprous oxide semiconductors and led to the discovery that this material is a P-type semiconductor and that its conductivity is greatly influenced by the oxygen pressure and heat treatment (38, 334), a fact which was later confirmed by thermoelectric power and Hall effect measurements (111a, 296).

The early theories (130, 258, 348) of contact rectification, based on "tunnel effect," are contradictory to Wagner's thermoelectric results and predict rectification in the wrong direction.

Wagner (337) was the first to state the rectifier equation as it is used today:

$$i = \text{const} \left\{ \exp\left(\frac{eV}{kT}\right) - 1 \right\} \quad (6)$$

where i is the current density, V is the voltage drop across the barrier, and the other letters have their usual significance. At room temperature $e/kT = \alpha$ is about 40 volts^{-1}.

The modern theory of contact rectification is due to Schottky (290) and Mott (251). Because of the difference in work function between the metal and the semiconductor, when they are brought together, there is a transient flow of electrons across the contact to establish equilibrium. A space charge region (barrier layer) is thus produced in the semiconductor (positive if electrons flow from an N type semiconductor into the metal), which gives rise to rectification. There are various types of rectification theories depending upon whether one is dealing with an artificial barrier, whether the field in the barrier is constant, or whether the thickness of the barrier is large or small, as compared to the mean free path. Mott assumed that the current is carried by thermally excited electrons which pass *over*, not *through* the potential barrier. He assumed a constant field intensity in the barrier (no space charge). On the other hand, Schottky took space charge into account and considered the collisions that occur in the barrier if it is thick in comparison with the mean free path (natural boundary layer, diffusion theory). If the boundary is thin in comparison with the mean free path, no collisions take place in the boundary (diode theory)[18] (61, 120, 166).

---

[18] Mean free path estimates in germanium and silicon (17) indicate that this condition is fulfilled for a metal-germanium contact of the "mixer-type" (28) and very nearly fulfilled for a metal-silicon contact.

With the development of microwave technique, point contact rectifiers again became important. PbS, which is an effective substance for detecting radio frequencies, seems to fail in the meter wavelength region, whereas silicon-metal combinations have been used down to wavelengths of the order of 1 mm (207). Since silicon is difficult to purify (28) and since its development was brought to high perfection by the British laboratories and the Bell Telephone Laboratories (28), the groups at Sperry (28) and Purdue (17) started the development of the germanium rectifier based on the use of relatively pure material with properties controlled by the addition of known amounts of impurity.

The rectifying action of the germanium-metal contact had been known for some time (55, 240), but only the development of the germanium diode and the high back voltage rectifier, together with the development of reproducibly uniform material, made its application of interest to the electrical industry.

Experimental rectifier characteristics usually show deviations from the theoretical curves[19] (10, 28, 196, 290b, 351). Contact rectification on N type germanium seems to be independent of the work function of the metal (17, 319). Germanium-germanium contacts exhibit the back direction characteristic, indicating that surface states produce a space charge layer at the surface of each piece (45, 57).

The germanium diode (17, 28, 58) has found a large number of industrial applications and was the starting point for the development of the transistor.

Photoconductivity

In 1873, W. Smith (311) discovered that illumination produces an increase in the electrical conductivity of selenium.[20] This observation was followed in 1885 by Bidwell's (62) discovery of the photoconductivity of silver sulfide and in 1887 by Arrhenius' (40) discovery of the photosensitivity of the silver halides. In 1904 Joffe (12) and Roentgen (277) found that rock salt, colored by irradiation with x-rays or radium rays, becomes photoconducting by visible light. This discovery became later, through the classical investigations of Pohl and collaborators (8, 9, 22), of the utmost importance in the development of an understanding of electronic conductivity in solids. Photoconductivity in silver iodide was carefully investigated by Scholl (287) in 1905 and Wilson (349) in 1907. In 1910 Rudert (279) investigated the photoconductivity of CuI. Volmer (336) discovered in 1915 photoconductivity in iodine, the mercury halides, and cinnabar (HgS).

---

[19] Some P-N boundary rectifiers seem to follow the ideal rectifier equation with $\alpha = e/kT$ (25, 126).

[20] The results of investigations in such substances as selenium can still be characterized by what Siemens wrote about them in 1876: "effects which are extremely complicated and unpredictable."

In 1917 Case (88) made a search for photoconducting properties in a large number of minerals and found photoconductivity in various substances, including Ag_2S (argentite), MoS_2 (molybdenite), and Cu_2O (cuprite). His most important contribution was the discovery of high photosensitivity in thallous sulfide which was oxidized and which he called thalofide (89). The fact that the oxygen-treated thallous sulfide became highly photosensitive made thalofide the prototype of a number of photoconductors such as PbS, PbSe, and PbTe which have been of increasing theoretical and practical interest in recent years because of their sensitivity in the infrared. Thalofide has been investigated in recent years by Cashman (90) and particularly by v. Hippel and collaborators (172). Photosensitization with oxygen does not change the crystal structure,[21] and one may assume that oxygen, dissolved in the thallous sulfide, makes it a P type conductor. This is borne out by thermoelectric power measurements (172, 174).

Investigations of numerous sulfides (MoS_2, Ag_2S, Bi_2S_3, and more complicated compounds) were carried out by Coblentz (96, 97) and his collaborators at the Bureau of Standards, but these investigations, like most of the early studies, were handicapped by the fact that the material was not in the form of uniform single crystals. Coblentz found that the photoconductivity is greatest in the wavelength region in which the absorption increases sharply. It is interesting that in MoS_2 no effect was found for wavelengths beyond 2.5 μ, whereas the optical absorption from 1.5 μ out seems to be constant. This may be an absorption by free carriers (8), whereas the sharply increasing absorption below 1 μ may be due to the main absorption band of the lattice releasing photoconductive carriers. In high-conductivity samples of MoS_2, photoconductivity is observable only at low temperatures (with increasing dark resistance). Coblentz recognized that high thermal emf, rectification, and photoconductivity are found together.

The photoconductivity of cuprous oxide (Cu_2O) was discovered by Pfund (271) in 1916. This investigation is of particular interest, since the material was not a natural crystal, but Cu_2O plates were produced by heating purest Cu in air at 900° C, thus starting the modern technique of artificially produced photoconductors (for single crystal growth of Cu_2O see 238). In Cu_2O the maximum in photoconductivity occurs at 0.63 μ (rise of absorption band). There is also another maximum found at 0.80 μ, due perhaps to impurities (286). Photoconductivity is found also from 1.7 to 4 μ (~ 0.3 ev). This photoconductivity is connected with the excess oxygen taken up.[21a] Hall effect and conductivity measurements by Engelhard (115) for the dark

---

[21] Complete oxidation leads to Tl_2SO_2 an insulator. This is quite similar to the formation of PbO · $PbSO_4$ (lanarkite), also an insulator on complete oxidation of PbS.

[21a] Recent investigations of the absorption in Cu_2O in the infrared indicate electronic transitions from local impurity levels with energies between .08 and .15 ev. [Gross, E. F., and Karryef, N. A., *Doklady Akad. Nauk S.S.S.R.* **84**, 261–264 (1952).]

current and the photocurrent show the same mobility for thermally excited photoelectrons.

Gudden and Pohl (150, 154), guided by Fajans' ideas about the distortion of electron shells and polarizability, have shown that photoconductivity might be expected to occur in substances (elements or compounds) which have a high refractive index ($n > 2$). Guided by this principle,[22] investigations of a number of compounds revealed the existence of photoconductivity where it was formerly unknown. By using single crystals of insulators (diamond, zincblende, and cinnabar) the basic principles of photoconductivity were clarified. To avoid complications due to time effects, hysteresis, and other secondary phenomena, Gudden and Pohl (152, 153) used short periods of illumination and weakest intensities. Thus they were able to study the creation of current carriers by illumination as a primary phenomenon. The photocurrent started and disappeared upon illumination and darkening without inertia and without the hysteresis usually observed in photoconductors due to secondary effects of the illumination on the dark current carriers. The primary photocurrent is proportional to the absorbed light intensity (for a definite wavelength) (see also 163). With increasing applied field a saturation value of the current is reached; the total possible saturation value (in regions of small optical absorption)[23] is equal to $i = e(I/h\nu)$, where I is the light energy absorbed per second; h is Planck's constant; ν is the frequency of the absorbed light; and e is the electronic charge. From their observations Gudden and Pohl deduced the photoelectric origin of the conductivity produced in crystals by illumination. They were able to determine h from the slope of the i vs λ curve. To produce a photocurrent only a section of the crystal has to be illuminated; therefore negative carriers must migrate from the illuminated part to the anode, or "defects" migrate to the cathode. The Hall effect, discovered by Gudden and Pohl (153) in diamond, and later investigated by Lenz (234), showed definitely that in diamond electrons are the carriers, and their number as released by light was found to be temperature independent down to the temperature of liquid air.

Studies of the optical and electrical behavior of the alkali halides (22) have established many relations which are now known to be characteristic of semiconductors in general. It is possible by optical and electrical measurements to establish the difference between effects which are due to the pure alkali halides themselves (intrinsic), and, on the other hand, the effects which are due to lattice defects or color centers which can be produced in the

---

[22] See also T. S. Moss, Photoconductivity of the Elements, *Proc. Phys. Soc.*, **A64,** 590 (1951); J. G. N. Braithwaite, Infra-red Photoconductivity of Certain Valence Intermetallic Compounds, *Proc. Phys. Soc.* B64, 274 (1951).

[23] In regions of strong optical absorption the saturation value of the current is only a fraction of the calculated one.

material either by chemical means (exposure to metal vapor or by irradiation with electrons, x-rays, or nuclear particles.[24] (For the effect of excess halogen see reference 245.)

The crystal imperfections which produce coloration in the alkali halides, the so-called F centers, electrons trapped at an anion vacancy (65, 253, 289), may also be excited thermally, and therefore the thermal ionization of these centers can be studied. In KCl, to which excess K atoms were added (by heating in K vapor, about one excess K atom for 10^5 atoms), measurements of the thermally released electron current show (316) that each F center releases thermally one electron.

The carriers may be "trapped" by impurities, lattice defects, etc., and are then no longer effective for current flow; they remain in these traps until thermal or optical excitation provides the activation energy necessary for their release. If the average distance traveled before being trapped is w, then in insulators $n(x) = n_0 \exp(-x/w)$, where n_0 is the number of carriers released at the point of illumination and $n(x)$ is the number of carriers at a distance x away from the origin; w, the average range ("Schubweg"), is proportional to the field strength applied; w is about 10^{-8} cm in NaCl for a field of 1 v/cm and about 10^{16} F centers per cm^3; w depends on the imperfections, for example F centers available in which electrons may be trapped. Glaser (140, 141) has shown that, depending on N, the number of color centers, w per unit electric field in NaCl can vary by a factor of 100 at a given temperature in such a way that Nw is about constant.[25]

In a semiconductor for each photoelectrically produced electron reaching the anode, an electron drifts into place from the cathode, and this takes place during the whole decay time of the state excited by absorption of light. Therefore, owing to the absorption of one light quantum in a semiconductor, many electrons can flow through the crystal, giving multiplication (168, 274, 320a) of the photoelectric current. The secondary photocurrent phenomena are found to be slow, because they depend on a long lifetime of the excited state. This explanation of the secondary photocurrent makes it also clear that it depends on the type of electron traps and donors and their activation energies; it is strongly temperature dependent and can lead to the complex behavior, well known in impurity semiconductors.

Among the substances which have been of particular interest in recent

---

[24] In KCl single crystals the maximum number of F centers obtained by additive coloration is $\sim 3 \times 10^{-4}$. However, recently [R. Kaiser, Z. Physik **132**, 482 (1952)] by condensing thin films of KCl + K at low temperatures it was possible to obtain as high as 1.1% F centers; also aggregates and colloids were observed. In Ag halides, also condensed at low temperatures, [W. Kaiser, Z. Physik **132**, 497 (1952)] it was found that unoccupied defects can shift the absorption edge toward longer wavelength [an effect somewhat anticiptated by F. Seitz, Rev. Mod. Phys. **23**, 328 (1951)]. By simultaneously condensing the halide with Ag, Cu, Au, new absorptions are found.

[25] Assuming quantum yield η = constant; in general $\eta w t$ = constant.

years is CdS, which originally was investigated as greenockite, but has now been grown in single crystal form in the laboratory (132). Some luminescent CdS crystals show a remarkable "multiplication" of the electron current released by β-ray bombardment [26] by factors of the order of 10^5. The processes observed in insulating and semiconducting crystals showing photoconductivity are basic to the understanding of the crystal counter (175, 202, 320a).

Photovoltaic Cells

In 1839, Becquerel (54) discovered that, if one of two electrodes dipping into an electrolyte is illuminated, a potential difference between them is observed. This was the first electrolytic photoelement, and it was some years before further progress was made.

The first photoelement of the barrier layer type was the selenium cell produced by Adams and Day (35) in 1876. In this all electronic cell a photovoltaic emf, produced by light, was observed; and Siemens, in repeating the later experiments of Fritts (134), quite rightly recognized that this is "a completely new physical phenomenon of great scientific importance; the direct transformation of light energy into electrical energy" (307).

These cells of Fritts were also good rectifiers since he found a ratio of current for the two different directions of current flow of 1 to 200, a value comparable to that of modern dry-disk rectifiers. Also the method of illumination through a goldleaf electrode resembled the modern technique. Among the numerous interesting investigations carried out during this period were those of Kalischer (201), who was able by heat treatment to produce light sensitive cells at will and who with a telephone observed photocurrents produced by chopped light. (See also 239.)

The first observations made in single crystals, rather than in crystalline layers, were those of Bose (68), who used galena (PbS) and other detector minerals in contact with a metal.[27] This crystal photoeffect was rediscovered by Coblentz and collaborators (96).

Not until the photovoltaic effect was rediscovered in the Cu-Cu_2O rectifier by Grondahl and Geiger (146) did photoelements enter modern practice. Soon afterwards, Lange produced photoelements yielding currents of the order of milliamperes on exposure to sunlight (16, 215). In 1922 Sheldon and Geiger (304) discovered a photo-emf in argentite (Ag_2S). The whole problem became of increasing interest when Schottky, from theoretical considerations of unidirectional conduction in rectifiers (295). showed that a

---

[26] Lenz (234) has previously shown that electron bombardment increases the conductivity of diamond and has found by Hall effect measurement (234) that electrons are released. Recently electron bombardment conductivity has become of interest in connection with electronic designs (39, 205, 276).

[27] For recent investigations see A. F. Gibson, *Proc. Phys. Soc.* **B65**, 196 and 214 (1952).

photo-emf should be developed at a barrier layer. Numerous other investigations followed. Those of Schottky and the later investigations of Mott (252) have led to an understanding of the photovoltaic effect as it is observed in a barrier photocell. A large amount of work has also been carried out by investigators under A. Joffe (187).

Photovoltaic effects are of particular interest as observed at P-N boundaries in semiconductors, e.g., germanium detectors (58) or germanium P-N boundaries (56, 126) or in photovoltaic barriers in PbS layers (315).

Optical Properties

As in the problems of photoconductivity and their relation to lattice defects, the optical investigation of ionic crystals by Pohl and his collaborators (19, 22) opened the way for the understanding of the relation between optical and thermal activation energy and anticipated in many details results obtained later with semiconducting materials. (For a general discussion of optical problems in solid state physics see 24, p. 629 ff.)

As early as 1903, Koenigsberger (208) investigated the relation between the reflectivity and absorptivity at long wavelengths and the electrical conductivity of oxides and sulfides. His investigation showed that the smaller the electrical conductivity the larger is the ratio of the experimentally found absorption for long wavelengths to the theoretical value. However, if the conductivity is high, then the observed infrared absorption is smaller than the value calculated from the conductivity (15a).

Although the reflectivity of metals increases with increasing wavelength, the reflectivity for most of the semiconductors investigated remains approximately constant.

The greatest difficulty in interpreting these results is due to the fact that because of the strong dependence of conductivity of semiconductors on impurity content, it is not possible to draw conclusions regarding the optical properties, except when the same material and the same specimen is used for both the electrical and the optical investigations (15, 43). Since minerals, as found in nature, were often used in the early investigations, the experimental difficulties due to the small dimensions of the specimen were considerable. Sb_2S_3 (100, 208), FeS_2 (as marcasite and pyrite), MoS_2, Fe_2O_3, and PbS were investigated (208). The tentative absorption curve obtained for Sb_2S_3 was quite similar in form to the absorption curves found later for germanium and silicon.[28] From reflectivity measurements Koenigsberger and others deduced the refractive indices and the dielectric constants of various semiconductors (208). In recent years the photoconductive lead compounds have been studied extensively. Absorptivity vs wavelength curves for most semiconductors indicate a strong absorption band near the visible. A photon energy of 1.66 ev corresponds to a cutoff at 0.74 μ (100)

---

[28] See also Pohl and collaborators for absorption of alkali halides (22).

as observed for Sb_2S_3. In MoS_2 the absorption edge occurs at 1.0 μ, in Ag_2S at 1.3 μ (96). In the photoconductive lead compounds the cutoff of the main absorption is observed in PbS at 0.8 μ, in PbSe at 1.0 μ, and in PbTe at 1.5 μ. However, there is a tail band in these absorption spectra extending to 6 μ (138). These lead compounds have similar absorption spectra, but, with increasing atomic weight of the electronegative component, the curve shifts toward longer wavelengths (138).[29] The effect of temperature on the band edge, the maximum, and the tail band has been studied in these compounds.

GENERAL SURVEY OF SEMICONDUCTORS

A large number of substances, mostly oxides and sulfides, are known as semiconductors. Because of the great sensitivity to lattice imperfections and their effects on the electrical behavior of semiconductors, the preparation and the physical state even of the chemically pure semiconductors may influence their electrical behavior.

Semiconductors may be used in the form of single crystals of definite orientation as in the case of germanium, silicon, tellurium (see below, *Elementary Semiconductors, Particularly Germanium*), lead sulfide,[30] and lead telluride,[30] zinc oxide (135, 244; also 32, p. 172), rutile (TiO_2),[31] carborundum (84), and barium oxide (Sproull *et al.*, 32, p. 122).[32] In this form the semiconductor should approach the ideal behavior, but interstitials and lattice vacancies, as produced for example by thermal treatment, may change the electrical properties. Some semiconductors are used in the form of sintered materials (309) or ceramic bodies (Verwey, 32, p. 157). In this case the grain boundaries play a decisive role in determining the electrical behavior.

In many cases semiconducting films are used which may show a completely different behavior from the bulk material. The most important examples are the semiconducting and photoconducting films of PbS, PbSe, and PbTe which have been used widely in recent years as infrared detectors (171, 315).

---

[29] The recent discovery of the relatively high transparency in certain single crystals of PbS beyond 3 μ [W. Paul, D. A. Jones, and R. V. Jones, *Proc. Phys. Soc.* **B64**, 258 (1951)] has led to a more detailed investigation of the absorption spectra of PbSe and PbTe crystals grown in the laboratory [W. D. Lawson, *J. Appl. Phys.* **22**, 1444 (1951)]. Secondary absorption edges in PbS and PbTe are in agreement with photoconductivity limits. However, in PbSe the absorption edge occurs at 4.7 μ at room temperature and at 6.7 μ at liquid air temperature. This was confirmed by measuring the photoconductivity of a PbSe photodiode (point metal contact) [A. F. Gibson, W. D. Lawson, T. S. Moss, *Proc. Phys. Soc.* **A64**, 1054 (1951)]. See also M. A. Clark and R. J. Cashman, *Phys. Rev.* **85**, 1043 (1952).

[30] E. H. Putley, *Proc. Phys. Soc.* **B65**, 388 (1952).

[31] D. C. Cronemeyer, *Phys. Rev.* **87**, 876 (1952).

[32] E. M. Pell, *Phys. Rev.* **87**, 457 (1952).

In many such cases ordinary DC conductivity measurements are misleading and special methods have to be used. Measurements at higher frequency may give results differing from the DC pattern (91^{33}; see also Volger, 32, p. 166). For the investigation of powders the method of dielectric losses of the powder embedded in an insulator has been used (155, 335). Conductivity has also been measured by observing the lowering of the Q factor of a coil filled with semiconducting powder (Busch et al., 32, p. 188).

Depending on whether the conductivity is due to electrons or holes, we are talking of N or P type semiconductors. If the conductivity, depending on the type of impurity, can change from N to P or vice versa, following the chemical analogy used by Schottky (292), we may call these semiconductors amphoteric.

Conductivity, independent of impurities, methods of preparation, and identical at the same temperature for all samples, is the intrinsic conductivity corresponding to pure material. Mott and Gurney make this cautious statement: "Although intrinsic semiconductors are theoretically possible, it is doubtful whether any are known, though it has been claimed that cuprous oxide is an intrinsic semiconductor at high temperature" (19, p. 152).

It has been definitely established now that the elementary semiconductors Ge (17, 25), Si (17, 25), and Te (69, 243, 283)[34] at high temperature are intrinsic semiconductors.

We believe that intrinsic conductivity also exists in Cu_2O (199) and in PbS and PbTe (Chasmar et al., 32, p. 208; also Putley, loc. cit.) in spite of some discrepancies in the behavior of the high-temperature conductivity and Hall effect. There are indications that BaO single crystals may be intrinsic at high temperatures (Pell, loc. cit.). Similar claims have also been made for single crystals of rutile (Cronemeyer, loc. cit.).

An analysis of electronically conducting materials (133) was made in 1944 by Meyer (241). This led to a table of semiconducting substances which were brought up to 1950 (3). We have added some new results and are listing the various types.

SEMICONDUCTING SUBSTANCES

N Type

TiO_2	Nb_2O_5	CdS	Cs_2Se	(Au_2O_3)	Hg_2S	ZnF_2	$BaTiO_3$
V_2O_5	MoO_3 (3)	CdSe	BaO	(UO_3)	(red)	(101)	(3)
(CuO)	CdO	SnO_2	Ta_2O_5	Fe_2O_3	Hg_2S	Al_2O_3	$PbCrO_4$
ZnO	Ag_2S	Cs_2S	WO_3	U_3O_8	(black)	(162)	(142)
					Ti_2O_3	Cu_2O_3	Bi_2Se_3
					(ScN)	(SnSe)	(UO_3)

---

[33] See E. S. Rittner and F. Grace, Phys. Rev. **86**, 955 (1952).

[34] See also Fukuroi et al., Sci. Rept. RITUA, Vol. 1, p. 371, 1949; P. R. Aigrain, Compt. rend. **235**, 145 (1952).

P Type

Ag$_2$O	CoO	Cu$_2$O	(Cu$_2$Te)	SnS (36)	(Tl$_2$O)	Bi$_2$Te$_3$	MoO$_2$
Cr$_2$O$_3$	(Co$_3$O$_4$)	Cu$_2$S	(GeO)	Sb$_2$S$_3$	(Bi$_2$O$_3$)	Te (69)[a]	Hg$_2$O
MnO	NiO	(Cu$_2$Se)	SnO	CuI (236)	(Bi$_2$S$_3$)	Se	Pr$_2$O$_3$[b]

Amphoteric Semiconductors

SiC	(RuO$_2$)	(PbO)	PbTe	Mn$_3$O$_4$	U$_3$O$_4$	Si (17, 25)	Sn (86)
(Cr$_5$O$_9$)	(OsS$_2$)	PbS	UO$_2$	Co$_3$O$_4$	Tl$_2$S	Ge	(Mg$_2$Sn)
(Mn$_2$O$_3$)	IrO$_2$	PbSe		[c]	(172, 174)	(17, 25)	(67)

[a] See also Fukuroi *et al.*, Sci. Rept. *RITUA*, Vol. 1, p. 371, 1949.
[b] R. L. Martin, *Nature* 165, 202 (1950).
[c] For the group of partially substituted oxidic semiconductors see Verwey *et al.* (332, 333): valency-controlled semiconductors.

ELEMENTARY SEMICONDUCTORS, PARTICULARLY GERMANIUM

INTRODUCTION

The discussion up to this point has described the historical development of our knowledge of the phenomena characteristic of semiconductors. Most of these phenomena have been studied in detail in germanium and silicon (17, 25), either in bulk samples or at contacts with a metal, or at contacts between N and P type layers of material.

The following discussion concentrates upon the behavior of germanium semiconductors in bulk form and their electrical properties as influenced by chemical impurities (substitutional imperfections) and by lattice defects (interstitials and vacancies) produced either by heat treatment or by nucleon bombardment; the optical properties are also treated as functions of these imperfections.

The wave mechanical theory of electrical conduction in semiconductors, developed by Wilson (347) and others (19, 24, 25, 26a, 79) through use of the energy band model (323, 247), takes into account the fact, first pointed out by Gudden (148), that impurities and lattice imperfections producing localized energy states may be largely responsible for the electrical behavior at room temperature and below. It is possible to summarize these results by reference to the energy band picture (Fig. 3). In general there will be bands of energy of allowed states alternating with bands of forbidden states. In a crystal with N cells a band can accommodate $2N$ electrons. If all the states are filled, these full bands cannot contribute to the conductivity. A conduction electron must be able to go to nonlocalized empty states. In a metal this is always possible (Fig. 3C) since the conduction band is only partially filled; therefore such states exist. If the full band is followed by an empty band, then, depending on the width of the forbidden band separating the full and empty bands, an insulator or a semiconductor exists.

According to this picture, insulators and semiconductors differ only in the activation energy necessary to produce free carriers (Fig. 3A,B,D). Electrons in pure insulators have activation energies so great that at ordinary temper-

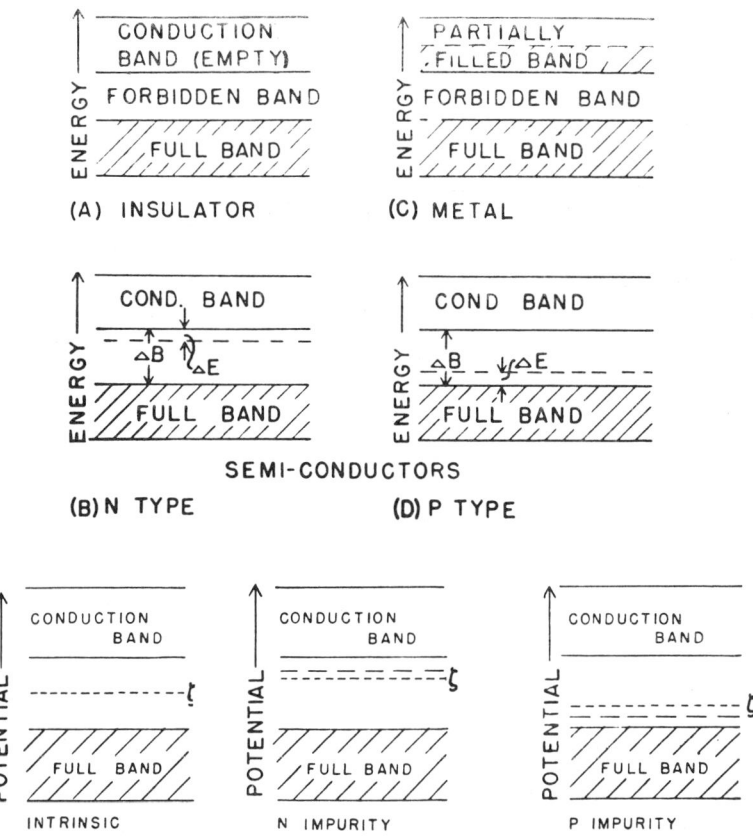

(E) POSITION OF ELECTRON CHEMICAL POTENTIAL IN SEMICONDUCTOR

Fig. 3. Energy level diagrams. In an insulator (A) the width of the forbidden band is so great (1 ev), that there is no appreciable transition of electrons from the full to the empty band at ordinary temperatures. A metal (C) is characterized by a partially filled band. The forbidden band width in a semiconductor (0.1-1.5 ev) is less than in an insulator and hence the former shows observable intrinsic conductivity at high temperatures. In the neighborhood of room temperature the conduction of a semiconductor is due to additional energy levels introduced by impurities or lattice defects. The impurity level in an N type semiconductor (B) is below the empty band with an activation energy ΔE comparable with kT; hence electrons are thermally excited to the conduction band. The P type semiconductor (D) has acceptor levels above the full band so that thermal excitation of electrons into these states produces holes in the full band.

The electron chemical potential (Fermi level) is about in the middle of the forbidden band in an intrinsic semiconductor (E); deviation from the center is due to difference between the effective masses of electrons and holes. The impurity levels and Fermi levels shown for impurity semiconductors are representative of germanium at 300°K with 10^{15} carriers per cm³. The position of the impurity level depends upon the amount and kind of impurity; the Fermi level depends on impurity content and temperature.

atures it is impossible to produce a detectable number of free carriers and therefore no electronic conduction is observed. On the other hand, in semiconductors this energy is smaller, and hence appreciable numbers of free carriers are produced by thermal excitation in the ordinary temperature range. If electrons are lifted from the full band into the conduction band, they leave empty spaces, "holes," in the distribution of the full band. Under the influence of a field, electrons within the band can now move into these empty spaces, or the holes seem to move in the opposite direction like a positive charge with a certain effective mass. Two cases must be considered in semiconductors: the activation energy involved may be the width of the "forbidden" band, determined by the structure and binding in the lattice itself (intrinsic semiconductor), or it may be the dissociation energy of an impurity or a lattice imperfection (impurity semiconductor).

It has been shown by Fowler (128) that it is possible to give a simple discussion of the pertinent facts based on the law of mass action and the dissociation equilibrium between bound and free electrons (343; see also 266, 292).

Equilibrium in an intrinsic semiconductor involves the transition:

Bound electron (full band) $\rightleftarrows$
Free electron (conduction band) + Free hole (full band)

Applying the law of mass action to holes and electrons one has (128)

$$n_1 n_2 = 32 h^{-6} (m_1^* m_2^*)^{3/2} (\pi k T)^3 e^{-\Delta B/kT} \qquad (7)$$

where n_1 and n_2 are electron and hole densities, respectively, m_1^* and m_2^* are electron and hole effective masses, and ΔB is the width of the energy gap between the full and conduction bands (the E_1 of *Conductivity* section). In true intrinsic conduction, $n_1 = n_2$, and thus

$$(n_1 n_2)^{1/2} = n' = A T^{3/2} e^{-\Delta B/2kT} \qquad (8)$$

where A is a constant depending on the material only through m_1^* and m_2^*. If there are either chemical impurities or lattice imperfections present, then, depending on type of imperfection, new energy levels (in the forbidden band) (182, 184, 310, 347) are created which can either give off electrons—donors— or take them up—acceptors.

The equilibrium condition in an impurity semiconductor with negligible intrinsic conduction corresponds to the transition:

Bound electron (donor level) $\rightleftarrows$
Free electron (conduction band) + Bound hole (donor level)

or

Bound electron (full band) $\rightleftarrows$
Bound electron (acceptor level) + Free hole (full band)

According to classical statistics the equation for either of the above conditions becomes

$$\frac{n^2}{N-n} = 2(2\pi mkT)^{3/2} h^{-3} e^{-\Delta E/kT} \qquad (9)^{34a}$$

where n is the carrier density, N the density of bound levels available, and ΔE the dissociation energy (the E_2 of *Conductivity* section).

Rigorously, the preceding treatment should be based upon the condition (180):

$$\begin{Bmatrix} \text{Density of free} \\ \text{electrons in} \\ \text{conduction band} \end{Bmatrix} + \begin{Bmatrix} \text{Density of} \\ \text{bound electrons} \\ \text{in acceptors} \end{Bmatrix} = \begin{Bmatrix} \text{Density of} \\ \text{bound holes} \\ \text{in donors} \end{Bmatrix} + \begin{Bmatrix} \text{Density of free} \\ \text{holes in full} \\ \text{band} \end{Bmatrix}$$

For the case of only N donor levels and n free electrons in the conduction band, one has:

$$N = \frac{N}{1 + \exp(-\Delta E - \zeta)/kT} + \frac{4\pi(2m)^{3/2}}{h^3} \int_0^\infty \frac{E^{1/2} dE}{1 + \exp(E-\zeta)/kT} \qquad (10)^{34a}$$

This reduced to Eq. (9) if ζ is less than about $-2kT$; with

$$\zeta = kT \ln\left(\frac{nh^3}{2(2\pi mkT)^{3/2}}\right) \qquad (11)$$

The energy parameter ζ in the Fermi distribution, also called the Fermi level, is the electron chemical potential and for the case of equilibrium has the same value in coexisting phases.

If $N \gg n$, Eq. (9) reduces to

$$n = N^{1/2} A' T^{3/4} e^{-\Delta E/2kT} \qquad (12)$$

The condition for this simplified equation is rarely satisfied in germanium, and thus calculations based upon the dissociation equation require use of the complete forms (9) or (10).

The carrier densities at room temperature to be expected in intrinsic semiconductors of various activation energies are represented in graphical form in Fig. 4. Upon setting $T = 300°K$ and $m_1^* = m_2^* =$ free electron mass, Eq. (8) becomes:

$$n' = 2.51 \times 10^{19} \exp(-\Delta B/0.0517); \Delta B \text{ in ev} \qquad (13)$$

One can see that the number of electrons at this temperature is small if the forbidden gap is of the order of one electron volt. For energy gaps which are only fractions of a volt, the electron density becomes larger and is in a range which can be observed easily. One sees that the preparation of intrinsic semiconductors i.e., highly pure semiconductors containing carrier densities due to lattice imperfections and chemical impurities that are negligible in comparison with the intrinsic electron density becomes a for-

---

[34a] It was not taken into account in equations (9) and (10) that the probability of occupation of "localized" and "non-localized" states will be different (19, p. 158; 25, p. 475; H. M. James Purdue Signal Corps Progress Rept., March 31, 1952).

midable chemical problem if the energy gap is greater than a few tenths of an electron volt.

In silicon the energy gap is 1.12 ev, and the number of intrinsic carriers at 300°K is about $10^{11}/cm^3$. This means that to prepare silicon having essentially intrinsic conduction at 300°K it is necessary (assuming one carrier per imperfection or impurity) to obtain material which contains not more than one lattice imperfection for about 10^{11} atoms.

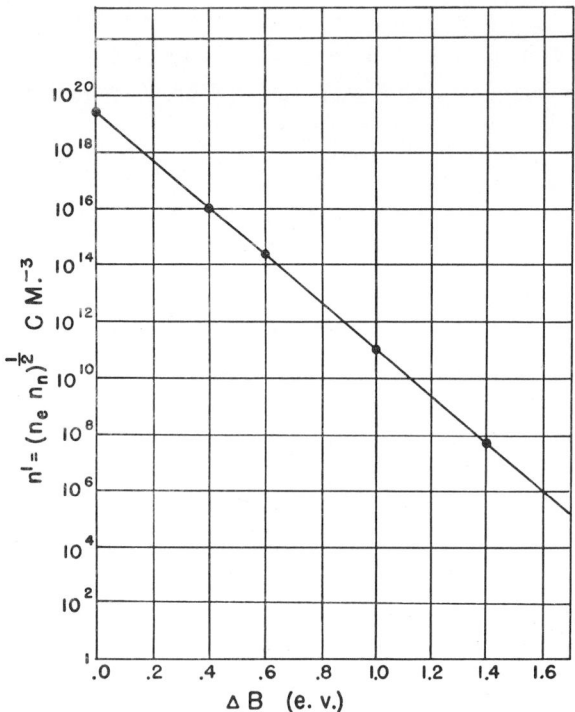

Fig. 4. Number of intrinsic carriers per cubic centimeter at 300°K as a function of the width of the forbidden energy band ΔB. Note that at 1 ev the carrier density is so small that it corresponds to a purification of the material from ionizable impurities to about one part in 10^{11}.

In germanium the energy gap is about 0.75 ev (17, 25, 222, 243, 320), and electrically pure material should contain less than one effective impurity in 10^8 atoms.

In tellurium with an energy gap of about 0.38 ev (14, 69, 243, 283), there must be not more than one effective impurity per 10^6 atoms.

In gray tin the energy gap is only 0.1 ev (86, 203) and therefore intrinsically conducting material contains not more than one effective impurity in about 10^4 atoms.

Electrical measurements are thus used as a very sensitive means of detecting ionizable impurities in semiconductors. In the high-resistivity intrinsic conductors like silicon and some oxides, one effective impurity atom in about 10^{11} lattice atoms can be detected. At low temperatures one detects even smaller impurity concentrations. Figure 5 shows measurements on a germanium sample in which one carrier per 10^{16} atoms is detected at 6°K. Generally, such purity can be checked only by electrical measurements and comparison, in some cases, with quantitative spectroscopic measurement (17) or radioactive indicator methods (267). Electrical measurements can only detect impurities contributing carriers and thus they cannot detect directly either neutral impurities or impurities of both signs compensating in their effects, except through their contribution to resistivity: scattering of current carriers.

It also becomes clear why the behavior of compounds is as complicated as was indicated above, since both deviation from stoichiometric composition and the addition of foreign atoms to the lattice may change the behavior of the semiconductor profoundly. For example, the thermo- and photo-electric behaviors of lead sulfide, lead selenide, and lead telluride are completely changed by heat treatment or the addition of oxygen (171; 32, p. 198; 315).

In the ideal semiconductor, i.e., one prepared so that (a) there are no chemical impurities in the lattice, (b) the sample is a single crystal, and (c) lattice imperfections (vacancies or interstitials) are absent, it should be possible to obtain a log resistivity vs reciprocal temperature curve which is linear throughout the whole temperature range with a slope characteristic for the intrinsic semiconductor.

Such ideal conditions cannot be fulfilled since there always are lattice imperfections. Bottom (69) was able to purify tellurium by successive distillations until single crystals were obtained of such uniformity that from room temperature up all the resistivity curves were identical. This material had imperfections to the extent of one in about 10^7 atoms. However, it was not possible to obtain Hall coefficients greater than 5000 cm^3/coulomb or impurity concentrations smaller than 10^{15}/cm^3. This may well be the number of imperfections frozen in during solidification.[35]

In the case of germanium the purification problem is more difficult. By starting with the purest obtainable germanium oxide [36] (it is important that arsenic be absent in this preparation), one obtains germanium powder by reducing it in hydrogen at about 650°C. The powder is melted in

---

[35] For a detailed discussion of the importance of defects see later sections on *Conductivity* and *Hall Effect* and on *Production of Lattice Defects and Their Influence on Electrical Properties*.

[36] This purification was worked out, starting in 1942, by the Eagle Picher Company Research Laboratories in collaboration with R. M. Whaley of the Purdue semiconductor group. Attempts to purify germanium were also carried out at Sperry (Woodyard). All commercial purifications follow essentially the same procedure (17, 28, 278, 325, 346).

vacuum or in helium, and, by properly cooling it or seeding the melt with a single oriented crystal, it is possible to obtain single crystals of germanium with resistivities approaching 60 ohm-cm at room temperature, the theo-

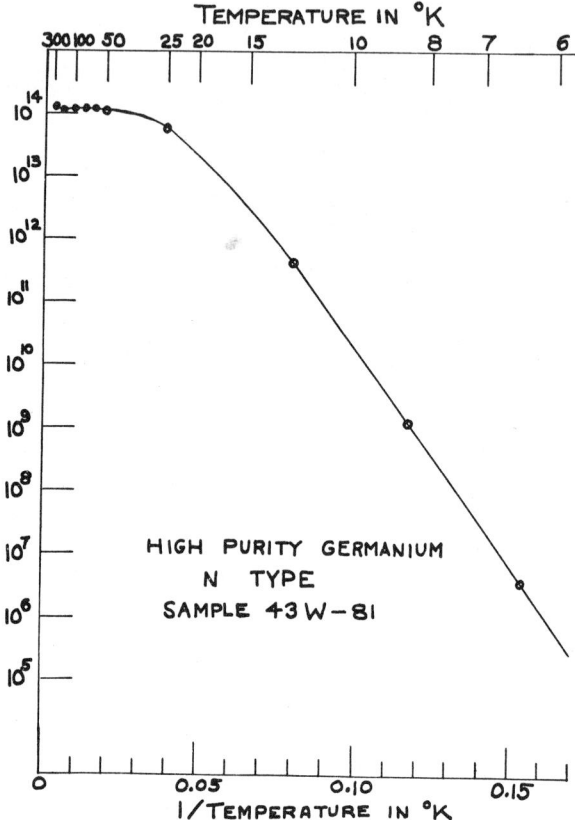

Fig. 5. The carrier density n (per cubic centimeter) of a high purity N type germanium sample as found from Hall effect measurements between approximately 6°K and about 300°K. The activation energy for this sample at low temperature has been calculated to be about 0.025 ev. (Measurements by Hung and Gliessmann.)

retically computed value of the room temperature intrinsic resistivity of germanium (278).[37]

In the case of silicon, the problem is indeed formidable, since silicon has a much higher melting point and has a tendency to attack the material in

---

[37] Recently introduced procedures (zone heating, combined with recrystallization technique and crystal growing with orientated seeds) have produced germanium of such purity that only at low temperatures can impurities be detected (B.T.L., G.E., R.C.A. and other laboratories).

which it is melted. As a consequence, material obtained so far has not been of much higher resistivity than that obtained for germanium, a value far below the theoretical value of about 2.4×10^5 ohm-cm expected at 300°K

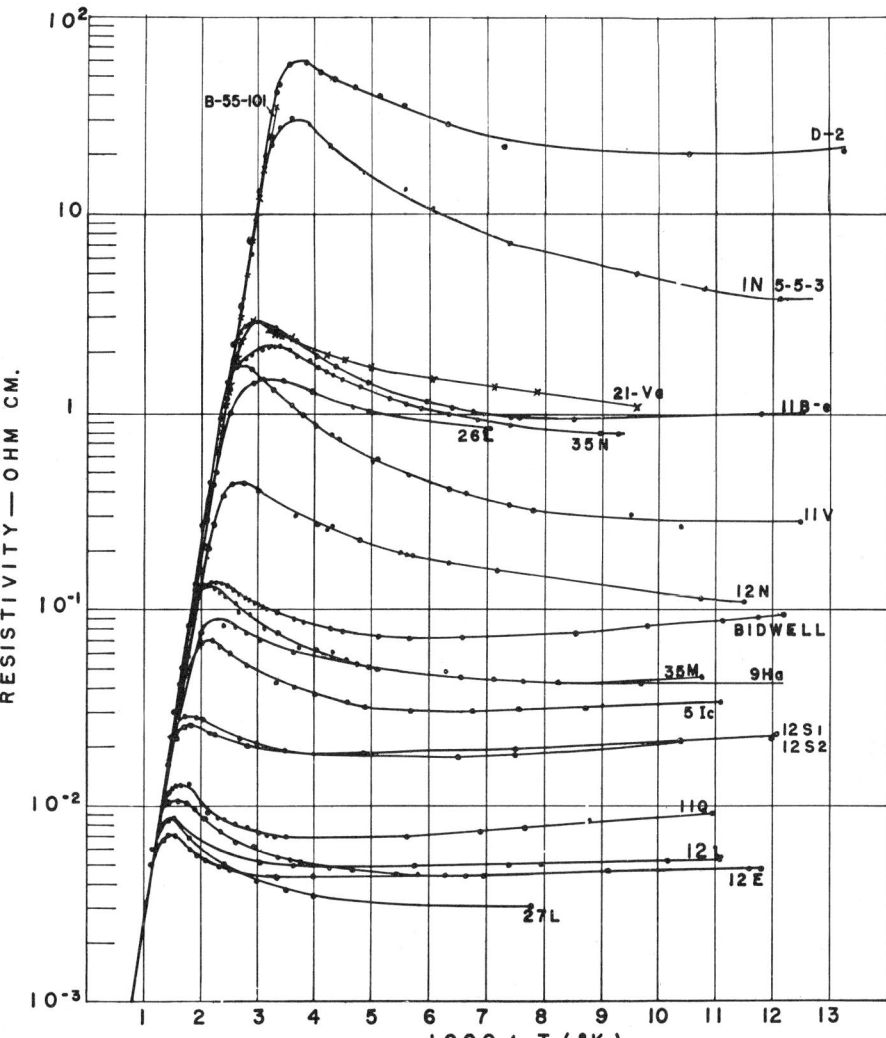

Fig. 6. Resistivity vs reciprocal of absolute temperature for typical germanium samples. Note the appearance of maxima between 300° and 600°K, the approach of all samples to a common intrinsic curve at high temperatures, and that the low resistivity curves show minima around 125°–200°K followed by resistivity increase as temperature decreases. (Measurements by Benko, Middleton, Miller, Scanlon, and Walerstein. Note that Bidwell's (63) sample fits well the character of these resistivity curves.)

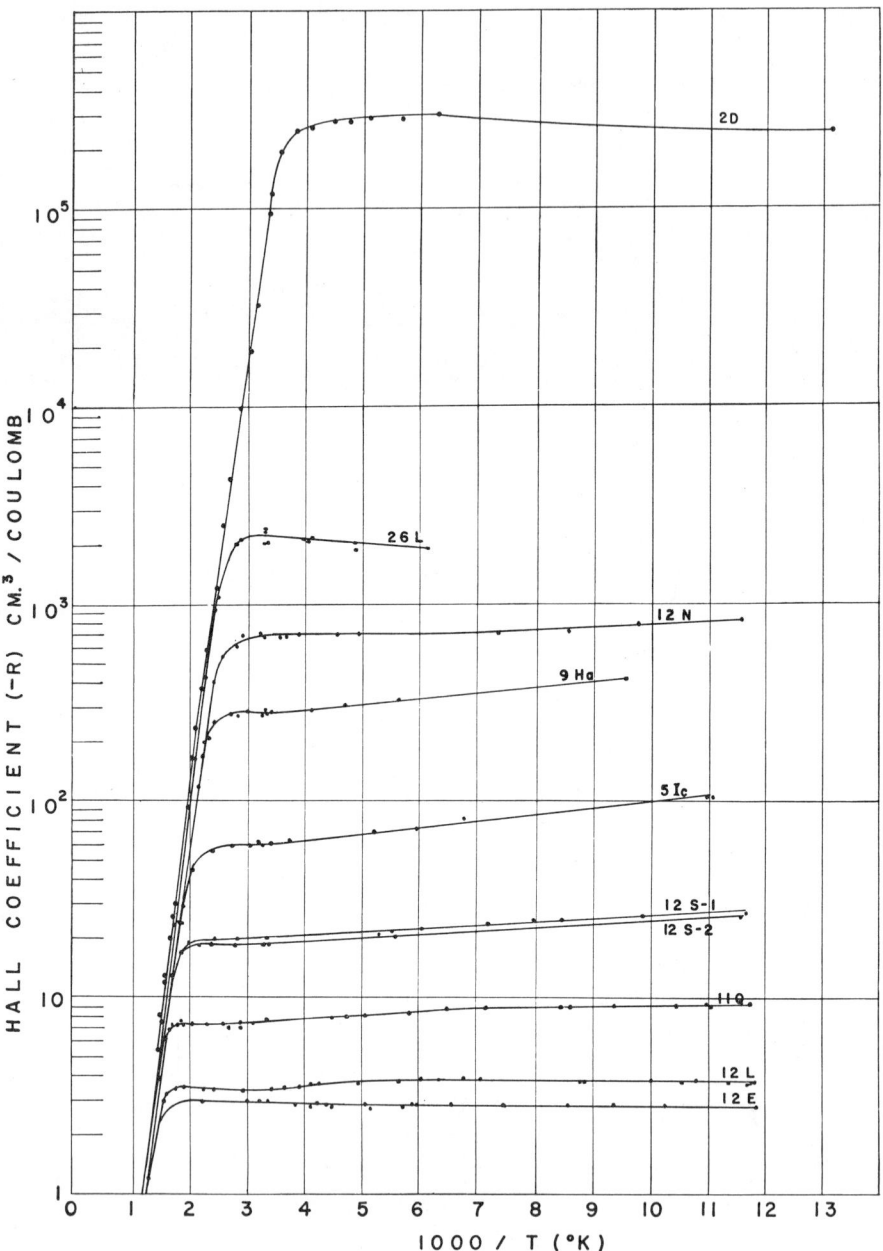

Fig. 7. Hall coefficient vs reciprocal of absolute temperature for typical N type germanium samples. Note the small low temperature slopes indicative of small activation energies of the impurities, the flat exhaustion region around 200°K–300°K, and the approach of all samples to a common intrinsic curve at high temperatures. (Measurements by Benko, Middleton, Miller, Scanlon, Walerstein.)

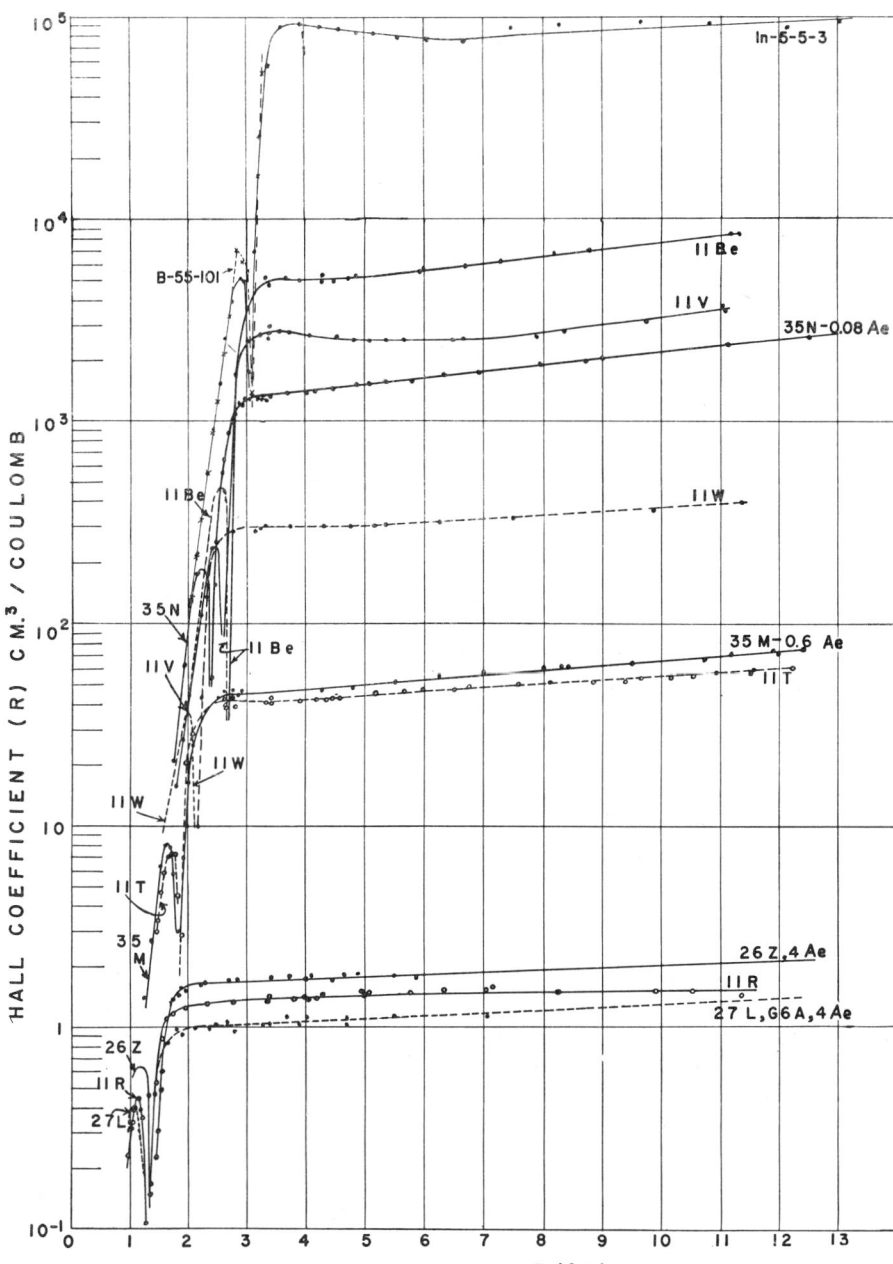

FIG. 8. Hall coefficient vs reciprocal of absolute temperature for typical P type germanium samples. The break in each curve represents its passage through zero. Note the small activation energies, the appearance of an exhaustion range, and the trend toward a common intrinsic curve after passing through the maxima. (Measurements by Middleton, Miller, Scanlon, Walerstein, and on two high resistivity single crystals by Benko.)

for pure silicon.[38] The metallurgical and chemical problem of silicon purification has been successfully attacked by the Bell Telephone Laboratories, the British groups (B.T.H. and G.E., Wembley) and Dupont. (For details see 28, 282.)[39]

As temperature rises, the intrinsic carrier concentrations increase exponentially and eventually mask the conductivity due to impurities and lattice imperfections. Such high-temperature intrinsic conduction has been experimentally observed in the elementary semiconductors silicon (17, 266), germanium (17, 46, 320), tin (86, 203), and tellurium (69, 283). (See Figs. 6, 7, and 8 and also Table 2, p. 92).

In the elementary semiconductors silicon and germanium, it is possible to produce N type conductivity by adding a small amount of any element from the fifth column of the periodic system, i.e., phosphorus, arsenic, and antimony. The addition of small quantities of any element from the third column of the periodic system, i.e., boron, aluminum, indium, and gallium, produces states which accept electrons and thus produce hole conduction (abnormal extrinsic semiconductor in Fowler's terminology or, briefly, P type) (17, 25, 282).

It is possible to prove that each impurity atom releases one carrier, and one only, in two different ways: (1) by adding radioactive impurities and counting the number of atoms taken up by the lattice and comparing this with the number of carriers present as determined from the Hall effect (267); (2) by activating the material in the nuclear reactor and producing by transmutation in the lattice a number of impurity centers which may be computed from the known flux of slow neutrons and the activation cross section of the isotope, and comparing this number with the number of carriers released as found from Hall effect measurements (93, 216). With both methods one finds, within the limits of experimental error, that the number of carriers is equal to the number of atoms added.

Conductivity and Hall Effect[40]

Figure 6 illustrates the temperature dependence of the resistivity of germanium impurity semiconductors. The curves of conductivity as a func-

---

[38] It has been possible, by deuteron bombardment with the Purdue cyclotron and neutron irradiation at Oak Ridge, to prepare P and N type samples of about 2×10^5 ohm-cm resistivity at 300°K, which means the free carriers have been removed, but the sample is now more disordered than before irradiation.

[39] Recently large single crystals of high purity silicon have been produced [G. K. Teal and E. Buehler, *Phys. Rev.* **87**, 190 (1952)]. V. Wartenberg succeeded in preparing purest Si single crystals of microscopic dimensions, transmitting light in the visible red [*Z. anorg. Chem,* **265**, 186 (1951)].

[40] This section and the next on thermoelectric power are based on experimental data by Middleton (232), Scanlon (283) and E. Benko, and reports by Johnson and Lark-Horovitz (14, 17).

tion of temperature indicate that at low temperature the conductivity depends on impurities added, but the curves are identical (intrinsic) at high temperature for all samples, whether N type or P type. Figures 7 and 8 show plots of Hall coefficient R vs $1/T$ for both N and P type samples. A close study of the Hall effect curves as functions of temperature indicates that four regions should be distinguished: the low-temperature region with a rather small slope of log R vs $1/T$ due to carriers released from the impurity centers; a flat region (exhaustion region) in which all impurities are ionized and the number of carriers is constant; a high-temperature transition region with a rather steep slope due to the release of intrinsic carriers (both electrons and holes); and, finally, at very high temperatures, the intrinsic region characterized by the common curve approached by all samples.

As temperature rises, the Hall coefficient of a P type sample goes from positive values through zero and becomes negative. This indicates that the mobility of the holes must be smaller than the mobility of the electrons. In both silicon and germanium this seems to be the case, and closer analysis also shows that the mobility ratio μ_1/μ_2 remains the same from sample to sample of the basic semiconductor and indicates but small temperature dependence.

The scattering of carriers due to lattice vibrations, as calculated by Bethe and Wilson (27, 31), leads to a mobility proportional to $T^{-3/2}$. Since the resistivity of an impurity type semiconductor is inversely proportional to carrier concentration and mobility, on the foregoing basis it follows that resistivity in the impurity range should follow the formula

$$\rho = DRT^{3/2} \quad (14)$$

where D is a constant of proportionality characteristic of the material (17).

However, as may be seen from Fig. 6, the observed resistivity curves usually do not fall off, with temperature dropping below the maximum position, in the manner predicted by Eq. (14). The equation is followed for a short temperature interval below the maximum and then the resistivity curve flattens and even rises in a number of cases. This behavior indicates that there must be, in addition to scattering by lattice vibrations, some other scattering mechanism effective in the impurity range. Furthermore, it follows that the new mechanism must lead to a mobility that decreases with dropping temperature, in contrast to the temperature behavior of the mobility due to lattice scattering.

Scattering by charged impurity ions was suggested by Lark-Horovitz as leading to the required mobility behavior. A formula for resistivity due to impurity scattering has been derived by Conwell and Weisskopf (98).

$$\rho_I = \frac{9 \times 10^{11} \, \pi^{3/2} e^2 m^{1/2} \ln\left(1 + \frac{36 \, \varepsilon^2_d k^2 T^2 d^2}{e^4}\right)}{2^{7/2} \, \varepsilon^2_d \, (kT)^{3/2}} \text{ ohm-cm} \quad (15)$$

where d is half of the average distance between impurity ions, ε_d is the dielectric constant,[41] and the other symbols are as defined previously.[42]

Inspection of the two resistivity expressions indicates that, although the resistivity due to lattice scattering is proportional to $T^{3/2}$, the resistivity due to impurity scattering changes as a negative power of T.

Since impurity scattering causes the temperature dependence opposite from lattice scattering, one might be able to explain the nature of the observed resistivity as a combination of these two types of scattering (see 17, 25, 216, 266).

It is clear that these two types of scattering are not the only ones present. Frequently not all of the impurity atoms added become ionized. It often happens that large numbers of impurity atoms remain either on grain boundaries or dispersed through the lattice as neutral atoms and so do not contribute carriers, but do contribute to the scattering of carriers. Therefore, one must also consider grain boundary (190) and neutral impurity scattering (116, 266) in a complete study of resistivity.

It has been found, especially for germanium samples, that one can obtain a good representation of the observed resistivity by considering it as the sum of resistivities due to lattice scattering and impurity scattering. In a first approximation this certainly is correct, but Shockley (as quoted in 266) has pointed out that there is some deviation from additivity of the two resistivities since l_L, the lattice mean free path, is energy independent, but l_I, the impurity mean free path, depends on the square of energy.

Schottky (private communication) and Jones (197) have also pointed out that the equation $\rho = \rho_L + \rho_I$ is inconsistent with the relation $1/l = 1/l_L + 1/l_I$ and that actually $\rho > \rho_L + \rho_I$. Johnson and Lark-Horovitz have, therefore, modified their argument (194) and taken $\rho_L + \rho_I = F\rho$, where F, a quantity less than unity and a function of ρ_I/ρ_L, is determined in the manner now described. If one assumes the Rutherford scattering model for evaluating ρ_I, the resistivity when only impurity scattering is present, then l_I is proportional to the square of the energy ε, $l_I = \alpha\varepsilon^2$. Since l_L is independent of energy, the reciprocal addition of mean free paths yields a combined mean free path:

$$l = \frac{l_L l_I}{l_L + l_I} = \frac{\alpha l_L \varepsilon^2}{l_L + \alpha\varepsilon^2} \tag{16}$$

When this expression is put into the usual expression for resistivity of a semiconductor, one obtains

---

[41] From this analysis we determined the dielectric constant in germanium as 16 in agreement with the value obtained later from reflectivity (221).

[42] An equivalent but somewhat different treatment of impurity scattering has been given by Bardeen (266).

$$\frac{1}{\rho} = \frac{8\pi e^2}{3m^2} \frac{nl_L}{kT} \left(\frac{m}{2\pi kT}\right)^{3/2} \int_0^\infty \frac{\varepsilon^3 \, e^{-\varepsilon/kT} \, d\varepsilon}{\varepsilon^2 + \frac{l_L}{\alpha}} \quad (17)$$

This is simplified by setting $\varepsilon/kT = x$, and relating l_L to ρ_L through

$$l_L = \frac{3m^2}{8\pi e^2} \frac{1}{kT} \left(\frac{2\pi kT}{m}\right)^{3/2} \frac{1}{n\rho_L} \quad (18)$$

Define $b^2 = l_L/(\alpha k^2 T^2) = 6\rho_I/\rho_L$. These relations combine to give

$$\frac{\rho_L}{\rho} = \int_0^\infty \frac{x^3 e^{-x} dx}{x^2 + b^2} \quad (19)$$

and

$$F = \frac{\rho_L + \rho_I}{\rho} = \left(1 + \frac{b^2}{6}\right)\left(1 - b^2 \int_0^\infty \frac{xe^{-x} \, dx}{x^2 + b^2}\right) \quad (20)$$

The integral in this equation has been evaluated for a series of b^2 values (see Table 1). The result is that F is appreciably less than unity except at the two limits $\rho_I = 0$, and $\rho_L = 0$, and that F varies only slightly as ρ_I/ρ_L varies between ¼ and 4. This latter fact probably accounts for the result that simple adding of ρ_I and ρ_L expressions accounts for the observed resistivity behavior of most germanium samples in the impurity range.

TABLE 1

$\frac{\rho_I}{\rho_I + \rho_L}$	$\frac{\rho_I}{\rho_L}$	b^2	F	r
0.0	0	0	1.00000	1.1781
.1	1/9	2/3	.79565	1.0289
.2	1/4	3/2	.73941	1.0411
.3	3/7	18/7	.71309	1.0650
.4	2/3	4	.7030	1.0962
.5	1	6	.7047	1.1348
.6	3/2	9	.7175	1.1833
.7	7/3	14	.7427	1.2475
.8	4	24	.7849	1.3399
.9	9	54	.8560	1.4963
1.0			1.0000	1.9328

The relative importance of lattice scattering and impurity scattering depends upon the temperature and the carrier concentration. Figure 9 compares, at 300°K, the resistivity due to lattice scattering with that due to impurity scattering and shows how their ratio varies with carrier density.

The ρ_L curve is plotted for three different mobility values. It is obvious that impurity scattering must be considered at 300°K if the carrier density

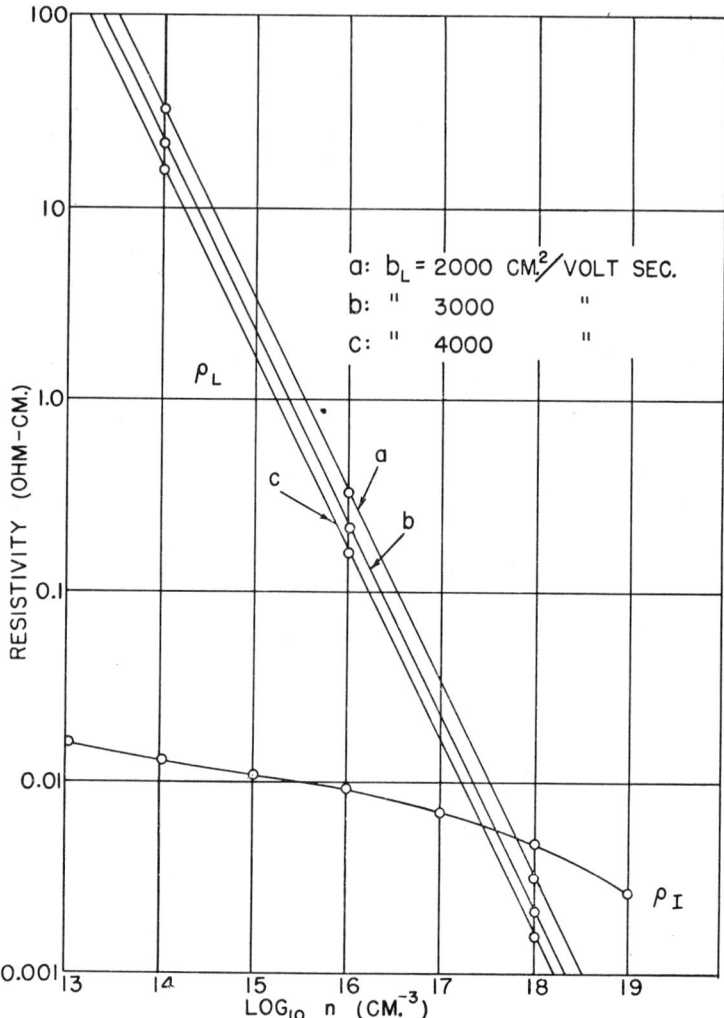

Fig. 9. Logarithmic plot of room temperature values of resistivity due to lattice scattering, ρ_L, calculated for various lattice mobilities, and resistivity due to impurity scattering, ρ_I, as functions of concentration of carriers per cubic centimeter. Note that, for the range of mobilities considered, impurity resistivity is greater than lattice resistivity when the carrier density approaches $10^{18}/cm^3$. Impurity resistivity cannot be neglected for carrier density as low as $10^{16}/cm^3$. At lower temperatures the ratio of the two types of resistivity is shifted in favor of the impurity resistivity.

n exceeds about $10^{16}/cm^3$. For a given n, ρ_I/ρ_L increases as T decreases and approaches zero as T increases.

The dependence of ρ_I upon energy influences the Hall coefficient as well as the resistivity. When only one kind of carrier is present and the magnetic field is relatively weak, the Hall coefficient is given by

$$|R| = \frac{3}{4\pi me} \frac{\int_0^\infty l^2 \frac{\partial f_0}{\partial \varepsilon} v^2 dv}{\left(\int_0^\infty vl \frac{\partial f_0}{\partial \varepsilon} v^2 dv\right)^2} \qquad (21)$$

Upon replacing l by $l_L l_I/(l_L + l_I)$, one gets

$$r = \frac{|R|}{(1/ne)} = \frac{\pi^{1/2}}{48} \frac{(b^2 + 6)^2}{F^2} \int_0^\infty \frac{x^{9/2} e^{-x} dx}{(x^2 + b^2)^2} \qquad (22)$$

The numerical quantity r (see Table 1) replaces the factor $3\pi/8$ in the usual Hall expression $R = 3\pi/(8ne)$.

At high temperatures, i.e., in the transition and intrinsic ranges, both electrons and holes are present in appreciable numbers. However, impurity scattering is negligible, and thus one can treat mobility as proportional to $T^{-3/2}$. The Hall coefficient and conductivity are given, respectively, by Eq. (3) and by

$$\sigma = n_1 e \mu_1 + n_2 e \mu_2 \qquad (23)$$

These equations may be solved for c, the ratio of electron to hole mobility, as follows (17):

N Type
$n_1 = n_2 + N$
$\mu_1 = B_1 T^{-3/2}$

P Type
$n_2 = n_1 + N$
$\mu_2 = B_2 T^{-3/2}$

$$1 - \frac{1}{c} = \frac{\frac{-8eR}{3\pi}\left(\frac{\sigma}{\mu_1 e}\right)^2 - N}{\frac{\sigma}{\mu_1 e} - N} \qquad c - 1 = \frac{N - \frac{8eR}{3\pi}\left(\frac{\sigma}{\mu_2 e}\right)^2}{\frac{\sigma}{\mu_2 e} - N} \qquad (24)^{42a}$$

In Eq. (24) N is the number of carriers at exhaustion and thus represents the contribution from impurity atoms; it is assumed that all ionizable impurities are ionized at exhaustion temperatures and so there will be no further increase from this source as temperature increases. The factors B_1 and B_2 are found from impurity range data. With c determined, one can calculate n_1 and n_2 by use of the conductivity equation.

The method described here in detail is only one of several for determining the mobility ratio c. Other methods include (1) the ratio of the maximum in the Hall curve,[43] when it reverses sign to the Hall value at exhaustion, and

---

[42a] If the temperature dependence of the mobility of holes is different from that of electrons, these simple formulas do not hold, and c is temperature dependent.

[43] R. M. Baum, M. S. Thesis, Purdue University, 1951; H. Fritzsche, *Science* **115**, 571 (1952).

(2) the measurement of drift mobilities (306)[44] for holes and electrons. The fact that the value of c for a given material depends to some extent upon whether it is calculated from high temperature or room temperature data suggests that c may have a slow variation with temperature.

The temperature dependence of n_1 and n_2 permits the calculation of ΔB and A in the dissociation Eq. (8). From these values one can estimate the intrinsic carrier density and resistivity at any desired temperature. Table 2 presents a summary of such calculations for the elementary semiconductors silicon, germanium, and tellurium; the mobility ratio values used in the table are based upon the use of Eq. (24).

TABLE 2

	Si	Ge	Te
c	3.0[a]	1.5	1.5[b]
μ_1 at 300°K cm²/volt-sec	290	3300	830
μ_2 at 300°K cm²/volt-sec	96	2200	540
ΔB (ev)	1.12	0.75	0.38
A (cm^{-3} deg$^{-3/2}$)	2.8×10^{16}	7.2×10^{15}	1.7×10^{15}
n' at 300°K (cm^{-3})	6.7×10^{10}	1.9×10^{13}	5.8×10^{15}
Intrinsic ρ at 300°K (ohm-cm)	2.4×10^5	60	1.0

[a] Direct measurements of the drift mobility in single crystals of silicon [J. R. Haynes and W. C. Westphal, *Phys. Rev.* **85**, 680 (1952)] give $\mu_1 \sim 1200$ and $\mu_2 \sim 250$.
[b] Calculated (V. A. Johnson) from the first reversal of Hall effect. Calculation from the Hall maximum (H. Fritzsche) leads to the value $c = 1.7$ with $\mu_1 = 920$. (See also P. R. Aigrain *et al.*, *Compt. rend.* **235**, 145–6 (1952); Fukuroi *et al*, (Sci. Rept. *RITUA*, Vol. 1, p. 371, 1949.)

The ΔB values given in Table 2 represent the width of the forbidden energy gap at 0°K. Analysis of optical data, electrical conductivity data, elastic constants, and thermoelectric power data indicates that this energy gap can be written as

$$\Delta B = B_0 + aT \quad (25)$$

The values of B_0 and a are presented in Table 3.

Mobilities and mean free paths determined from the combination of Hall effect and conductivity data may vary greatly from sample to sample because of variation in experimental conditions (25). This is particularly important because, as has been shown by the Bell Telephone Laboratories research group, it is possible to observe directly the drift mobility of carriers in a semiconductor (306). When these mobilities are compared with the ones determined from Hall effect and conductivity, it is found that the Hall mobilities are usually smaller and show more variation from one sample to another.

---

[44] See also J. R. Haynes and W. C. Westphal, *Phys. Rev.* **85**, 680 (1952).

The fact that impurity scattering introduces a mean free path which depends upon energy, as described earlier in this section, influences the calculation of mobility from conductivity and Hall data. It can be shown that the consideration of the ratios F and r leads to larger values of μ_L than are obtained without their consideration. Since $\mu_L = (ne\rho_L)^{-1}$ and $R = r/(ne)$ then

$$\mu_L = \frac{R}{r\rho_L} \quad (26)$$

The experimental resistivity curve of a sample is the ρ curve, and from its analysis one obtains two curves, which may be termed the ρ_L' and ρ_I'

TABLE 3

Element	B_0(ev)	a(ev/°K)	Type of Data	Ref.
Si	≈1.12	-5×10^{-4}	Optical	52
		-3.0×10^{-4}	Resistivity	266
		-3.58×10^{-4}	Elastic constants	124
Ge	≈0.75	-5.4×10^{-4}	Optical	52
		-1.06×10^{-4}	Elastic constants	124
		-1.1×10^{-4}	Resistivity	191
		-1.0×10^{-4}	Drift mobilities	47
		-4×10^{-4}	Thermoelectric power	a
		-1×10^{-4}	Optical	b
Te	≈0.36	$+2.0 \times 10^{-4}$	Optical	250
		$+1.8 \times 10^{-4}$	Resistivity	c

[a] Johnson and Lark-Horovitz, Purdue Prog. Rept., Jan. 1–Mar. 31, 1952, p. 22.
[b] Haynes and Briggs, *Phys. Rev.* **86**, 647A (1952).
[c] Fan and Johnson, Purdue Prog. Rept., Sept. 1–Nov. 30, 1950, p. 23.

curves. These curves are determined as the curves of appropriate temperature behavior that add arithmetically to give the ρ curve. Since $\rho_L + \rho_I = F\rho$, $\rho_L' + \rho_I' = \rho$, and since F is constant to within 4% as temperature change causes the ρ_I/ρ_L ratio to vary from $\frac{1}{4}$ to 4, it is usually a good approximation to take $\rho_L = F\rho_L'$, $\rho_I = F\rho_I'$, and $\rho_L/\rho_I = \rho_L'/\rho_I'$. Thus, in terms of ρ_L, one calculates the mobility as

$$\mu_L = \frac{R}{rF\rho_L'} \quad (27)$$

Since rF may be as low as 0.75, in comparison with the usual $3\pi/8 = 1.18$, μ_L values may be as much as 1.6 times greater than the values computed without considering the combination of lattice and impurity scattering.

μ_L was estimated in the manner just described for five single crystal specimens of N type germanium, and the results are indicated in Table 4.

TABLE 4

(All values at room temperature)[a]

Sample	R (cm³/coulomb)	ρ (ohm-cm)	$\mu = \dfrac{3\pi\,\rho}{8R}$ (cm²/volt-sec)	Corrected μ_L (cm²/v-sec)
43W-81	−59200	14.1	3570	3690
44B-1	−1050	0.292	3050	3960
44B-6	−1050	0.415	2150	2640
45J-3	−52	0.0368	1200	2190
43W-3	−43000	11.5	3170	3280

[a] Hall and resistivity values measured by Hung and Gliessman, Purdue Prog. Rept., Mar. 1, 1949–May 31, 1950.

The average value from the last column is

$$\mu_L\,(300°K) = 3150 \pm 600 \text{ cm}^2/\text{volt-sec}$$

for N type germanium. These mobility values are substantially higher than those found from $8R/(3\pi\rho)$, and the values for the relatively pure samples are in rather good agreement with the drift mobilities (306).

Baum and Hung (Baum Thesis, Purdue, 1951) have measured mobilities of germanium single crystal samples converted from N to P type by heat treatment. Their results are summarized in Table 5. It is interesting to note again the high values of mobilities close to drift mobility.

Calculations up to this point are based on the assumption that the electron gas in the semiconductor is nondegenerate and obeys classical Maxwell-Boltzmann statistics. This is equivalent to saying that the number of electrons in the conduction band is negligible compared to the available levels which they may occupy. However, degeneracy occurs at low tem-

TABLE 5

Sample	ρ (ohm-cm)	R (cm³/coulomb)	μ (cm²/v-sec.)	N/cm³
44B22				
Before	6.80			3.4×10^{14}
After	1.40	+3280	2000	2.2×10^{15}
44B3				
Before	3.60			6.4×10^{14}
After	1.20	+3600	2550	2.0×10^{15}
45J20				
Before	4.16	−17600	3600	4.2×10^{14}
After	1.68	+4830	2440	1.6×10^{15}

peratures in semiconductors having high carrier concentrations and very small values of the activation energy ΔE. Quantum, or Fermi-Dirac, statistics must be employed in calculations pertaining to such degenerate samples.

An approximate rule for the existence of degeneracy is the definition of degeneracy temperature

$$T_d = 4.20 \times 10^{-11}\, n^{2/3}\, {}^\circ\text{K} \tag{28}$$

below which quantum statistics must be used to avoid substantial error (193, 305). The Hall curve of a germanium sample which becomes degenerate at relatively high temperatures (100–150°K) is observed to be quite flat at low temperature.[45] The resistivity curve is also relatively flat. Germanium samples showing degenerate behavior were selected at Purdue for low temperature measurements, and measurements on them were made down through the liquid hydrogen and liquid helium ranges at Carnegie Institute of Technology (118), at the Naval Research Laboratory,[46] and since 1949 at Purdue (177, 179).

The observed resistivity has been explained (193) upon the following basis: (1) lattice scattering becomes negligible compared to impurity scattering as degeneracy is approached and reached; (2) the resistivity due to impurity scattering is calculated on the basis of Rutherford scattering of carriers by impurity ions with quantum statistics employed in the derivation. The resistivity calculated in this way agrees well with experiment.[47] The theoretical resistivity expression approaches a residual resistivity as a limit:

$$\rho_R = \frac{6270}{n^{1/3}}\ \text{ohm-cm} \tag{29}$$

where n is the constant low-temperature carrier density computed from the flat Hall curve.

Johnson and Shipley (Purdue Thesis, 1952) calculated the isothermal Hall coefficient for a degenerate semiconductor with impurity scattering only, using Fermi statistics. The quantity r defined by Eq. (22) drops

---

[45] Similar results were also obtained on silicon at Bell Telephone Laboratories (266).

[46] We are indebted to I. Estermann at Carnegie Institute of Technology and Mr. Ambrose of the Naval Research Laboratory for making such measurements for us.

[47] At high temperatures degeneracy is not reached, as has been shown by Johnson and Lark-Horovitz. The question of whether a degenerate sample may become nondegenerate again as absolute zero is approached (180) has been discussed in correspondence with Ehrenberg, Schottky, and ter Haar. In a detailed analysis Schottky has pointed out (April 23, 1952) that, if ΔE has the values required to give the flat Hall curves observed for the degenerate samples, it is not at all likely that the Fermi level near absolute zero will drop low enough to cause the sample to return to nondegeneracy. He has discussed the possibility of a band due to electron interaction.

from about 1.9 to 1.0 as the quantity ζ/kT changes from -2 to $+15$. (ζ represents the chemical potential or Fermi level.) It should be noted that the change of r must be employed in analyzing a Hall curve at low temperatures.[48]

Semiconductor theory predicts that in a nondegenerate sample the carrier concentration decreases exponentially with decreasing temperature corresponding to an increase in Hall effect. However, it was found (177) in many cases at low temperatures that the Hall effect goes through a maximum and in some cases decreases by a factor of more than 100 as the temperature is reduced further. The resistivity, however, approaches a saturation value. In the temperature region above that at which the anomaly occurs, experiment and semiconductor theory agree and the resistivity can be accounted for by the scattering from both acceptor and donor levels (178). Hung pointed out (179) that the anomaly in the resistivity and the Hall curves can be understood by assuming conductivity in an impurity band (185) with mobilities varying between 10^{-4} and 100 cm^2/v-sec as impurity concentration is increased. At low temperatures all the electrons move in the impurity band, but at room temperature all are assumed to be excited into the conduction band. While this theory gives good agreement with experiment, it is difficult to understand the formation of an impurity band due to interaction between impurity states at concentrations of about 10^{15}/cc.

Sometimes other low temperature anomalies, such as multiple reversal of Hall effect, are observed in samples with barriers or oxide layers. These anomalies usually disappear on annealing.

In tellurium an anomaly in the Hall effect has been known for some time. In spite of careful purification (69, 283) and the detailed investigation of its electrical properties, (69, 87, 158, 160, 285, 350), the behavior of tellurium is not yet completely understood.[49] The early results of Wold (350) on the anomalous Hall effect (double reversal: low-temperature P type, medium-temperature N type, high-temperature P type again) cannot be due to lack of purification, since Bottom found that two reversals occur in the Hall coefficient vs temperature curve only in pure samples containing less than 1.5×10^{17} carriers per cubic centimeter. This behavior is not understandable on the basis of the semiconductor model previously discussed. Scanlon and Middleton (243, 283) found that impure samples do not show any reversal of sign in Hall effect. But Bottom (69) has shown that, whereas the reversal of the Hall effect at 230°C seems to be characteristic of the material itself, the low-temperature reversal depends on the number of impurity centers in the sample. He found an empirical relation

---

[48] Finlayson, Johnson, and Shipley, *Phys. Rev.* **87**, 1141, (1952).

[49] Fritzsche has clarified the anomalous Hall effect behavior in tellurium by assuming a lattice defect concentration proportional to exp $(-W/kT)$ where $W \approx 0.70$ ev [*Science* **115**, 571 (1952)]. See also section on lattice defects.

between the Hall constant at exhaustion R_{ex} and the (lower) reversal temperature T_r:

$$\log R_{ex} = a/T_r + b$$

So far it has not been possible to produce at low temperature, i.e., in the impurity range, any tellurium which is N type. In view of the heat treatment effects observed in germanium and silicon to be discussed later in the section on lattice defects it seems that the production of lattice defects, particularly vacancies, may be responsible in tellurium for the production of only P type conduction (160). The method of production of tellurium, involving rapid quenching from the melting point, may be responsible for this behavior. This also may explain why it is not possible to obtain tellurium with less than 10^{15} acceptors per cubic centimeter (69), why structure investigations indicate some crystal deformation, and why the density of the material as determined by various investigators shows large fluctuations (261, 285).

MEASUREMENT AND THEORY OF THERMOELECTRIC POWER IN GERMANIUM SEMICONDUCTORS

Middleton and Scanlon (243, 283), following preliminary investigations by Lark-Horovitz, Miller, and Walerstein (17, 222), have measured thermoelectric power as a function of temperature of a number of germanium samples, both N type and P type. Figure 10 shows typical thermoelectric power curves.

Fowler and others (81, 128, 305) have shown that the thermoelectric power of an impurity semiconductor is expressible in terms of its chemical potential (or Fermi level) ζ:

$$Q = \pm \frac{1}{e}\left(2k - \frac{\zeta}{T}\right) \tag{30}$$

if it is assumed that the mean free path is independent of energy and where the sign of Q is the sign of the carrier in the convention used in this paper. The chemical potential of an impurity semiconductor obeying classical statistics is determined by the carrier density:

$$\zeta = kT \ln\left\{\frac{nh^3}{2\,(2\pi m^* kT)^{3/2}}\right\} \tag{31}$$

If one takes the effective mass m^* as equal to the free electron mass m_0 and relates n to the Hall coefficient R by $n = 3\pi/(8eR)$, then the thermoelectric power of an impurity semiconductor may be written (17, 192):

$$Q = \pm \frac{k}{e}\left\{\ln\,(RT^{3/2}) - 5.32\right\} \tag{32}$$

If one takes $n = r/(eR)$, where r depends upon the proportion of impurity scattering present (see preceding section), allows for a difference between

m^* and m_0, and considers that the mean free path is dependent upon energy whenever impurity scattering is present, then a more complete form is obtained for the thermoelectric power of an impurity semiconductor.[50]

$$Q = \pm \frac{k}{e} \left\{ \ln (RT^{3/2}) - \ln r - 5.16 + 1.5 \ln (m^*/m_0) + Q' \right\} \quad (33)$$

The quantity Q' is related to the ratio of impurity resistivity to the sum

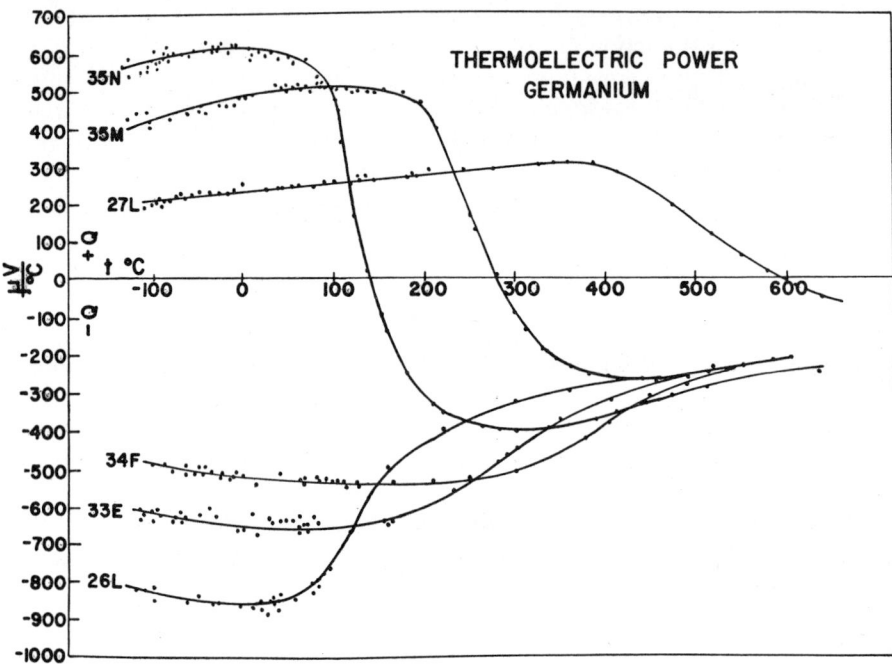

Fig. 10. Thermoelectric power in microvolts per °C as a function of temperature for various P type (35N, 35M, 27L; all aluminum-doped) and N type (34F, 33E, 26L; antimony-doped) germanium samples. (Measured by Middleton and Scanlon.) Note the similarity in the shape of the P type curves and their reversal of sign at various temperatures, equivalent to conversion from P type to N type behavior. Compare these curves with Table 7 which gives the calculated values of the temperatures of maxima and reversal points. The higher the impurity content, the lower the thermoelectric power at room temperature and below (impurity range).

of impurity and lattice resistivities as indicated by the values of Table 6. Both the complete expression and the shorter formula have been used for calculating the low-temperature (intrinsic conduction negligible) portion of the theoretical thermoelectric power curves shown in Figs. 11 and 12.

---

[50] V. A. Johnson and K. Lark-Horovitz, Purdue Prog. Rept., Dec. 1, 1950–Feb. 28, 1951, p. 5.

TABLE 6

$\dfrac{\rho_I}{\rho_I + \rho_L}$	0.0	0.1	0.2	0.3	0.4	0.5
Q'	0.000	0.379	0.558	0.698	0.809	0.947
$\dfrac{\rho_I}{\rho_I + \rho_L}$	0.6	0.7	0.8	0.9	1.0	
Q'	1.065	1.199	1.380	1.639	2.000	

For carriers of both signs in appreciable number, the expression of Fowler may be generalized to[51]

$$Q = \frac{-2k}{e} \tanh z + \frac{1}{eT}\left(\zeta + \frac{\Delta B}{2}\right) - \frac{\Delta B}{2eT}\tanh z \qquad (34)$$

where ΔB is the width of the forbidden energy gap between filled and conduction bands and

$$z = \frac{\zeta}{kT} + \frac{\Delta B}{2kT} + \frac{1}{2}\ln\left\{\frac{cm_1^{3/2}}{m_2^{3/2}}\right\} \qquad (35)$$

in which c is the ratio of electron mobility to hole mobility and m_1 and m_2 are, respectively, the effective masses of electrons and holes. The chemical potential may be related to n_1, the electron density, and n_2, the hole density, to yield the thermoelectric power in an alternate expression (17, 192):

$$Q = \frac{-k}{e}\frac{1}{n_1 c + n_2}\left[2(n_1 c - n_2) - n_1 c \ln\left\{\frac{n_1 h^3}{2(2\pi m_1 kT)^{3/2}}\right\}\right.$$
$$\left. + n_2 \ln\left\{\frac{n_2 h^3}{2(2\pi m_2 kT)^{3/2}}\right\}\right] \qquad (36)$$

This expression was used in calculating the high-temperature portion of the curves in Figs. 11 and 12. As may be seen from these figures, the theoretical curves agree with the observed points as well as can be expected in view of the scatter of the experimental data.

For a purely intrinsic semiconductor such that $n_1 = n_2$, the equations given above reduce to

$$Q = \frac{-k}{e}\frac{(c-1)}{(c+1)}\left\{\frac{\Delta B}{2kT} + 2\right\} \qquad (37)$$

If one allows for the approximately linear temperature variation of the forbidden energy gap by writing $\Delta B = B_0 + aT$, then the preceding equation becomes

$$Q = \frac{-k}{e}\frac{(c-1)}{(c+1)}\left\{\frac{B_0}{2kT} + 2 + \frac{a}{2k}\right\} \qquad (38)$$

---

[51] V. A. Johnson and K. Lark-Horovitz, Purdue Prog. Rept., Jan. 1–Mar. 31, 1952, p. 22.

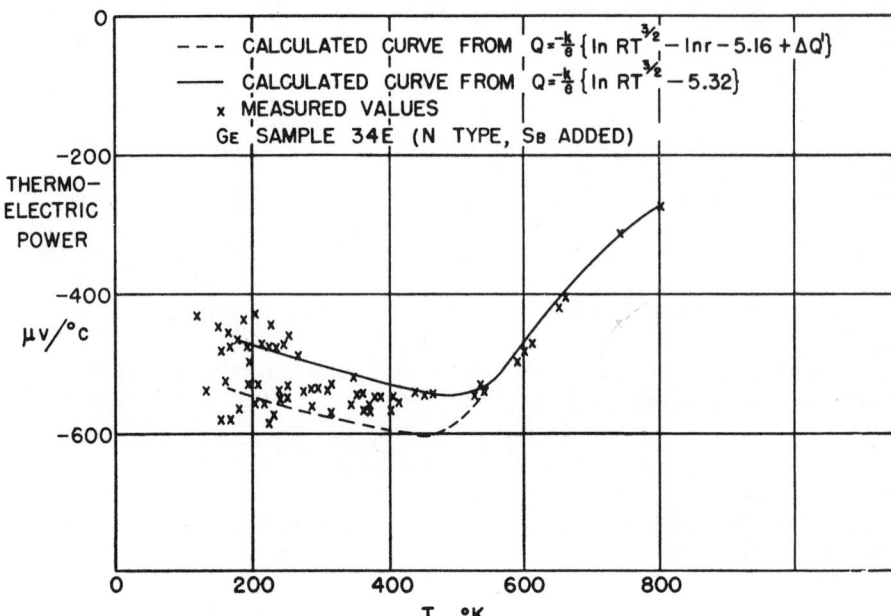

FIG. 11. Thermoelectric power of an N type germanium sample (antimony-doped) as a function of temperature. The observed values were measured by Middleton and Scanlon; the theoretical curve was calculated by use of the formulas 32, 33, 36 (Johnson and Lark-Horovitz). Note the agreement between theory and experiment at high temperatures and the scattering at low temperatures, probably due to difficulties in measurement of small temperature differences.

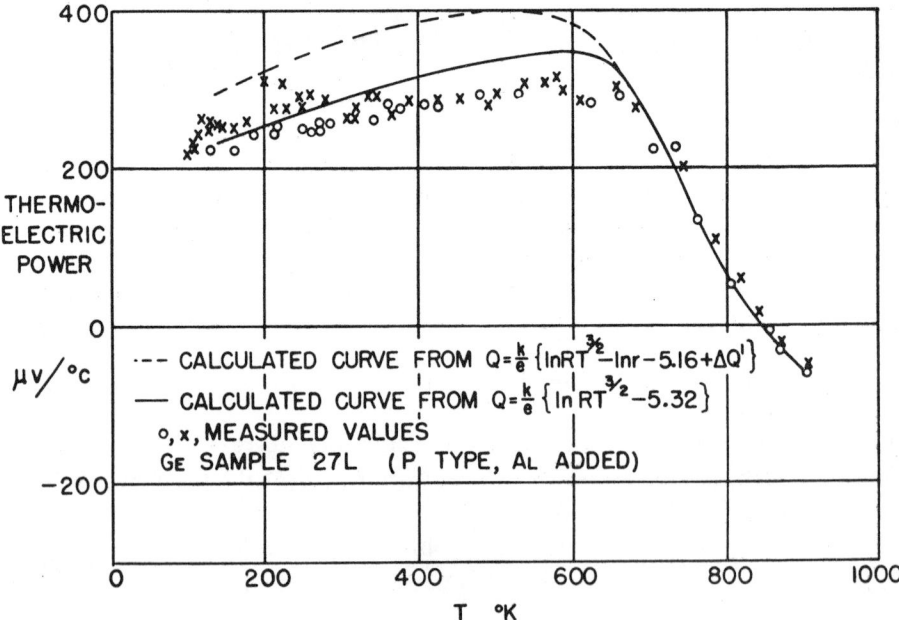

FIG. 12. Thermoelectric power for a P type (aluminum-doped) germanium sample as a function of temperature (Middleton and Scanlon). As in Fig. 11, theory and experiment agree well at high temperature and scattering at low temperature.

This means that, on a plot of Q vs T, the thermoelectric power curves of all samples, of a given material, such as germanium, approach a common hyperbola at high temperatures. The high-temperature measurements of Middleton and Scanlon on three comparatively pure but polycrystalline N type germanium samples yield for this hyperbolic expression:

$$Q = \frac{-k}{e}\left(\frac{2430}{T} - 0.34\right) \qquad (39)$$

From these numbers one obtains a high-temperature c value of about 3 and an a value of about -4×10^{-4} ev/°K, which is within the range of values mentioned in the preceding section.

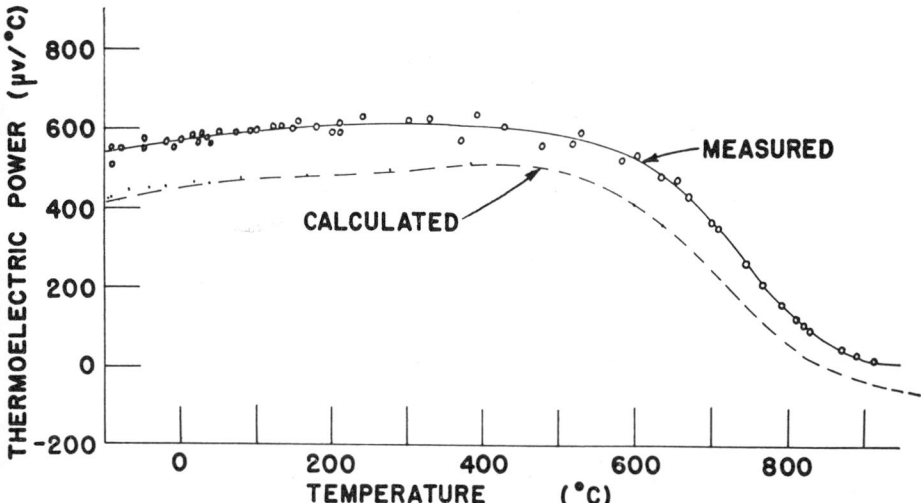

FIG. 13. Thermoelectric power of a P type silicon sample as a function of temperature. (Measurements by Middleton and Scanlon, calculations from Eqs. 33 and 36). Note the difference between theory and experiment, possibly due to the presence of scattering in addition to lattice and impurity scattering, as indicated by analysis of the resistivity curve.

By varying the amount of a given impurity introduced into the same semiconductor it is possible to obtain a family of curves (Fig. 10) which shows the dependence of thermoelectric power behavior upon impurity content.

The effect produced by a change in impurity concentration is most readily seen in the changes in position of maxima, minima, and zero points of the thermoelectric power curves. Table 7 indicates how these observed points on the curves of Fig. 10 agree with the corresponding points calculated by putting observed Hall data for the samples into the equations for thermoelectric power. Therefore, one can say that it is possible to predict the behavior of thermoelectric power in a semiconductor of small activation energy from its Hall effect and conductivity. For semiconductors of high activation energy there are deviations from this simple behavior (101, 174).

TABLE 7 Comparison between Calculated and Measured Thermoelectric Power Curves for Germanium Samples

Sample	Temperature of Occurrence of:					
	Maximum magnitude of T. E. P.		T. E. P. = 0 (P type only)		Neg. minimum of T. E. P. (P type only)	
	Meas.	Calc.	Meas.	Calc.	Meas.	Calc.
35N	−23°C	−13°C	144°C	136°C	284°C	281°C
35M	102°	106°	281°	288°	404°	410°
27L	310°	304°	580°	569°	Not reached	
26L	−7°	2°				
33E	116°	107°				
34F	175°	200°				

It should be pointed out that the preceding equation for the impurity range is not applicable at very low temperatures. One would expect the thermoelectric power to approach zero as the temperature approaches 0°K. This has been shown to be the case by Frederikse,[52] who measured the thermoelectric power of two N type germanium samples from 10° K upward. He found that at low temperatures the thermoelectric power can be represented empirically by

$$Q = \text{const} \times T \qquad (40)$$

We have also measured the thermoelectric power in silicon (Fig. 13) and tellurium. In both cases agreement between theory and experiment is less striking than it was in germanium. In both cases analysis of the resistivity curves indicates strong contributions of grain boundary scattering and defect scattering to the resistivity.

PRODUCTION OF LATTICE DEFECTS AND THEIR INFLUENCE ON ELECTRICAL PROPERTIES

In some of the early measurements of Hall effect and resistivity of P type germanium samples made at Purdue (243), it was observed that a comparatively high activation energy initially existed at low temperatures (Fig. 14);[53] the accompanying Hall effect measurements indicated that the samples were P type and sometimes showed low-temperature anomalies (multiple

---

[52] H. P. R. Frederikse, *Phys. Rev.* **86**, 647 (1952).

[53] In this case calculations based on the dissociation equation gives $\Delta E \sim 0.3$ ev, far higher than in chemically doped material. (However gold doped Ge gives activation energies of this order of magnitude.)

reversal of the Hall effect). The activation energy decreased markedly when the sample was heat treated for some hours at about 500°C. The samples apparently became P type by rapid cooling from high temperatures and were found to be N type after heat treatment. To explain this behavior we assumed that lattice defects (131, 289, 294) frozen in by rapid solidification produce acceptor states and that these are removed by heat treatment. (For systematic heat treatment experiments see 17, 281, 325, 346.)

A more detailed investigation has been carried out by Taylor (325). To investigate the effect of heat treatment, the samples were held in a partial vacuum ($\sim 10^{-2}$ mm Hg) at elevated temperatures for some time and then quenched in a bath of ethyl alcohol. The resistivity was observed as a function of the heat treatment temperature, and it was found that the resistivity of a P type sample decreases with increasing heat treatment temperature, and the resistivity of an N type sample first increases, reaches a maximum, and then decreases again (Fig. 15), indicating that acceptor states are formed which first remove electrons from the conduction band and then increase hole conduction. If one anneals the sample at 450–500°C for a period of 6 to 24 hours until no more changes in resistivity are observed, it is possible to restore the original properties of the material.[54]

The number of ionized impurities originally present determines whether conversion from N to P type takes place. If one plots the logarithm of the number of carriers per unit volume (determined only approximately by Taylor, using an empirical relation between Hall effect and resistivity) introduced into germanium by quenching from elevated temperatures against $1/T_q$ (T_q, the quenching temperature), a straight line is obtained. This indicates that if each defect produces one carrier, the number of defects N_D as a function of lattice atoms N_L is given by (19)

$$N_D = N_L \exp(-A/kT) \exp(-B/k) \qquad (41)$$

where A is 1.8 ev/atom and B is -5×10^{-4} ev/atom/°K or $\exp(-B/k) \cong 330$ (taking the energy of formation of defects $= W = A + BT$ (Fan) (325).[55]

---

[54] However, lifetime measurements of holes injected into quenched and then annealed material which is reconverted to N type and in which the original resistivity has been restored by heat treatment indicate that the lifetime is still appreciably shortened [D. Navon, R. Bray, H. Y. Fan, *Bull. Am. Phys. Soc.* 27, No. 4, 14, (1952)].

[55] Studies by C. S. Fuller and W. van Roesbroeck [*Phys. Rev.* 85, 678 (1952)] lead to a similar expression, but their numerical values differ from Taylor's ($A = 1.2$ ev, $\exp\{-B/k\} = 0.053$). C. Goldberg (Westinghouse Sci. paper 1659, May, 1952) used rapid quenching into oil and measured the Hall effect on single crystals of N and P type Ge before and after quenching. His data fit an equation of the type used by Taylor with $A = 1.45 \pm 0.05$ ev and $\exp\{-B/k\} = 0.96 \pm 0.05$. The interpretations of the three investigators are based on different concepts, and additional experiments will be necessary to decide between these various possibilities. The effects of copper impurities in the thermal conversion of germanium have been emphasized recently [W. P. Slichter and F. D. Kolb, *Phys. Rev.* 87, 527 (1952); also C. S. Fuller and J. D. Struthers, *Phys. Rev.* 87, 526 (1952)].

A particular type of quenching process is apparently going on in the deposition of thin films of germanium, produced either by evaporation[56] or by dissociation of GeH_4 on a suitable heated substrate.[57] In both cases only P type films are obtained. The electrical properties (Hall effect, resistivity, mobility) indicate strong disordering, and we believe that the

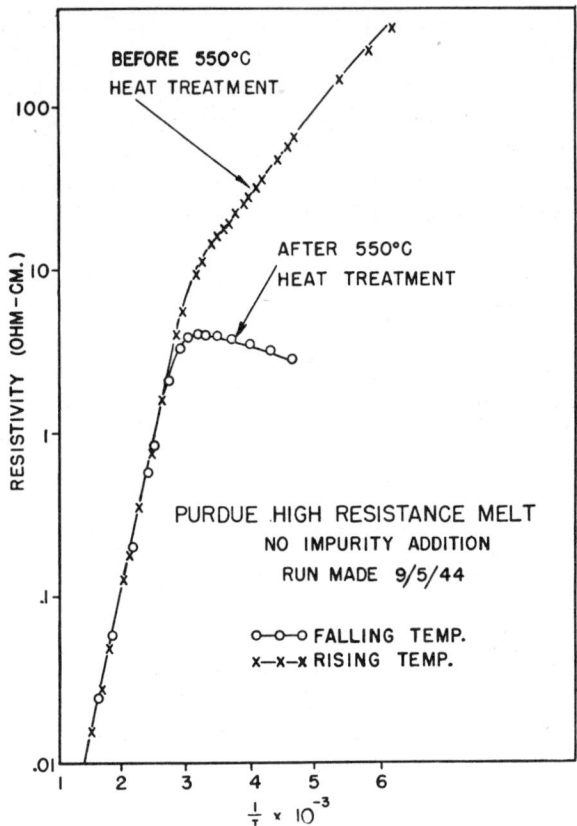

FIG. 14. Resistivity as a function of temperature (log ρ vs $1/T$) of high-resistance germanium before and after heat treatment (Middleton, 1944). (Hall effect indicates sample before heat treatment was P type; after heat treatment N type.)

P type character of the films is due to acceptors, introduced by defect production. This is also in agreement with recent observations[58] on the P type character of a polish layer on N type germanium.

The first semiconductor in which heat treatment and quenching effects

---

[56] *Phys. Rev.* **82,** 762 (1951).

[57] Purdue Prog. Rept., February 1951 to date.

[58] C. A. Hogarth and T. W. Granville, *Proc. Phys. Soc.* (*London*) **E64,** 993 (1951).

were observed was tellurium (160). It was first pointed out by Schottky[59] that in a lattice like tellurium a production of defects might be inherent in the preparation of the material, thus producing what he called a new type of "intrinsic defect conductivity." In such a semiconductor at high temperatures the energy to produce carriers is not given by the work necessary to produce an electron-hole pair in the basic lattice (intrinsic conductivity), but by the energetic relations which correspond to the formation or dissociation of a defect plus carrier. This idea has been formulated inde-

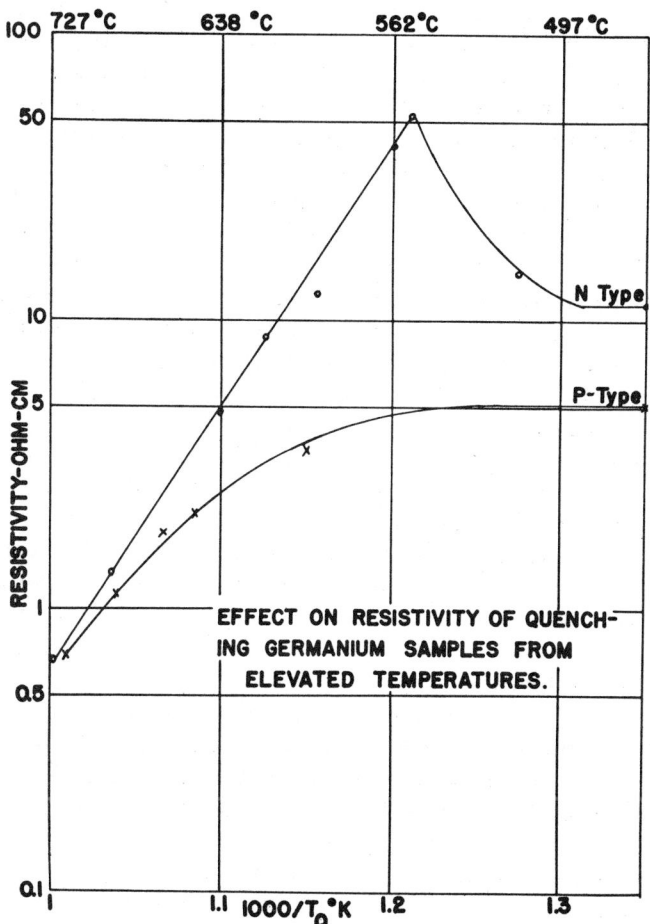

FIG. 15. The effect on room temperature conductivity of quenching germanium samples from various elevated temperatures, plotted as logarithm of room temperature conductivity vs reciprocal of the absolute temperature at which quenching occurs. (Measured by W. E. Taylor.)

---

[59] Private communication, February, 1950.

pendently by H. Fritzsche,[60] assuming that the number of defects N_D follows an equation of the type (41), and that the total number of carriers

$$N = N_{ex} + N_D$$

N_{ex} is the number of impurities at exhaustion and N is constant in a temperature range where N_D is negligible, but follows (41) at high temperatures. Fritzsche assumed that $\Delta B = 0.36$ ev and the energy A in (41) is 0.72 ev.

Quenching and heat treatment experiments are of particular interest in view of the experiments carried out to produce defects by the passage of fast particles through semiconductors. If fast neutrons, alpha particles, deuterons, or even high-energy electrons pass through a solid, atoms are displaced in the lattice, (Wigner effect, 83, 301), and lattice imperfections, both vacancies and interstitials, are formed (217). They produce localized states (186), which in turn can act as donors and acceptors and thus profoundly change the electrical behavior of the semiconductor.

Radiation effects on solids have been known for a long time. The fatigue of scintillating zinc sulfide produced by α-particle bombardment was first analyzed by Rutherford (60, 280). The isotropization of crystals[61] with radioactive inclusions described by Muegge (213, 254) indicates that the passage of heavy particles profoundly alters the lattice of a solid.

It was expected, therefore, that the great sensitivity of pure semiconductors to imperfections would make radiation effects easily observable as electrical changes in the material. One would expect that a heavy particle passing through matter might, by elastic collision, produce a displacement resulting in a lattic vacancy and an interstitial atom. If vacancies and interstitials are equal in number and equally effective in either accepting or donating carriers, their electrical effect would be an increase in the scattering and a consequent decrease in the mean free path and a corresponding decrease in conductivity. On the other hand, a net excess of either type of carrier can be measured electrically by the Hall effect and may result in large changes in the conductivity.

In addition to lattice disordering, transmutations may be produced by nucleon bombardment. Since one is interested primarily in the lattice defects produced by displacements, and not in transmutations, it is convenient to use particles which have such a small cross section for transmutation that this effect can be neglected. Deuterons and α-particles have such a small cross section for transmutation that, in the time necessary to produce effects due to displacements, the transmutation effect can be neglected. The use of charged particles has the advantage that it is possible

---

[60] *Science* **115.** 571, (1952).

[61] The crystals retain their form, but show liquid type disorder (metamict) (80) both by optical methods and x-ray diffraction (331). The disorder can be "healed out" by heat treatment (213). Recently such an effect has been produced by α-particles from radon in some organic material. (B. Stech, *Z. Naturforsch.* **7a,** 175 (1952).

to produce well-defined beams and thus to concentrate the radiation effect in a definite region. On the other hand, it is more difficult, except in the case of some naturally radioactive sources (73, 104, 106), to control the influx of the number of particles to the target as carefully as one can in the nuclear reactor using neutrons. Charged particles affect a definite area only, and their interaction is limited to the range of the particle in the material. On

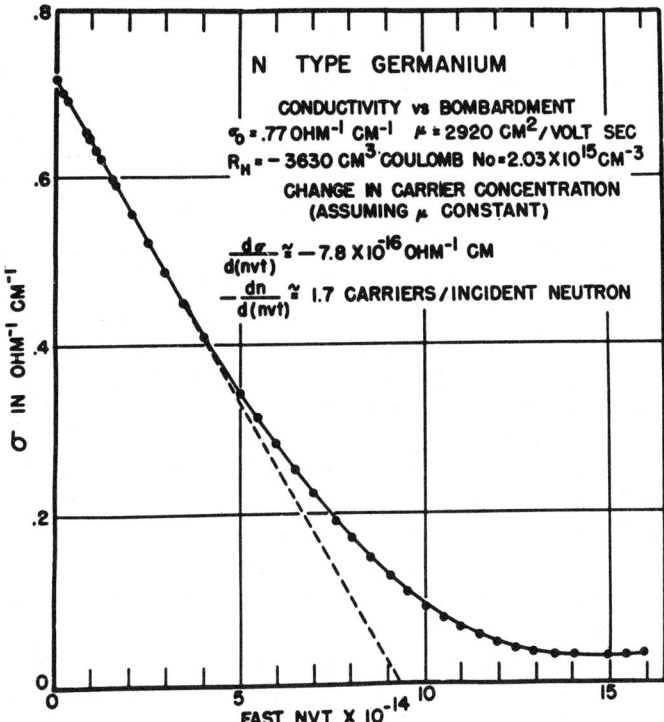

FIG. 16. Conductivity of N type germanium as a function of bombardment. The abscissa gives the number of incident fast neutrons per sec, per cm² and the slope indicates the number of carriers removed per incident neutron. Note that the minimum is reached only after about twice as many neutrons have fallen on the sample as there were electrons originally removed. (Measured by W. E. Johnson and K. Lark-Horovitz, analyzed by J. H. Crawford.)

the other hand, with fast neutrons bulk samples of material of considerable thickness can be irradiated.

The first orienting experiment carried out with the Purdue cyclotron (219) indicated at once that deuteron irradiation affects the resistivity and Hall effect in germanium semiconductors in a very pronounced and easily measurable way. It was found that N type germanium becomes converted to P type and that the conductivity of P type germanium increases with increasing bombardment and seems to reach saturation. Similar effects were observed with α-particles from polonium. Subsequent heat treatment experiments showed that these resistivity effects are due to a disordering which is reversible, since most of the defects can be healed out at about 500°C (104, 106).

It is obvious from the Hall effect curves observed immediately after bombardment (first obtained by Davis, 104) and after proper heat treatment that fast neutrons produce levels different from those due to chemical substitution.

Figure 16 (93, 94) was obtained in material which has been bombarded in the nuclear reactor with fast and slow neutrons. After bombardment, and before heat treatment, the fast neutron effect is dominant and different from the effect of substitutional impurities.

The deuteron bombardment experiments also indicate that, after a certain number of displacements have been produced in P type germanium, upon further bombardment saturation is reached (104). This is due in part to self-healing under the conditions of bombardment and in part to the fact (186, 232) that when both donor and acceptor levels are introduced, as is evidently the case here, the Fermi level should approach a definite position upon prolonged bombardment, and therefore conductivity will no longer change at a fixed temperature.

It is thus possible to describe the effect on carrier concentration due to irradiation of germanium by a shift in the Fermi level. In N type material there are states near the conduction band and so the Fermi level is located in the upper part of the forbidden band (Fig. 3E), but when the material becomes P type, the Fermi level has shifted close to the edge of the full band[62] (232, 233).

The analysis of an irradiation curve[63] and Hall effect curve such as is represented in Figs. 16 and 17 makes it clear that a distribution of energy levels, rather than a single level, is produced by the defects and that these levels are much deeper than the ones produced by the chemical impurities. This analysis (102, 103) indicates that at the start of bombardment of an N type germanium sample the change in conductivity is extremely rapid until a certain conductivity minimum is reached. Beyond this minimum the material (as tested by Hall effect, thermal emf, or rectification) is P type.

---

[62] Depending on the initial concentration of impurities, it is possible to shift the Fermi level by bombardment either down or up and as a consequence to produce respectively an increase or decrease in the conductivity of P type germanium. [Such effects have been observed recently with neutron irradiation: *Phys. Rev.* **85**, 730A (1952); **84**, 861 (1951); **86**, 641 (1952). Similar results have also been observed in deuteron bombardment: *Phys. Rev.* **86**, 643 (1952)].

[63] A similar analysis was carried out for α-particle irradiation of germanium and indicates agreement with theoretical predictions (73).

There is a pronounced difference in slope on the two sides of the minimum. This becomes understandable if one considers that the donors due to interstitials and the acceptors due to vacancies have much higher activation energies than chemical impurities. One assumes that the donor levels are depressed far below the conduction band near to the top of the full band, and that the acceptor levels are raised somewhat above them (217). It is clear that as long as there are electrons available in the conduction band, these electrons will fall into the empty acceptor states until the material has reached a maximum resistivity. After conversion to P type the conductivity increases again, because now electrons from the full band can be lifted into acceptor states. This, however, is possible only if there is enough energy available; and hence the production of holes will not take place as rapidly as the removal of electrons when the material was N type. In the conductivity vs irradiation curves the ratio between the slopes before and after passage through the minimum conductivity (Fig. 16) indicates that the acceptor states are closer to the full band than to the conduction band.[64]

In agreement with this observation it was found that, in the case of P type germanium, the number of carriers added per incident neutron is smaller than the number of carriers removed from N type material: on the average about 3.2 carriers [65] are removed per incident neutron at about 300°K as compared to $\sim$ 0.7 added at the same temperature. Also the conductivity change in P type material is temperature dependent, as is expected from these considerations.

The minimum energy necessary to displace a germanium atom from its lattice site can be determined by using electrons instead of α-particles or deuterons. M. M. Mills pointed out in a discussion that fast electrons have enough energy to displace an atom from its lattice site. One can therefore expect to observe conversion of germanium from N to P type by fast electrons as well as by the passage of heavy particles.

The use of electrons has the advantage that by varying the energy (E) of

---

[64] H. M. James and K. Lark-Horovitz [*Z. phys. Chem.* **198**, 107 (1951)] have developed a more detailed model of the lattice defects produced by bombardment of silicon and germanium. Analysis of observations on bombarded germanium indicates that the second electron ionization energy of germanium interstitials is of the order of the forbidden band width, 0.75 ev, and that the second hole ionization is low, of the order of 0.2 ev. It then appears that in bombarded germanium the Fermi level is determined by the equilibrium between single and doubly ionized interstitials and vacancies. In bombarded silicon most of the interstitials and vacancies will be singly ionized. This model is now being tested with neutron [*Phys. Rev.* **86**, 641 (1952)] and deuteron bombardment [*Phys. Rev.* **86**, 643 (1952)].

[65] This is of the order of magnitude calculated by G. E. Evans assuming a reasonable scattering cross section of fast neutrons in germanium. In deuteron bombardment Forster *et al.* [*Phys. Rev.* **86**, 643 (1952)] 11 electrons are removed in N type germanium per incident deuteron and 3 holes in high-conductivity P type germanium. In P type silicon 31 holes are removed per incident deuteron.

the incoming particles the threshold energy T can be determined, below which such displacements cannot take place.

$$T = \frac{2Em}{M}\left(\frac{E}{mc^2} + 2\right) \qquad (42)$$

where m is the electronic rest mass, M the mass of the germanium atom, and c the velocity of light.

In the case of germanium this effect can be established in a simple way. N type germanium was bombarded so that in one and the same piece of germanium adjacent areas were exposed successively to electron beams of known energies between 2 Mev and 0.5 Mev, and some areas remained unexposed. Since a point contact of metal on N type germanium gives a rectifier characteristic which is inverted if the material converts to P type, one simply probes the irradiated and unirradiated portions of the sample and thus determines at what energy conversion sets in. The experimentally determined threshold energy below which no change was observed is about 0.67 Mev, indicating that the energy (Wigner energy) required to move a germanium atom from its position in the lattice must be about 30 ev (206, 217).

The use of charged particles and their interaction with the target in a well-defined area makes it possible to produce by bombardment in N type germanium a P type region adjacent to an area shielded during bombardment and remaining N type. Therefore P-N barriers (104, 106, 126) can be produced at will and with a predetermined depth. By using deuterons or α-particles of various energies, it is possible to obtain P type layers of different thicknesses on top of the N type layers with as sharp an intersection as the straggling of the range of the particles permits.

A P-N boundary acts as a rectifier. Electrical equilibrium is established through the establishment of a double layer of charge at the boundary only when the chemical potential in the two materials becomes the same. The resulting potential barrier maintains the required relatively large concentration of electrons in the N type material, despite the low concentration of electrons in the adjacent P type material, and similarly maintains the greater concentration of holes in the P type material. If the positive and negative terminals of a battery are connected to the P and N type materials, respectively, the barrier is lowered, and electrons tend to flow from the N type material, where they are abundant, into the P type material. Similarly, holes flow in the opposite direction. If the terminals of the battery are reversed, then the barrier is heightened and electrons tend to flow from the P type material, where they are relatively scarce, into the N type material. Similarly, holes flow from the N type into the P type material. It is evident that the greater current will flow in the former case, in which the voltage is applied in the "forward" direction. The characteristic of the rectifier is almost ideal (25, 126) since it has the temperature dependence predicted by theory.

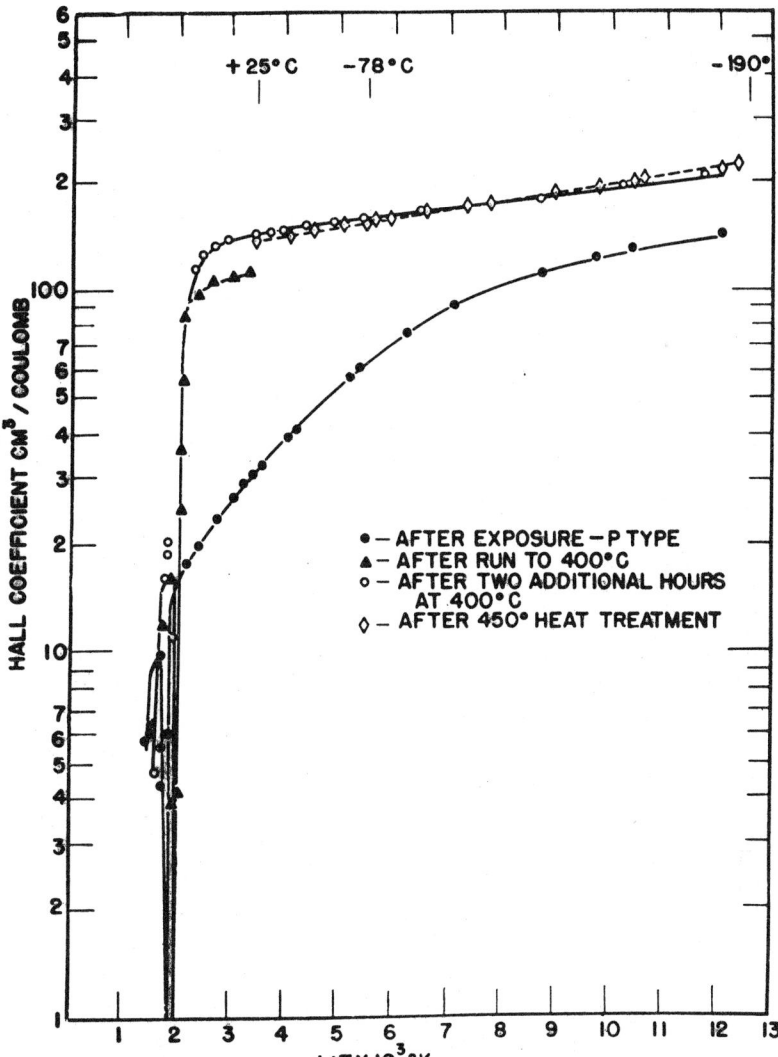

FIG. 17. Hall coefficient of a germanium sample after bombardment by fast and slow neutrons in a nuclear reactor. (Measured by J. Cleland.) The sample was N type before exposure and P type afterwards. Note the sharp curvature in the Hall curve taken immediately after exposure. Heating at 450°C returns the sample to typical N type behavior with an activation energy equal to about 0.03 ev.

It is also possible to produce such P-N barriers by irradiating one part of a sample of germanium while shielding the remainder. P-N barriers in germanium and silicon are not only rectifiers (25, 126) but also photoconductive and photovoltaic elements (25, 56, 59). Moreover, it has been possible to show that these barriers act as crystal counters (237, 260).

It is possible to start with material of preselected conductivity, then to bombard it to form a P-N barrier, and thus to produce practically any of the nonohmic behaviors which have been discussed above for various semiconductor-metal combinations. One has the advantage of dealing with a rectifying and photovoltaic junction between two semiconductors of the same basic material which have well-defined carrier densities and mobilities.

In the study of these photovoltaic P-N barriers in germanium, Fan and Becker noticed that the effect of light was so large that it was difficult to understand why the unilluminated part of the material did not short-circuit the illuminated part. The investigation of these optical properties has turned out to be of both practical and theoretical importance and interest.

Optical Properties of Germanium and Silicon

The optical properties of semiconductors, as already mentioned, have been investigated before, but unfortunately many of the samples investigated optically were not the same samples that were investigated electrically, with the exception of the insulating crystals selected by Gudden and Pohl for which optical, photoelectric, and electrical properties were measured (8, 9, 22). Hence correlation between the number of carriers and their mobility and optical properties was usually not possible. The optical investigation of nonpolar semiconducting samples for which both the Hall effect and the conductivity were known was carried out on germanium and silicon semiconductors (221). The reflectivity of silicon and germanium, for both N and P type samples, was measured with respect to aluminum with a NaCl spectrometer in the range 0.6 μ to 12 μ and with residual rays in selected regions up to 152 μ.

Although the reflectivity of metals increases with increasing wavelengths, the reflectivities of silicon and germanium samples of medium conductivity remain approximately constant throughout the range of the spectrum investigated, $\sim 36\%$ for germanium and $\sim 30\%$ for silicon. The values for silicon are in agreement with observations made by former investigators (181). From these measurements it was concluded that the dielectric constant of silicon is ~ 12 (256) and the dielectric constant of germanium 16, in agreement with the value of the dielectric constant deduced from the interpretation of the resistivity contribution due to impurity scattering (17, 192).

Reflectivity measurements on both N type and P type high conductivity germanium reveal a peak in the reflectivity curve at about 117 μ, an observation which is difficult to understand (124). For this range of conductivity, corresponding to about 10^{18} carriers per cubic centimeter or more, additional experiments will be necessary to establish definitely the long wavelength behavior and its temperature dependence.

The very high infrared transmissivity of germanium and silicon was discovered by Becker and Fan (51) in the investigation of the P-N boundary photovoltaic effect (51, 52, 53). The study of semiconductors of various conductivities has shown that absorptivity is very small and that the nearly pure material is "transparent" in a wide range of the infrared spectrum [up to 12 μ as investigated at Purdue and up to 25 μ as investigated by Burstein et al. (82) and Briggs (78)].

The high transparency makes it possible to measure absorption through bulk material, instead of using thin films. The measurements are thus free of the effect of surface layers, which is such a serious drawback in most optical measurements. This becomes clear if one compares the work on germanium films, as carried on by various investigators (72, 259), with the reflectivity of the bulk samples.[66]

To test transmissivity photographically, a few seconds exposure to a white light source through a silicon window gave intense blackening on IZ photographic plates (sensitized to be responsive out to 1.3 μ). Exposures up to 36 hours with germanium did not give any detectable blackening (53), consistent with the cutoff difference corresponding to the different width of the energy gap ΔB, 1.12 ev in silicon and 0.75 ev in germanium. To test whether absorption as a function of thickness obeys exponential attenuation, the 1.083 μ line of a helium discharge tube was used as a source, silicon as absorber, and a germanium P-N boundary as a receiver. A Gaertner L-235 infrared spectrometer using a single NaCl prism was employed. Special tests for scattering indicated that the error due to scattering of light of less than about 1.5 μ was not more than 2% at 10 μ. Carefully polished samples selected for uniformity by electrical tests were studied. Bulk silicon up to several millimeters in thickness has been used to determine the percentage transmission as a function of thickness at various wavelengths. Samples were prepared in such a way that the surface reflectivity was the same in every case. The absorption, determined from the slope of the log percentage transmission vs sample thickness plot, varies exponentially with thickness; and the reflectivity, 31%, determined by the intercept of the graph extrapolated to zero thickness, taking into account multiple reflections in the sample, agrees with the values obtained by direct measurement.

In germanium the absorption increases strongly for wavelengths shorter than 2 μ. For measurements at shorter wavelengths very thin samples have been used. For wavelengths above 2 μ the absorption is so small that samples several centimeters thick have to be used for accurate measurements. Samples of resistivity 5 ohm-cm and 0.015 ohm-cm have been used,

---

[66] O'Bryan (259), using Ge film, found in the visible a reflectivity of $\sim 36\%$. Bottom, using Thornhill's films at Purdue, found for a Ge film deposited on optically polished silica 37%, but with the same light (Na) 45% for the bulk material. Direct measurement of reflectivity of bulk material (221) (0.6–0.8 μ) gave 47–51%, decreasing to 35% above 2 μ. Brattain and Briggs's work on film led to a reflectivity from 52% at 0.4 μ to 39% above 5 μ (72).

and the reflectivity deduced from these observations is the same for both samples in this spectral range.

It is important to correlate absorption with the presence of free carriers. Since germanium semiconductor samples can be prepared to vary in resistivity from about 0.001 ohm-cm to about 30 ohm-cm, silicon semiconductors from about the same low resistivity to considerably higher resistivities, and tellurium from 0.001 ohm-cm to 0.5 ohm-cm, it is possible to investigate the optical properties as functions of the number of carriers and their mobility.

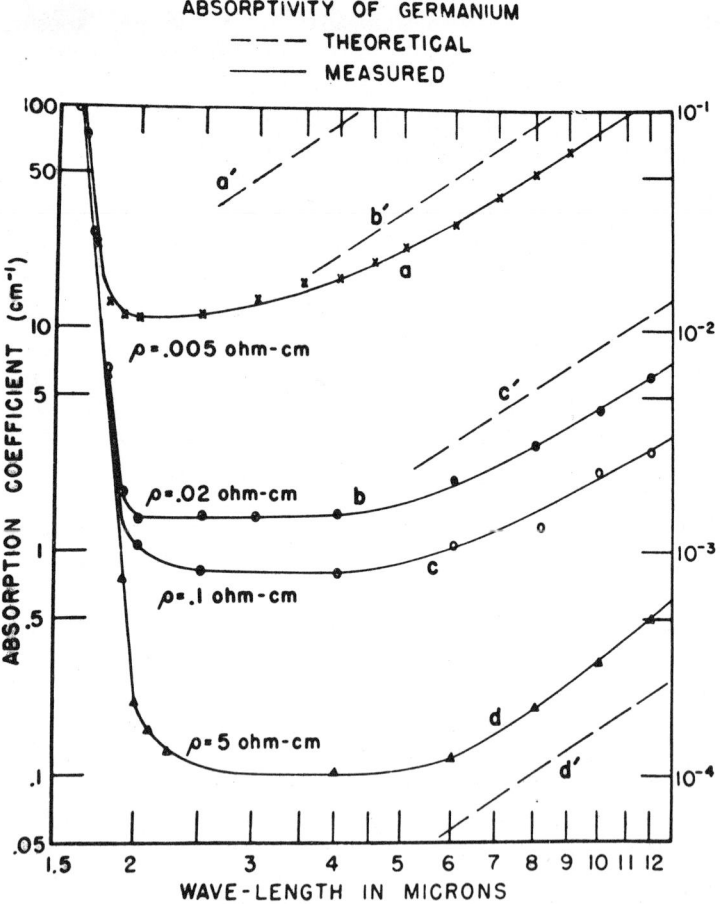

Fig. 18. Absorptivity of germanium as a function of wavelength for samples of various resistivities. Full curves with vertical scale on left are experimental. Broken lines with scale on right are calculated curves derived from the numbers of free carriers. Observe the considerable differences between theoretically predicted and observed absorptivity (a' and a, etc., belong to the same sample). (Measured by M. Becker.)

Depending upon whether the photon energy $h\nu$ is greater or smaller than the width of the forbidden gap, ΔB, ($h\nu \gtrless \Delta B$), excitation can occur from the full band to the conduction band or impurities may be excited.

The theory of absorption by free carriers has been formulated quantum mechanically by Kronig (212). The theory predicts proportionality of absorption to the number of carriers and gives for the near infrared (collision frequency < radiation frequency) proportionality to the square of the wavelength. It is possible to express this dependence in the following form

$$\mu = \frac{1}{\pi n c^3} \left(\frac{e}{m^*}\right)^2 \frac{\lambda^2}{\mu_e^2 \rho} \tag{43}$$

In the equation μ is the absorption coefficient (cm^{-1}), λ the wavelength (cm), μ_e the electronic lattice mobility (cm^2/v-sec), ρ the resistivity (ohm-cm), m^* the effective mass, and n the index of refraction. Fan has also considered absorption by neutral impurities due to photoionization (32). Figure 18 is a plot of the absorptivity of germanium as a function of wavelength in microns. The dotted lines indicate the value calculated from theory. The observed absorptivities are higher by several orders of magnitude than the theoretically predicted ones, and there is only an approximate correlation between the absorptivity and the number of carriers. However, it is important to note that all the curves approach the same limiting value at the short wavelength edge.

Similar results are obtained for silicon as shown in Fig. 19. Most of the P type silicon shows absorptivities proportional to the square of the wavelength, and the experimentally found absorption is only 5 to 10 times higher than that theoretically predicted.

In semiconductors both the number of carriers and the mobility are functions of temperature. Since they can vary in marked manner from sample to sample — for example, degenerate samples may show little or no variation of number of carriers or mobility over a wide temperature range and some nondegenerate samples show very drastic changes—it is particularly interesting to investigate the optical properties of these materials as functions of temperature. Such experiments have been carried out by Becker and Fan (52, 53). When the number of carriers remained constant, the absorptivity decreased with decreasing temperature. On the other hand, in a nondegenerate sample of silicon the number of free carriers decreased thirty times as the temperature decreased from 300 to 100°K; but the absorption increased, indicating a strong influence of neutral impurities.

The question of how far free carriers influence absorption can be investigated not only by the effect of temperature changes and the presence of undissociated or dissociated impurity centers, but also by the investigation of nucleon bombarded material. It has been found that neutron or deuteron bombardment of silicon produces an increase in resistivity both in P type and N type samples (217). Heat treatment experiments by Davis (104) have shown that probably very deep lying levels (about 0.7 ev) are created. It was, therefore, of interest to study nucleon bombarded silicon and its absorptivity, since in this case the number of free carriers was considerably reduced, whereas the number of impurity centers was increased. Such an experiment would give an indication about the importance of free carriers and the effect of impurity centers on transmissivity.

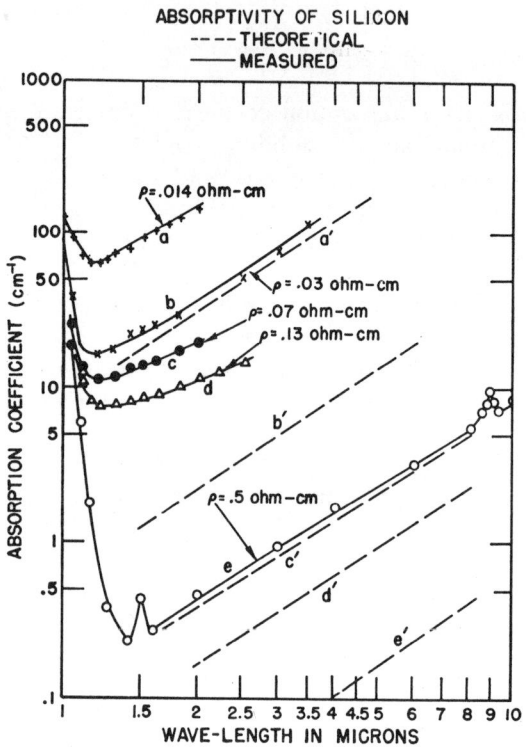

FIG. 19. Absorptivity of silicon. The solid lines represent values measured for samples of various resistivities; the dashed lines represent values calculated from the free-carrier theory. Small letters (a, a', etc.) denote corresponding pairs of measured and calculated lines. Observe the absorption lines at 1.7 μ and 8 μ on the highest resistivity sample; the first one may be due to defects produced by quenching similar to defects produced by nucleon bombardment. (Measured by M. Becker.)

Such measurements were carried out (218) and, as shown in Fig. 20, a definite new absorption was observed at about 1.8 μ. Even more important is the fact, also shown in Fig. 20, that beyond this level the transmissivity has been considerably increased, and the absorption edge shifts to larger wavelengths with bombardment. Recent experiments (52) show that the new absorption band increases with increasing bombardment and sharpens with decreasing temperatures, a behavior which is quite similar to the absorption in F centers (22) and indicates transition between localized states.

The absorption band at 9 μ which is clearly observed in this sample was found earlier in silicon samples by both Burstein (82) and Briggs (78). At Purdue this absorption peak was found in all samples which have absorptivity low enough to permit measurements to be made up to and beyond 9 μ. Furthermore, this peak corresponds to a *constant* additional absorptivity in all samples and must therefore be a property of the lattice itself.

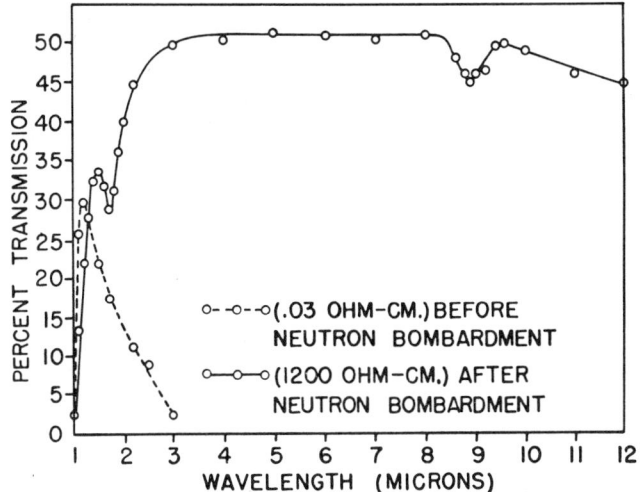

Fig. 20. Per cent transmission of a P type silicon sample, before and after neutron bombardment, as a function of wavelength. Note the relative opacity of the sample beyond 3 μ before bombardment and its transparency after bombardment and the appearance of a new absorption line at 1.7 μ. Also note the absorption at 9 μ, a characteristic of the silicon lattice. (Measurements by M. Becker.)

The measurements on both silicon and germanium have recently been extended at Purdue to longer wavelengths, beyond 12 μ.[67] For high-resistivity (small concentration of free carriers) germanium and silicon which showed very small absorptivity up to 12 μ a number of absorption bands are found at longer wavelengths. For germanium there is a strong peak at $1/\lambda = 340$ cm^{-1} and weaker peaks at 410, 540, and 650 cm^{-1}. For silicon the strongest peak is at 610 cm^{-1}, with weaker peaks at 520, 720, 890, and 1110 cm^{-1}, the last one being the absorption band at about 9 μ found previously. Similar results have been reported recently by Lord.[68]

---

[67] R. J. Collins and H. Y. Fan, Purdue Prog. Rept., Dec. 31, 1951 and Mar. 31, 1952; see also *Bull. Am. Phys. Soc.* **27**, Mar. 20, 1952.

[68] R C. Lord, *Phys. Rev.* **85**, 140 (1952).

The absorption in N type germanium of different resistivities has been investigated up to 110 µ with the vacuum grating of the University of Michigan. The absorption in samples of low resistivities (high carrier concentrations) can be resolved into two components. One is the band structure observed in high-resistivity samples and the other rises smoothly with increasing wavelength approximately proportional to λ^2. The latter component in different samples is roughly proportional to the conductivity of the sample, indicating that it is an absorption due to free electrons. The first component with a band structure is common to samples of all resistivities; therefore it is apparently a lattice absorption. It is interesting to note that the frequency of the most prominent peak in both silicon and germanium is close to the fundamental frequency of the lattice vibrations, calculated from the Nagendra Nath-Born relation[69] using the known elastic constants.

Summarizing, one can say that these experiments indicate that germanium and silicon are transparent in a wide range in the infrared beyond the absorption edge. The sharp increase in absorption at an energy corresponding to the energy of the forbidden gap indicates that one is dealing with a transition from the full to the conduction band.

Absorption does depend, in the region $\lambda > hc/\Delta B$, on the presence of free carriers, but is many orders of magnitude greater than can be calculated either on the basis of the conventional Drude-Kronig theory (212) or a quantum mechanical modification proposed by Fan and Fröhlich (32).

Just how far the band width can be changed by impurity content will be of particular interest in the further study of the optical properties.

The fact that in these semiconductors one deals with optical material of very high dielectric constant and high refractive index makes it possible to use them effectively for optical elements in the infrared region. Because of the high refractive index and transparency it is possible to make lenses, prisms, and filters (78) which may be useful in the infrared regions since they are not attacked by moisture or the atmosphere. The F number (relative aperture) of a germanium lens would be about 4 times that of a glass lens of the same geometry.

Conclusions

The preceding discussion has shown that germanium semiconductors follow, in most respects, the predictions of modern semiconductor theory. It is possible to produce imperfections of predictable properties in the material and to understand their behavior, particularly that of the chemically introduced substitutional imperfections.

It is possible to change the type of conduction from N type to P type either by heat treatment, by irradiation with nucleons or by changing the chemical impurities through transmutation. Resistivity and thermo-

---

[69] M. Born, *Nature* **157**, 582 (1946).

electric power can be predicted from a knowledge of the number of carriers as a function of temperature and their scattering mechanism.

However, there are a number of problems which at the present time are not completely understood and which will engage the attention of both experimental and theoretical physicists for some time to come.

The field dependence of the Hall effect, the magneto resistance, and the low-temperature conductivity show anomalies not explainable by present theory.

The mean free path calculations, and particularly the calculation of the effective mean free path when one deals with scattering mechanisms involving different energy dependence of mean free path (293) present a problem which needs further consideration.

Optical absorption as a function of carrier (impurity) concentration and of lattice imperfections is incompletely understood. The optical absorption, small as it is, is by orders of magnitude larger, particularly in germanium, than the values which have been calculated on the basis of free carriers. The direct correlation between optical behavior and the number of imperfections has not been established.

The interaction of imperfections, particularly the dependence of the activation energy on the number of impurity centers, has not yet been explained satisfactorily (see 32, Mott, p. 1, Castellan and Seitz, p. 8) (see also 273). Additional theoretical and experimental investigations will be necessary to clarify this point (see also 33, Henisch, p. 41). It seems that the production of controlled numbers of impurities *in situ* by transmutation in a nuclear reactor may be a particularly important tool for experimental study of the dependence of the activation energy on the number of impurities present.

The low-temperature behavior of degenerate samples, as well as of purer materials, cannot be understood on the basis of current semiconductor theory (117, 179, 180).

At the present time it is possible to estimate the distribution of energy levels due to lattice defects produced by displacement (vacancies and interstitials) only from the experiments. Such large perturbations cannot be treated by the usual methods, and even the application of the band picture to this process is doubtful.

Finally, it should be remembered that the question of the energy transfer across the energy gap, in both radiative and nonradiative processes, is a problem which has scarcely been attacked, although it has both practical and theoretical importance for the simple elementary semiconductors, such as we have discussed here, as well as for the more complicated ones which play a large part in practical applications.

It is not possible so far to predict from the atomic structure and the lattice structure the correct value of the energy gap observed in diamond, silicon, or germanium. While, in principle, methods can be applied which would allow such a calculation, the complications involved in the determination of the energy band structure are so great that there is a large discrepancy between calculated and experimental values (256).

ACKNOWLEDGMENTS

The work on germanium was first supported by an NDRC contract, (1942 to 1945), and later by a Signal Corps and an AEC contract.

The author is indebted to his colleagues, H. Y. Fan and H. M. James for many helpful discussions and criticism, but particularly to V. A. Johnson for her help in preparing the illustrative material for this article and for the use of some unpublished material.

REFERENCES

Selected Texts and Review Articles

1. Baedeker, K., *Die elektrischen Erscheinungen in metallischen Leitern*. Vieweg, 1911.
2. Brillouin, L., *Die Quantenstatistik*, Berlin, 1931.
3. Busch, G., Elektronenleitung in nicht-Metallen. *Z. angew. Math. u. Physik* **1**, 3, 81 (1950).
4. Campbell, L. L., *Galvanomagnetic and Thermomagnetic Effects*. Longmans, London, 1923.
5. Fowler, R. H., *Statistical Mechanics*. Cambridge, 1936.
6. Fröhlich, H. *Elektronentheorie der Metalle*. Berlin, 1936.
7. Grondahl, L. O., The Copper-Cuprous Oxide Rectifier. *Revs. Mod. Phys.* **5**, 141 (1933).
8. Gudden, B., *Ergeb. exakt. Naturw.* **3**, 143 (1924); **13**, 223 (1934); (General Review of electronic phenomena in non-metals).
9. Gudden, B., *Lichtelektrische Erscheinungen*. Springer, Berlin, 1928.
10. Henisch, H. K., *Metal Rectifiers*. Oxford Press, 1949.
11. Hughes, A. L., Photoconductivity in Crystals. *Revs. Mod. Phys.* **8**, 294 (1936).
12. Joffè, A. F., *Actualités sci. et. ind.* **87**, (1933); **114**, (1934).
13. Joffè, A. F., *The Physics of Crystals*. McGraw-Hill, New York, 1928.
14. Johnson, V. A., and Lark-Horovitz, K., Second Final Report Including Subject Analysis of Purdue Quarterly Reports. Signal Corps Contract No. W 36-039-SC 3200, December 1948.
15. (a) Koenigsberger, J., *Jahrb. Radiaakt. u. Elektronik* **4**, 158 (1907); (b) *ibid*. **11**, 84 (1914); (c) *Graetz' Handb. Elektr. u. Magn.* **3**, 597 (1920).
16. Lange, B., *Die Photoelemente*. Leipzig, 1936.
17. Lark-Horovitz, K., Preparation of Semiconductors and Development of Crystal Rectifiers. NDRC Rept. 14-585, March 1942 to November 1945.
18. Meissner, W., Elektronenleitung, etc. *Handb. d. Experimentalphys.* Vol. 11, Pt. 2, 1935.
19. Mott, N. F., and Gurney, R. W., *Electronic Processes in Ionic Crystals*. Oxford Press, 1940.
20. Nix, F. C., Photoconductivity. *Revs. Mod. Phys.* **4**, 723 (1932).
21. Nordheim, L. W., *Mueller Pouillet* IV, Pt. 4, 243 (1933).
22. Pohl, R. W., *Phys. Z.* **39**, 36 (1938).
23. Pringsheim, P., *Fluorescence and Phosphorescence*. Interscience, New York, 1949.
24. Seitz, F., *The Modern Theory of Solids*. McGraw-Hill, New York, 1940.
25. Shockley, W., *Electrons and Holes in Semiconductors*. New York, 1950.
26. Slater, J. C., Electronic Structure of Metals. *Revs. Mod. Phys.* **6**, 209–80 (1934).
26a. Slater, J. C., *Quantum Theory of Matter*. McGraw-Hill, New York, 1951.
27. Sommerfeld, A., and Bethe, H., *Handb. d. Physik*, Vol. 24, Pt. 2, 333-620, 1933.
28. Torrey, H. C., and Whitmer, C. A., *Crystal Rectifiers*. McGraw-Hill, New York, 1948.
29. Wiedemann, G., *Elektrizitaet*. Vol. II, p. 230 ff., 1894.
30. Wilson, A. H., *Semiconductors and Metals*. Cambridge Press, 1939.

31. Wilson, A. H., *The Theory of Metals*. London, 1936.
32. *Semiconducting Materials*, Butterworth Publication, London, 1951.
33. Schottky-Festband, *Z. physik. Chem.* **198**, (1951).

Journal References

34. Abt, A., *Ann. Physik* **2**, 266 (1900).
35. Adams, W. G., and Day, R. E., *Proc. Roy. Soc. (London)* **25**, 113 (1877).
36. Anderson, J. S., and Morton, M. C., *Nature* **155**, 112 (1945); *Trans. Faraday Soc.* **43**, 185 (1947).
37. Andrews, J. P., *Proc. Phys. Soc. (London)* **59**, 990–98 (1947).
38. Angello, S. J., *Phys. Rev.* **62**, 371-77 (1942).
39. Ansbacher, A., and Ehrenberg, W., *Nature* **164**, 144 (1949).
40. Arrhenius, S., *Wien, Ber.* **96**, 831 (1887).
41. Auwers, O. v., *Z. Physik* **93**, 90-91 (1934).
42. Baedeker, K., *Ann. Physik* **22**, 749 (1907).
43. Baedeker, K., *Ann. Physik* **29**, 566 (1909).
44. Baedeker, K., *Physik. Z.* **13**, 1080 (1912).
45. Bardeen, J., *Phys. Rev.* **71**, 717-27 (1947).
46. Bardeen, J., and Brattain, W., *Phys. Rev.* **75**, 1216 (1949).
47. Bardeen, J., and Shockley, W. S., *Phys. Rev.* **80**, 72 (1950).
48. Bauer, G., *Ann. Physik* **30**, 433 (1937).
49. Baumbach, H. H., and Wagner, C., *Z. physik. Chem.* B**22**, 199 (1933).
50. Becker, J. A., Green, C. B., and Pearson, G. L., *Elec. Eng.* **65**, 711 (1946).
51. Becker, M., and Fan, H. Y., (a) *Phys. Rev.* **75**, 1631 (1949); (b) **76**, 1530-31 (1949).
52. Becker, M., Ph.D. Thesis, Purdue University, 1951.
53. Becker, M., and Fan, H. Y., Purdue Signal Corps Prog. Rept., Contract W36-039-38151, May, 1949, and subsequent reports.
54. Becquerel, A. H., *Compt. rend.* **9**, 144, 561 (1839).
55. Benedicks, C., *Intern. Z. Metallog.* **7**, 225-37 (1915).
56. Benzer, S., *Phys. Rev.* **72**, 1267 (1947).
57. Benzer, S., *Phys. Rev.* **71**, 141 (1947).
58. Benzer, S., *J. Applied Phys.* **20**, 804 (1949).
59. Benzer, S., *Phys. Rev.* **73**, 1256 (1948).
60. Berndt, G., *Radioactive Leuchtfarben.* Vieweg, 1920.
61. Bethe, H. A., R L Report No. 43-12 (1942).
62. Bidwell, S., *Phil. Mag.* **20**, 178, (1885).
63. Bidwell, C. C., *Phys. Rev.* **19**, 447 (1922).
64. Bloch, F., *Z. Physik* **52**, 555 (1928); **59**, 208 (1930).
65. de Boer, J. H., *Rec. trav. chim.* **56**, 301 (1937).
66. de Boer, J. H., and Verwey, E. J. W., *Proc. Phys. Soc. (London)* **49**, 59-73 (1937).
67. Boltaks, B. I., *J. Tech. Phys.* (U. S. S. R.) **20**, 180 (1950).
68. Bose, J. C., U. S. Patent 755840, 1904.
69. Bottom, V., PhD. Thesis, Purdue University, 1949; *Phys. Rev.* **74**, 1218 (1948); **75**, 1310 (1949).
70. Brandes, H., *Elektrotech. Z.* **27**, 1015 (1906).
71. Brattain, W. H., *Phys. Rev.* **72**, 345 (1947).
72. Brattain, W. H., and Briggs, H. B., *Phys. Rev.* **75**, 1705 (1949).
73. Brattain, W. H., and Pearson, G. L., *Phys. Rev.* **78**, 646 (1950).
74. Braun, A., and Busch, G., *Helv. Phys. Acta* **18**, 251 (1945).
75. Braun, F., (a) *Ann. Phys. Pogg.* **153**, 556 (1874); (b) *Ann. Phys. Wied.***1**, 95 (1877); (c) **4**, 476 (1878); (d) **19**, 340 (1888).
76. Braun, F., *Elektrotech. Z.* **27**, 1199 (1906).
77. Bray, R., Ph.D. Thesis, Purdue University, 1949.

78. Briggs, H. B., *Phys. Rev.* **77**, 287 (1950).
79. Brillouin, L., *Compt. rend.* **191**, 198, 292 (1930); *J. phys.* (VII) **1**, 377 (1930).
80. Brögger, W. C., *Z. Krist.* **25**, 427 (1896).
81. Bronstein, M., *Physik. Z. Sov.* **2**, 28 (1932).
82. Burstein, E., and Oberly, J. J., *Phys. Rev.* **78**, 642 (1950).
83. Burton, M., *Ann. Rev. Phys. Chem.*, p. 113 (1950).
84. Busch, G., *Helv. Phys. Acta* **19**, 167, 463 (1946).
85. Busch, G., *Helv. Phys. Acta* **19**, 189 (1946).
86. Busch, G., *Helv. Phys. Acta* **23**, 528 (1950).
87. Cartwright, C. H., *Ann. Physik* **18**, 656 (1933); *Phys. Rev.* **49**, 443 (1936). Cartwright, C. H., and Haberfeld-Schwarz, H., *Proc. Roy. Soc. (London)* **A148**, 648 (1935).
88. Case, T. W., *Phys. Rev.* **9**, 305 (1917).
89. Case, T. W., *Phys. Rev.* **15**, 289 (1920).
90. Cashman, R., NDRC Rept. 16.4-6 (1943).
91. Chasmar, R. P., *Nature* **161**, 281 (1948).
92. Chasmar, R. P., et al., *International Conference Semiconductors*. Butterworth, London, 1951.
93. Cleland, J., Lark-Horovitz, K., and Pigg, J. C., *Phys. Rev.* **78**, 814 (1950).
94. Cleland, J. C., Thesis, Purdue University.
95. Coblentz, W. W., *Bull. Bureau Standards* **7**, 197 (1911).
96. Coblentz, W. W., and Emerson, W. B., *Wash. Acad. Sci.* **7**, 525 (1917); *Natl. Bureau Standards* (U. S.) *Sci. Papers* 322, 486.
97. Coblentz, W. W., *Phys. Rev.* **13**, 154 (1919); **14**, 534 (1920); numerous *Bull. Bur. Standards* 1919 ff.
98. Conwell, E., and Weisskopf, V. F., *Phys. Rev.* **69**, 258 (1946); **77**, 388 (1950).
99. Corbino, O. M., *Physik. Z.* **12**, 561, 842 (1911).
100. Crandall, I. B., *Phys. Rev.* **2**, 343 (1913).
101. Crawford, J. H., Ph.D. Thesis, University of North Carolina, 1949; Crawford, J. H., and Williams, F., *J. Chem. Phys.* **18**, 775 (1950).
102. Crawford, J. H., and Lark-Horovitz, K., *Phys. Rev.* **78**, 815 (1950); **79**, 889 (1950).
103. Crawford, J. H., Cleland, J. W., Lark-Horovitz, K., Pigg, J. C., and Young, F. M., *Phys. Rev.* **84**, 861 (1951).
104. Davis, R. E., M. S. Thesis, Purdue University, 1950.
105. Davis, R. E., Johnson, W. E., Lark-Horovitz, K., and Siegel, S., *Phys. Rev.* **74**, 1255 (1948); AECD Rept. 2054, June, 1948.
106. Davis, R. E., and Lark-Horovitz, K., Signal Corps Prog. Rept., Contract No. W36-039-32020, Nov. 1, 1947–Jan. 31, 1948; Aug.–Oct., 1948.
107. Davydov, B., and Gurevich, B., *J. Phys. U. S. S. R.* **7(a)**, 138 (1943).
108. Dilworth, C. C., *Proc. Phys. Soc. (London)* **60**, 315 (1948).
109. Drude, P., *Ann. Physik* **1**, 566 (1900); **3**, 369, 869 (1900); **7**, 687 (1902); **14**, 936 (1904).
110. Dunaev, Y. A., *Compt. rend. acad. sci. U. R. S. S.* **55**, 21 (1947).
111. Dünwald, H., and Wagner, C., (a) *Z. physik. Chem.* **B17**, 467 (1932); (b) **B22**, 212 (1933).
112. Dunwoody, H. H. C., U.S. Patent 837616, 1906.
113. Eisenmann, L., *Ann. Physik* **38**, 121 (1940).
114. Engelhard, E., and Gudden, B., *Z. Physik* **70**, 701 (1931).
115. Engelhard, E., *Ann. Physik* (5) **17**, 501 (1933).
116. Erginsoy, C., *Phys. Rev.* **79**, 1013 (1950).
117. Erginsoy, C., *Phys. Rev.* **80**, 1104 (1950).
118. Estermann, I., Foner, A., and Randall, J. A., *Phys. Rev.* **71**, 484 (1947); **72**, 530 (1947). Estermann, I., Foner, A., and Zimmerman, J. E., *Phys. Rev.* **75**, 1631 (1949).
119. Ettenreich, R., *Physik. Z.* **21**, 211 (1920).
120. Fan, H. Y., *Phys. Rev.* **61**, 365 (1942); **62**, 388 (1942); **74**, 1505 (1948).
121. Fan, H. Y., *Phys. Rev.* **75**, 1631 (1949).

122. Fan, H. Y., Signal Corps Prog. Rept. for June 1–Aug. 31, 1949, Contract No. W36-039-SC-38151.
123. Fan, H. Y., Signal Corps Prog. Rept., Mar. 1–May 31, 1949, Contract No. W36-039-SC-38151.
124. Fan, H. Y., *Phys. Rev.* **78**, 808 (1950); **82**, 900 (1951).
125. Faraday, M., *Exptl. Research* [Series iv], ¶ 433-39, 1833; *Pogg. Ann.* **31**, 241 (1984).
126. Forster, J. F., and Lark-Horovitz, K., *Purdue Signal Corps Prog. Rept.*, Aug. 1948.
127. Forster, J. F., *Purdue Signal Corps Prog. Rept.*, Dec. 1951.
128. Fowler, R. H., *Proc. Roy. Soc. (London)* **A140**, 505 (1933).
129. Frank, F. C., *Trans. Faraday Soc.* **33**, 513 (1937).
130. Frenkel, J., and Joffe, A., *Physik. Z. Sowjetunion* **1**, 61 (1932).
131. Frenkel, J., *Z. Physik.* **35**, 652 (1926).
132. Frerichs, R., *Phys. Rev.* **72**, 594 (1947).
133. Friederich, E., *Z. Physik.* **31**, 813 (1925).
134. Fritts, C. E., *Am. J. Sci.* (3) **26**, 465 (1883).
135. Fritsch, O., *Ann. Physik.* **22**, 375 (1935).
136. Froehlich, H., and Mott, N. F., *Proc. Roy. Soc. (London)* **160**, 230 (1937); **171**, 496 (1939).
137. Gans, R., *Ann. Physik.* **20**, 293 (1906).
138. Gibson, A. F., *Proc. Phys. Soc. (London)* **B63**, 756 (1950).
139. Gisolf, J. H., *Ann. Physik.* **1**, 3 (1947).
140. Glaser, G., and Lehfeldt, W., *Nachr. Ges. Wiss. Göttingen* **3**, 91 (1936).
141. Glaser, G., *Nachr. Ges. Wiss. Göttingen* **3**, 31 (1937).
142. Goldman, J., and Lawson, A. W., *Phys. Rev.* **64**, 11 (1943).
143. Goldsmith, G. J., and Lark-Horovitz, K., *Phys. Rev.* **75**, 526 (1949).
144. Gottstein, G., *Ann. Physik* **43**, 1079 (1914).
145. Grondahl, L. O., *Phys. Rev.* **27**, 813 (1926); *Science* **64**, 306 (1926).
146. Grondahl, L. O., and Geiger, P. H., *Trans. A.I.E.E.* **46**, 357 (1927).
147. Gudden, B., and Pohl, R. W., *Z. Physik* **7**, 65 (1921).
148. Gudden, B., *Erlangen Ber.* **62**, 289 (1930).
149. Gudden, B., *Physik Z.* **32**, 825 (1931).
150. Gudden, B., and Pohl, R. W., *Naturwissenschaften* **11**, 348 (1923).
151. Gudden, B., and Pohl, R. W., *Z. Physik* **16**, 42 (1923).
152. Gudden, B., and Pohl, R. W., *Z. Physik* **16**, 170 (1923).
153. Gudden, B., and Pohl, R. W., *Z. Physik* **21**, 1 (1924).
154. Gudden, B., and Schottky, W., *Z. tech. Physik* **16**, 323-27 (1935).
155. Guillery, P., *Physik Z.* **32**, 891 (1931); *Ann. Physik* **14**, 216-20 (1932).
156. Guinchant, J., *Compt. rend.* **134**, 1224 (1902).
157. Gundermann, J., Hauffe, K., and Wagner, C., *Z. phys. Chem.* **B37**, 148-54 (1937).
158. Guntz, A., and Broniewski, W., *Compt. rend.* **147**, 1474 (1908); **148**, 204 (1909).
159. Hahn, E. E., Russell, B. R., Miller, P. H., *Phys. Rev.* **75**, 1631 (1949).
160. Haken, W., *Ann. Physik* **32**, 291 (1910).
161. Hall, E. H., *Am. J. Math.* **2**, 287 (1879); *Am. J. Sci.* **19**, 200 (1880).
162. Hauffe, K., *Ann. Phys.* **8**, 201 (1950).
163. Hecht, K., *Z. Physik* **77**, 235 (1932).
164. Heisenberg, W., *Ann. Physik* **10**, 888 (1931).
165. Henninger, F. P., *Ann. Physik* **28**, 245 (1937); *Physik Z.* **39**, 216-24 (1938).
166. Herzfeld, K. F., NDRC Rept. 14-286, Purdue, 1944.
167. Hevesy, G. V. *Danske Videnskab. Selskab.* **3**, 12 (1921); *Naturwissenschaften* **13**, 225 (1925).
168. Hilsch, R., and Pohl, R. W., *Z. tech. Physik* **16**, 338 (1935).
169. Hilsch, R., and Pohl, R. W., *Z. Physik* **108**, 55 (1937).
170. Hilsch, R., and Pohl, R. W., *Z. Physik* **111**, 399 (1938); **112**, 252 (1939).

171. Hintenberger, H., *Z. Physik* **119**, 1 (1942); *Z. Naturforsch.* **1**, 13 (1946).
172. Hippel, A. v., *et al.*, *J. Chem. Phys.* **14**, 355 (1946). Hippel, A. v., and Rittner, E. S., *J. Chem. Phys.* **14**, 370 (1946).
173. Hittorf, W., *Pogg. Ann.* **84**, 1 (1851).
174. Hochberg, B. M., and Sominski, M. J., *Phys. Z. Sowjetunion* **13**, 198 (1938).
175. Hofstadter, R., *Nucleonics* **4**, April, p. 2; May, p. 29 (1949).
176. Hogarth, C. A., *Nature*, **161**, 60 (1948); *Phil. Mag.* **40**, 273 (1949).
177. Hung, C. S., and Gliessman, J. R., *Phys. Rev.* **79**, 726 (1950).
178. Hung, C. S., and Johnson, V. A., *Phys. Rev.* **79**, 535 (1950).
179. Hung, C. S., *Phys. Rev.* **79**, 727 (1950).
180. Hutner, R. A., Rittner, E. S., and DuPre, F. K., *Philips Research Repts.* **5**, 188 (1950).
181. Ingersoll, L. R. *Astrophys.*, *J.* **32**, 265 (1910).
182. James, H. M., *Phys. Rev.* **74**, 1218 (1948).
183. James, H. M., *Phys. Rev.* **76**, 1602 (1949).
184. James, H. M., *Phys. Rev.* **76**, 1611 (1949).
185. James, H. M., and Ginzbarg, A. S., *Phys. Rev.* **77**, 749 (1950).
186. James, H. M., and Lark-Horovitz, K., *Z. phys. Chem.* **198**, 107 (1951).
187. Joffè, A., and Joffè, A. V., *Z. Physik* **82**, 754 (1933).
188. Joffè, A. V., *J. Phys. U.S.S.R.* **10**, 49-60 (1946).
189. Joffè, A. V., and Joffè, A. F., *J. Exp. Theoret. Phys. U.S.S.R.* **9**, 1451-58 (1939); *J. Phys. U.S.S.R.* **2**, 283 (1940).
190. Johnson, V. A., Purdue U. Prog. Rept., Aug. 1949, p. 26.
191. Johnson, V. A., and Fan, H. Y., *Phys. Rev.* **79**, 899 (1950).
192. Johnson, V. A., and Lark-Horovitz, K., *Phys. Rev.* **69**, 258 (1946).
193. Johnson, V. A., and Lark-Horovitz, K., *Phys. Rev.* **71**, 374, 909 (1947); **71**, 483 (1947).
194. Johnson, V. A., and Lark-Horovitz, K., *Phys. Rev.* **79**, 176 (1950); **82**, 977 (1951.)
195. Johnson W. E., and Lark-Horovitz, K., NEPA Rept. 1178-IER-23; *Phys. Rev.* **76**, 442-43 (1949).
196. Johnson, V. A., Yearian, H. J., and Smith, R. N., *J. Applied Phys.* **21**, 283 (1950).
197. Jones, H., *Phys. Rev.* **81**, 149 (1951).
198. Jost, W., *J. Chem. Phys.* **1**, 466 (1933).
199. Juse, W. P., and Kurtschatow, B. W., *Physik. Z. Sowjetunion* **2**, 454-67 (1932).
200. Kalabuchow, N., *J. Physik U.S.S.R* **10**, 61 (1946).
201. Kalischer, S., *Wied. Ann.* **35**, 397 (1888).
202. Kallmann, H., and Warminsky, R., *Ann. Physik* **4**, 69 (1948).
203. Kendall, J. T., *Proc. Phys. Soc. (London)* **63**, 821 (1950).
204. Klaiber, F., *Ann. Physik* **3**, 229 (1929).
205. Klontz, E., and Lark-Horovitz, K., Signal Corps Contract W36-039-32020, Aug.–Oct. 1948, Prog. Rept.
206. Klontz, E., and Lark-Horovitz, K., Signal Corps Contract W36-039-SC-38151, Dec. 1949–Feb. 1950, Prog. Rept.; *Phys. Rev.* **82**, 763 (1951).
207. Klumb, H., *Physik Z.* **40**, 640 (1939).
208. Koenigsberger, J., *Physik Z.* **4**, 495 (1903); *Ann. Physik* **43**, 1205 (1914).
209. Koenigsberger, J., and Schilling, K., *Ann. Physik* **32**, 179–229 (1910).
210. Koenigsberger, J., and Gottstein, G., *Physik Z.* **14**, 232 (1913); *Ann Physik* **46**, 446 (1915).
211. Kowalenko, V., *Tech. Phys. U.S.S.R.* **5**, 789 (1938).
212. Kronig, R. de L., *Proc. Roy. Soc. (London)* **133**, 255 (1931); *Nature* **133**, 211 (1934).
213. Kuestner, H., *Z. Physik* **10**, 41 (1922).
214. Kurtschatow, I. V., *et al.*, *Phys. Z. Sowjetunion* **7**, 129 (1935); Kurtschatow, I. V., and Sinelnikov, K. D., *Phys. Z. Sowjetunion* **1**, 23-41 (1932).
215. Lange, B., *Physik Z.* **32**, 850-56 (1931).
216. Lark-Horovitz, K., *Elec. Eng.* **68**, 1047 (1949).

217. Lark-Horovitz, K., *Nucleon-Irradiated Semiconductors*, International Conference on Semi-conductors, Butterworth, London, 1951.
218. Lark-Horovitz, K., Becker, M., Davis, R. E., and Fan H. Y., *Phys. Rev.* **78,** 334 (1950).
219. Lark-Horovitz, K., Bleuler, E., Davis, R. E., and Tendam, D., *Phys. Rev.* **73,** 1256 (1948).
220. Lark-Horovitz, K., and Johnson, V. A., *Phys. Rev.* **69,** 258 (1946).
221. Lark-Horovitz, K., and Meissner, K. W., *Phys. Rev.* **76,** 1530 (1949).
222. Lark-Horovitz, K., Middleton, A. E., Miller, E. P., Scanlon, W. W., and Walerstein, I., *Phys. Rev.* **69,** 258 (1946).
223. Larmor, J., *Ether and Matter*, Cambridge University Press, Cambridge, 1900, p. 301 ff.
224. Le Blanc, M., and Sachse, H., *Z. Elektrochem.* **32,** 204 (1926).
225. Le Blanc, M., and Sachse, H., *Ber. Verhandl Sächs. Akad. Wiss. Leipzig, Math. phys. Klasse* **82,** 133, 153 (1930).
226. Le Blanc, M., and Sachse, H., *Ann. Physik* (5) **11,** 727 (1931).
227. Le Blanc, M., and Sachse, H., *Physik Z.* **32,** 887 (1931).
228. Le Blanc, M., Sachse, H., and Schoepel, H., *Ann. Physik* **17,** 334 (1933).
229. Lehfeldt, W., *Z. Physik* **85,** 717 (1933).
230. Lehfeldt, W., *Göttingen Nachr.* **2,** 171–86 (1935).
231. Lehovec, K., *Phys. Rev.* **74,** 463 (1948).
232. Lehman, G., *Phys. Rev.* **81,** 321 (1951).
233. Lehman, G., and James, H. M., *Semiconducting Materials*. Butterworth Scientific Publication, London, 1951, pp. 171–77.
234. Lenz, H., *Ann. Physik* **82,** 775 (1927); *Physik Z.* **25,** 435 (1924); *Ann. Physik* **77,** 449 (1925).
235. Lorentz, H. A., *Konink. Akad. Wetenschap. Amsterdam, Proc.* **7,** 438, 565 (1905).
236. Maurer, R. J., *J. Chem. Phys.* **13,** 321 (1945).
237. McKay, K. G., *Phys. Rev.* **74,** 1606-07 (1948); **76,** 1537 (1949).
238. Menzel, E., *Z. anorg. Chem.* **256,** 49 (1948).
239. Merritt, E., *Phys. Rev.* **23,** 555 (1924).
240. Merritt, E., *Proc. Natl. Acad.* **11,** 743 (1925).
241. Meyer, W., *Z. Elektrochem.* **50,** 274–90 (1944).
242. Meyer, W., and Neldel, H., *Z. tech. Physik* **18,** 588 (1937).
243. Middleton, A. E., Ph.D. Thesis, Purdue University, 1945.
244. Miller, P. H., *Phys. Rev.* **60,** 890 (1941); Miller, P. H., and Hahn, E. E., *Phys. Rev.* **78,** 349 (1950); Miller, P. H., *Semiconducting Materials*. Butterworth Scientific Publications, London, (1951).
245. Mollwo, E., *Göttingen Nachr.* **20,** 215 (1935).
246. Mollwo, E., and Stöckmann, F., *Ann. Physik* **3,** 223 (1948).
247. Morse, P. M., *Phys. Rev.* **35,** 1310 (1930).
248. Morton, M. G., *Trans. Faraday Soc.* **43,** 194 (1947).
249. Moss, T. S., *Proc. Phys. Soc. (London)* **A62,** 264 (1949); **B63,** 167 (1950).
250. Moss, T. S., *Phys. Rev.* **79,** 1011 (1950).
251. Mott, N. F., *Proc. Roy. Soc. (London)* **A171,** 27 (1939).
252. Mott, N. F., *Proc. Roy. Soc. (London)* **A171,** 281 (1939).
253. Mott, N. F., and Gurney, R. W., *Trans. Faraday Soc.* **34,** 506 (1938).
254. Muegge, O., *Zentr. Mineral. Geol.* **1922,** 721, 752 (1922).
255. Mueser, H., *Z. Naturforsch.* **5a,** 18 (1950).
256. Mullaney, J. F., *Phys. Rev.* **66,** 326 (1944).
257. Nelson, J. B., and McKee, J. H., *Nature* **158,** 753 (1946).
258. Nordheim, L. W., *Z. Physik* **75,** 434 (1932).
259. O'Bryan, H. M., *J. Optical Soc. Am.* **26,** 122 (1936).
260. Orman, C., Fan, H. Y., Goldsmith, G. J., and Lark-Horovitz, K., *Phys. Rev.* **78,** 646 (1950).

261. Orndoff, J. D., Thesis, Purdue University, 1944; *Phys. Rev.* **65**, 348 (1944).
262. Pauli, W., *Z. Physik* **41**, 81 (1927).
263. Pauling, L., *Nature* **161**, 1019 (1948); *Physica* **15**, 23 (1949); *Proc. Roy. Soc. (London)* **A196**, 343 (1949).
264. Pearson, G. L., *Phys. Rev.* **76**, 179 (1949).
265. Pearson, G. L., *Elec. Eng.* **66**, 638 (1947).
266. Pearson, G. L., and Bardeen, J., *Phys. Rev.* **75**, 865 (1949).
267. Pearson, G. L., Struthers, J. D., and Theurer, H. C., *Phys. Rev.* **75**, 344 (1949).
268. Peierls, R., *Z. Physik* **53**, 255 (1928); *Ann. Physik* **4**, 121 (1930).
269. Peierls, R., *Ergeb. Exakt. Naturw.* **11**, 264, 314, 319 (1932).
270. Pfestorf, G. *Ann. Physik* **81**, 906 (1926).
271. Pfund, A. H., *Phys. Rev.* **7**, 289 (1916).
272. Pierce, G. W., *Phys. Rev.* **25**, 31 (1907); **28**, 153 (1909).
273. Pincherle, L., *Proc. Phys. Soc. (London)* **A64**, 603 (1950).
274. Pohl, R. W., and Stöckmann, F., *Ann. Physik* **6**, 89 (1949).
275. Ringer, W., and Welker, H., *Z. Naturforsch.* **3a**, 20 (1948).
276. Rittner, E. S., *Phys. Rev.* **73**, 1212 (1948).
277. Roentgen, W. C., and Joffé, A., *Ann. Physik* **64**, 1 (1921).
278. Roth, L. M., and Taylor, W. E., *Proc. I.R.E.* **40**, 1338 (1952).
279. Rudert, G., *Ann. Physik* **31**, 559 (1910).
280. Rutherford, E., *Proc. Roy. Soc. (London)* **83**, 561 (1910).
281. Scaff, J. H., and Theurer, H. C., *J. Metals* **191**, 59 (1951).
282. Scaff, J. H., Theurer, H. C., and Schumacher, E. E., *Metals Trans.* **185**, 383 (1949).
283. Scanlon, W. W., Thesis, Purdue University, 1948; *Phys. Rev.* **72**, 530 (1947); **73**, 125 (1948).
284. Schleede, A., and Buggisch, H., *Jahrb. drahtl. Teleg.* **30**, 190 (1937).
285. Schmid, E., and Staffelbach, F., *Ann. Physik* **29**, 273 (1937).
286. Schoenwald, E., *Ann. Physik* **15**, 395 (1932).
287. Scholl, H., *Ann. Physik* **16**, 193, 417 (1905).
288. Schottky, W., *Z. Physik* **14**, 63 (1923); *Physik. Z.* **32**, 833 (1931).
289. Schottky, W., *Z. physik Chem.* **B29**, 335 (1935).
290. Schottky, W., *Z. Physik* **113**, 367 (1939); *Siemens Ver.* **XVIII**, 1-68 (1939); *Z. Physik* **118**, 539 (1942).
291. Schottky, W., and Duhme, E., *Naturwissenschaften* **16**, 735 (1930).
292. Schottky, W., *Z. Elektrochem.* **45**, 33 (1939).
293. Schottky, W., *Ann. Physik* **6**, 193 (1949).
294. Schottky, W., *Naturwissenschaften* **23**, 656 (1935).
295. Schottky, W., *Z. tech. Physik* **11**, 458 (1930); *Physik Z.* **31**, 913 (1930); **32**, 833 (1931).
296. Schottky, W., and Waibel, F., *Physik Z.* **36**, 912 (1935).
297. Schulze, A., *Physik Z.* **31**, 1062 (1930); *Z. tech. Physik* **11**, 443 (1930).
298. Schuster, A., *Phil. Mag.* **48**, 251 (1874).
299. Seebeck, T. J., *Berlin Akad.* p. 265 (1822-23); reprinted in Ostwald's Klassiker, No. 70, Leipzig, 1895.
300. Seitz, F., *J. Applied Phys.* **16**, 553 (1945).
301. Seitz, F., *Faraday Soc.* **1949**, 271, No. 5.
302. Seitz, F., *Phys. Rev.* **73**, 549 (1948).
303. Seitz, F., and Johnson, R. P., *J. Applied Phys.* **8**, 84, 186, 246 (1937).
304. Sheldon, H. H., and Geiger, P. H., *Proc. Natl. Acad. Sci. U.S.* **8**, 161 (1922); *Phys. Rev.* **22**, 461 (1921).
305. Shifrin, K. C., *J. Phys. U.S.S.R.* **8**, 242 (1944); *J. Tech. Phys. (U.S.S.R.)* **14**, 40–42 (1944).
306. Shockley, W., Pearson, G. L., and Haynes, J. R., *Phys. Rev.* **78**, 295 (1950).
307. Siemens, W., *Beibl. Ann. Physik* **10**, 115 (1886); *Berlin Akad.* **8**, 147 (1885).

308. Simpson, O., *Nature* **160,** 791 (1947).
309. Skaupy, F., *Z. Physik* **1,** 259 (1920).
310. Slater, J. C., *Phys. Rev.* **76,** 1592 (1949).
311. Smith, W., *Am. J. Sci.* **5,** 301 (1873).
312. Sommerfeld, A., *Z. Physik* **47,** 1, 43 (1928).
313. Sommerfeld, A., and Frank, N. H., *Revs. Mod. Phys.* **3,** 1 (1931).
314. Somerville, A. A., *Phys. Rev.* **33,** 77 (1911).
315. Sosnowski, L., *Phys. Rev.* **72,** 641 (1947). Sosnowski, L., Soole, B. W., and Starkiewicz, J., *Nature* **160,** 471 (1947). Starkiewicz, J., Sosnowski, L., and Simpson O., *Nature* **158,** 28 (1946).
316. Stasiw, O., *Nachr. Ges. Wiss. Göttingen* 261 (1932).
317. Stasiw, O., *Nachr. Ges. Wiss. Göttingen* 199 (1935).
318. Steinberg, K., *Ann. Physik* **35,** 1009 (1911).
319. Stephens, W. E., Serin, B., and Meyerhof, W. E., *Phys. Rev.* **69,** 42 (1946).
320. Stoeckmann, F., *Naturwissenschaften* (a) **36,** 82 (1949); (b) **37,** 85, 105, 523 (1950).
321. Streintz, F., *Physik. Z.* **4,** 106 (1902).
322. Strock, L. W., *Z. Krist.* **93,** 285 (1936).
323. Strutt, M. J. O., *Ann. Physik* **86,** 319 (1928).
324. Szekely, A., *Wien, Ber.* **127,** 719 (1918).
325. Taylor, W. E., Ph.D. Thesis, Purdue University, June 1950; Signal Corps Spec. Rept., Contract W36-039-sc-38151, 1950.
326. Tissot, C., *Jb. draht. Telg. u. Telf.* **2,** 115 (1908).
327. Tubandt, C., *Z. Elektrochem.* **26,** 358 (1920).
328. Tubandt, C., et al., *Z. Elektrochem.* **37,** 589 (1931).
329. Tubandt, C., et al., *Z. Elektrochem.* **39,** 227 (1933).
330. Tubandt, C., and Reinhold, H., *Z. phys. Chem.* **24,** 22 (1934).
331. Vegard, L., *Phil. Mag.* **32,** 65 (1916).
332. Verwey, E. J. W., *Nederland. Tijdschr. Natuurk.* **14,** 205 (1948).
333. Verwey, E. J. W., Haaijman, P. W., and Romeijn, F. C., *Chem. Weekblad* **44,** 705 (1948); *J. Chem. Phys.* **15,** 181 (1947). Verwey, E. J. W., Haaijman, P. W., Romeijn, F. C., and van Oosterhout, G. W., *Philips Research Rept.* **5,** 173 (1950).
334. Vogt, W., *Ann. Physik* **7,** 183 (1930).
335. Voelkl, A., *Ann. Physik* **14,** 193 (1932).
336. Volmer, M., *Z. Elektrochem.* **21,** 113 (1915).
337. Wagner, C., *Physik. Z.* **32,** 641 (1931).
338. Wagner, C., *Z. physik. Chem.* **B21,** 42 (1933).
339. Wagner, C., *Z. physik. Chem.* **B22,** 181 (1933).
340. Wagner, C., *Z. physik. Chem.* **B22,** 195 (1933).
341. Wagner, C., *Z. Physik. Chem.* **B22,** 469 (1933).
342. Wagner, C., Baumbach, H. V., and Dunwald, H., *Z. physik. Chem.* **B22,** 226 (1933).
343. Wagner, C., and Schottky, W., *Z. physik. Chem.* **B11,** 163 (1930).
344. Wagner, C., *Trans. Faraday Soc.* **34,** 851 (1938).
345. Wartenberg, H. V., *Ber. deut. physik. Ges.* **12,** 105 (1910).
346. Whaley, R. M., NDRC Purdue Prog. Rept., Oct. 1945.
347. Wilson, A. H., *Proc. Roy. Soc. (London)* **A133,** 458 (1931); **134,** 277 (1931).
348. Wilson, A. H., *Proc. Roy. Soc. (London)* **A136,** 487 (1932).
349. Wilson, W., *Ann. Physik* **23,** 107 (1907).
350. Wold, P. S., *Phys. Rev.* **7,** 169 (1916).
351. Yearian, H. J., *J. Applied Phys.* **21,** 214 (1950).
352. Zener, C., *Nature* **132,** 968 (1933).

FLOW OF ELECTRONS AND HOLES IN SEMICONDUCTORS*

J. BARDEEN †

Bell Telephone Laboratories, Murray Hill, New Jersey

Electrons in a semiconductor can carry current in two different ways: (1) by excess, or conduction, electrons which do not fit into the valence-bond structure of the solid, and (2) by defect electrons, or holes, which represent electrons missing from the valence bonds. While the former have the normal negative charge of an electron, the latter behave in all respects like positively charged particles. The two types of carriers are denoted by the terms conduction electrons (or simply electrons) and holes, respectively.‡

This paper is concerned with the flow of electricity in semiconductors under conditions in which appreciable numbers of both types of carriers are present. It is not intended to be a comprehensive review. We shall discuss several examples of flow in order to illustrate the principles involved. It is based mainly on studies, both experimental and theoretical, which have been carried out at the Bell Telephone Laboratories as a part of the research program on transistors. Although a number of groups in the Laboratories have contributed in important ways, the work which we shall describe has been largely centered in a group under the general direction of William Shockley (1). Germanium is the semiconductor used in most of this research.

The transistor is very roughly analogous to a vacuum tube. Flow of electrons in a semiconductor replaces flow in a vacuum. Carriers introduced at one electrode, called the emitter, are drawn to a second electrode, called the collector. These electrodes have functions similar to the cathode and plate of a vacuum tube. The analogy is crude because the nature of flow in semiconductors is quite different from flow in a vacuum. The notions with which we deal are similar to those used in discussions of flow in electrolytes and in gas discharges: mobilities and conduction by electric fields, recombination and lifetime, diffusion and space charge.

We shall first give a brief review of the properties of semiconductors and of the nature of rectifying barrier layers to show how the conductivity can be changed by current flow, then present some of the basic equations which

---

* Revisions have been made to bring the references up to date and an expanded section on *p-n* junctions has been included.

† Present address: Department of Physics, University of Illinois, Urbana, Illinois.

‡ The nature of the conduction of electricity in semiconductors is discussed in the paper of K. Lark-Horovitz.

determine the flow when both electrons and holes are present, and finally discuss several experiments and their interpretation in terms of the theory. These include experiments on the flow of holes in n-type germanium filaments, on the effect of added concentration of holes on the current-voltage characteristics of rectifying contacts and on p-n junctions.

The modern theory of conduction in solids is based on an application of quantum theory to the motion of electrons in the periodic field of the crystal lattice. Although we shall not be concerned with quantum ideas to any great extent, we shall give an outline of the theory as applied to semi-conductors. Figure 1 is an energy level diagram which shows the allowed levels for electrons in a crystal such as germanium or silicon. There is a continuous band of levels normally occupied by the electrons in the valence bonds, an energy gap in which there are no levels in the ideal crystal, and then another

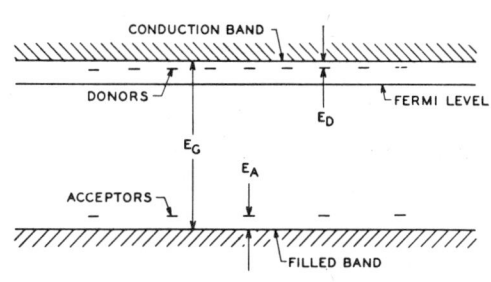

FIG. 1. Energy level diagram for a semiconductor such as germanium or silicon. Electrons in the valence bonds occupy the levels of the filled band. The energy gap, E_G, in which there are no allowed levels in the ideal crystal, is about 0.75 ev in germanium and 1.1 ev in silicon. Electrons with energies in the conduction band and missing electrons or holes in the filled band are mobile and contribute to the conductivity. Donor and acceptor levels are those of electrons localized at foreign atoms.

continuous band of levels, the conduction band, normally unoccupied. The energy gap in semiconductors is smaller than in insulators and is usually of the order of one electron volt.

At high temperatures, electrons can be thermally excited from the valence band to the conduction band, and electrons in the conduction band and the missing electrons or holes in the valence band both contribute to the conductivity. Conductivity of this type is called *intrinsic*. In most cases the conductivity at room temperature is due mainly to impurities. Impurities which contribute to the conductivity are of two types: (1) *donors*, which have energy levels a little below the conduction band, are normally neutral, and become positively ionized by thermal excitation of electrons to the conduction band, and (2) *acceptors*, which have energy levels a little above the valence band, are normally neutral, and become negatively charged when the levels are occupied by thermal excitation of electrons from the filled band. The energies involved are so small in germanium that practically all donors and acceptors are ionized at room temperature. A semiconductor with donor impurities conducts by electrons in the conduction band and is said to be n-type; one with acceptors conducts by holes in the filled band and is said to be *p*-type. If both types of impurities are present, electrons will be transferred from the donor levels to the lower-lying acceptor levels. The conductivity type then depends on which is in excess and under conditions of electrical neutrality the concentration (number per unit volume) of mobile carriers is equal to the difference between the concentrations of donor and acceptor ions.

The equilibrium concentration of the minority carrier is small but not entirely negligible. It can be shown that under equilibrium conditions the product of the electron and hole concentrations, which we denote by the symbols n and p, is independent of impurity concentrations and depends only on the temperature:

$$np = K = n_i^2 \qquad (1)$$

The product is equal to the square of the intrinsic concentration, n_i, and increases rapidly with temperature. In germanium, at room temperature, n_i is about $2.5 \times 10^{13}/\text{cm}^3$. If there is a large excess of donors so that n is large, p is correspondingly small.

Injection of Carriers at Point Contacts

It is possible to increase the concentration of carriers, and thus the conductivity, without introducing space charge by adding equal numbers of electrons and holes. It has long been known that this could be done by application of heat, giving intrinsic conductivity, or by light or other radiation, giving photoconductivity. In both cases electrons are raised from the filled band to the conduction band giving rise to an electron in the conduction band and a hole in the valence band. A third method, which is the one used in the transistor and is the one which is our main concern, is by current flow from an appropriate contact. An example is a metal-point or catwhisker contact to *n*-type germanium. Current flowing from the contact biased positively with respect to the germanium consists in large part of holes, that is, of carriers of opposite sign to those normally present in excess in the interior. The metal takes electrons directly from the filled band and the holes left behind flow into the germanium. The space charge of these added holes is compensated by an increase in the concentration of conduction electrons. That the holes introduced in this way actually flow into the body of the germanium has been demonstrated by experiments (2).

The nature of the flow, in which both conduction and diffusion play a role, is illustrated in a schematic way in Fig. 2. In the vicinity of the contact from which the current is flowing there is a high concentration of both holes and electrons. This high concentration is not due to a rise in temperature, but exists because current flowing across the rectifying barrier layer at the metal-germanium contact consists largely of holes. Both holes and electrons tend to diffuse away from the region of high concentration. The electric field is in such a direction as to move the holes away from the contact and electrons toward the contact. Thus, the diffusion and conduction currents add for holes and subtract for electrons. This makes it possible for the net hole current to be larger than the net electron current. It is estimated that the concentrations of electrons and holes in the immediate vicinity of the contact may be as much as 50 times as large as the normal concentration of conduction electrons in the germanium (3).

A point contact of this sort acts as a rectifier; current flows much more readily when the point is positive so as to inject holes into the n-type germanium than when the point is negative. In the high resistance direction, the current must consist of electrons flowing from the point to the conduction band of the germanium or of holes from the germanium to the

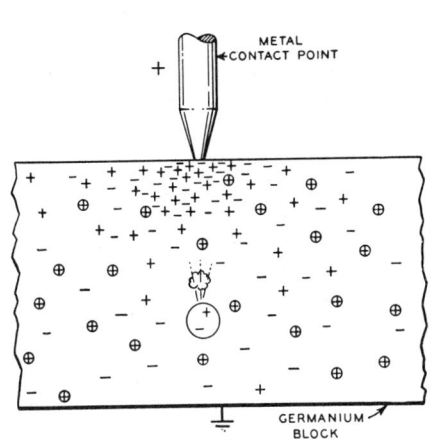

FIG. 2. Current flow from metal point to n type germanium block. Although the carriers normally present in the block are conduction electrons (−), a large part of the flow consists of holes, (+), flowing away from the point contact. The conductivity in the vicinity of the contact is enhanced by the injected holes and by the electrons which flow in to neutralize the space charge of the holes. The fixed charges of the donor ions are represented by ⊕. Electric fields and concentration gradients are both important in the flow in the vicinity of the contact. Holes introduced in this way have a relatively short lifetime as they tend to combine with electrons and disappear. One such process is illustrated schematically in the central part of the figure.

point where they are occupied by electrons from the metal. The current to the conduction band is small because there is a potential barrier at the junction which electrons from the metal must surmount to enter the germanium. The hole current is small because there are few holes present under equilibrium conditions in the n-type germanium.

In the point-contact transistor, there are two contacts in close proximity. One, called the *emitter*, is biased positively so as to inject holes into the germanium. The second, called the *collector*, is biased negatively, the direction of high resistance, with a relatively large voltage. There is a third large-area low-resistance contact to the germanium block, called the *base* electrode. Holes injected into the germanium by the emitter are attracted to the collector by the electric field produced by the collector current. The magnitude of the emitter current is determined mainly by the voltage between the emitter and base rather than by the voltage on the collector. A relatively small signal applied between emitter and base may be used to control a much larger power in the collector circuit. In the rough analogy with a vacuum tube, mentioned earlier, the emitter corresponds to the cathode, the collector to the plate, and the base electrode to the grid. This analogy is closer for the junction transistor, to be described later, than for the point-contact transistor.

Holes introduced at the emitter contact do not exist indefinitely in the n-type germanium, but tend to recombine with conduction electrons. In other words, electrons in the conduction band drop into the state from which an electron is missing in the filled band, and both conduction electron and hole disappear. At large distances from the contact, injected holes disappear and the current consists almost entirely of electrons. The lifetime of a hole in n-type germanium is relatively long, however, because it is difficult for the electron and hole to give up the energy of recombination to the lattice. It is believed that recombination occurs most frequently at lattice imperfections, either in the interior or at the surface of the crystal, where both an electron and a hole can be trapped. Lifetimes as long as 1000 microseconds have been observed in well-annealed single crystals.

At any temperature, electrons are continually being thermally excited from the valence band to the conduction band, creating electron-hole pairs. The rate of thermal generation of pairs increases rapidly with temperature. Under conditions of thermal equilibrium, there is a balance between electrons and holes recombining and thermal generation. The equilibrium concentration of minority carriers is equal to the product of the rate of generation, g (number of pairs/cm³/sec) and the lifetime, τ. For n-type material,

$$g\tau_p = p_n = n_i^2/n_n$$

or

$$g = n_i^2/n_n\tau_p \qquad (2)$$

A word in regard to notation: the subscript n on p_n and n_n is used to denote conductivity type; the subscript p on τ_p indicates that it refers to lifetime of holes.

Injected holes may become trapped in the lattice, forming sites with fixed positive charge. Electrical neutrality is then maintained by an increased concentration of conduction electrons. Acceptors, normally negatively charged, may be neutralized by capture of holes. These processes may play a role in the vicinity of the collector contact of a transistor (4).

THEORY OF FLOW OF HOLES AND ELECTRONS

Our picture of a semiconductor is then a medium in which there may be mobile carriers of both signs—conduction electrons and holes, and fixed charges of both signs—ionized donors and acceptors. Both diffusion and conduction are important in determining the flow of carriers. The concentrations of electrons and holes may be increased above the equilibrium values in various ways. There is then a tendency to reestablish equilibrium by recombination.

The electric field in a semiconductor is determined in part by local space charge. Rectifying barrier layers which occur between a metal and semiconductor or between two semiconductors are cases in which fields produced by space charge are very important. Such barrier layers may also occur between two regions of a semiconductor which differ in impurity content or at the free surface of a semiconductor. We shall discuss later the barrier separating regions of n- and p-type conductivity. Our main concern is about current flow in regions where there is substantial electrical neutrality. It is usually possible to treat the space-charge regions separately. In fact, they really require separate treatment because the currents which flow are usually determined by the concentrations at the boundaries of the regions rather than by what goes on within the barrier layer itself. Equations of the diode theory rather than diffusion theory should then be used (5).

Substantial electrical neutrality must exist throughout the volume of a semiconductor of any size. It is not difficult to show that an unbalance of charge extending over an appreciable distance would produce a very large field. There are about 10^{15} electrons/cm^3 in germanium with a resistivity of about 2 ohm-cm. A layer with a 10% unbalance of charge extending over a distance of 10^{-3} cm would give a field of more than 10^4 volts/cm and a potential drop of the order of 10 volts.

The influence of an electric field on the motion of carriers may be expressed in terms of mobility. Since electrons and holes suffer collisions with the lattice, they are not accelerated indefinitely but acquire a limiting velocity proportional to the electric field. The mobility is defined as the drift velocity in unit field. Values for germanium at room temperature are

    Electrons $\mu_n = 3500$ cm^2/volt sec
    Holes $\mu_p = 1700$ cm^2/volt sec

These values were determined by Haynes and Shockley (6) by a direct method which we shall describe later. Mobilities in germanium are higher than those of carriers in any other known substance except InSb.

Electrons and holes also flow as a result of concentration gradients. The diffusion coefficients, D_n and D_p, for electrons and holes may be expressed in terms of the mobilities μ_n and μ_p by Einstein's relation:

$$D_n = \mu_n kT/e \qquad D_p = \mu_p kT/e \tag{3}$$

in which k is Boltzmann's constant, T is the absolute temperature, and e is the magnitude of the electronic charge. The corresponding electrical conductivities are:

$$\sigma_n = ne\mu_n \qquad \sigma_p = pe\mu_p \tag{4}$$

where n and p are the concentrations of concentrations electrons and holes. The expressions for the current densities, i_n and i_p:

$$i_n = \sigma_n E + eD_n \text{ grad } n \tag{5a}$$
$$i_p = \sigma_p E - eD_p \text{ grad } p \tag{5b}$$

may be written in the form

$$i_n = \mu_n (enE + kT \text{ grad } n) \quad (6)$$
$$i_p = \mu_p (epE - kT \text{ grad } p) \quad (7)$$

In these equations, E is the electric field strength and the other symbols have been defined above. The first term in each case comes from conduction and the second from diffusion.

The electron and hole currents are not conserved independently. Electrons and holes can be lost by trapping and by recombination. If trapping is neglected, the steady state conservation equation is:

$$\text{div } i_n = -\text{div } i_p = e \times \text{ net rate of recombination} \quad (8)$$

The total current density, $i = i_n + i_p$, does satisfy the conservation equation,

$$\text{div } i = 0 \quad (9)$$

The final equation required is that which gives electrical neutrality. In germanium the donors and acceptors are completely ionized at room temperature, so that

$$p - n + N_D - N_A = 0 \quad (10)$$

If the donors and acceptors are not completely ionized, further equations which show how the degree of ionization is related to the concentrations of carriers are required.

These are the basic equations which determine the flow. It is sometimes useful to rewrite them in a different form by introducing the diffusion potentials:

$$\zeta_n = -(kT/e) \log (n/n_0) \qquad \zeta_p = (kT/e) \log (p/p_0) \quad (11)$$

for electrons and holes, respectively. In terms of these, and the electric potential,

$$E = -\text{grad } V \quad (12)$$

the expressions for the electron and the hole current densities are:*

$$i_n = -\sigma_n \text{ grad } (V + \zeta_n) \qquad i_p = -\sigma_p \text{ grad } (V + \zeta_p) \quad (13)$$

As kT/e is equal to 0.025 volt at room temperature, a large difference in concentration is required to produce a diffusion potential of more than a few hundredths of a volt.

In equilibrium, there is no current flow, so that $i_n = i_p = 0$,

$$n = n_0 \exp (eV/kT) \qquad p = p_0 \exp (-eV/kT) \quad (14)$$

The product pn is independent of V and depends only on the temperature. Variations in V may occur between regions of a semiconductor with differing impurity content. We shall discuss later the characteristics of a junction between p- and n-type regions.

---

* The sum, $V + \zeta$, is the electrochemical potential of the electrons and has been called the "imref" by Shockley.

These equations have been applied to a number of situations and compared with experiment. Some examples will now be discussed.

Flow in Filaments

One of the most fruitful studies has been that of flow in filaments of germanium (1). The germanium is in the form of a single crystal of small cross section, perhaps a few mils on a side. As the flow is essentially in one

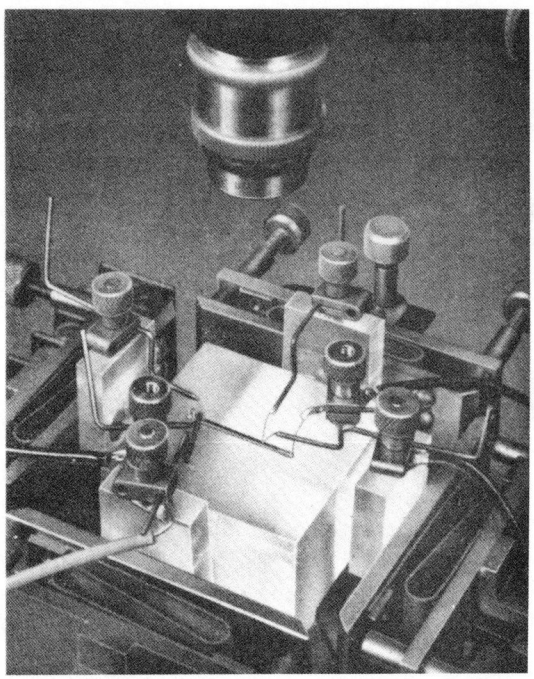

Fig. 3. Experimental arrangement used by Haynes for the study of the flow of holes in an n type germanium filament. Positions of the point contacts on the filament can be adjusted by micromanipulator screws. The microscope objective indicates the scale. The electrical circuits used for measurements of mobility are shown in Fig. 4.

dimension, the analysis of the data is comparatively simple. The small cross section allows the possibility of applying relatively large fields without overheating. Large area electrodes are placed at the ends of the filament and micromanipulator adjustments are used to place metal point contacts at various places along its length. Point contacts are used as potential probes as well as for current leads. A crystal used by Haynes is shown in Fig. 3.

Haynes and Shockley (6) have made accurate measurements of mobility by injecting pulses of holes from a point contact near one end of a filament of n-type germanium (or electrons in p-type germanium) and measuring the time taken for the pulse to be swept down the filament by a field applied along its length. Similarly, mobilities of electrons are measured by injecting them into p-type filaments. The experimental arrangement is shown schematically in Fig. 4. The field acting along the length of the filament can be changed by varying the voltage applied between the two ends. A square current pulse is applied to the emitter contact. A large fraction of this current consists of holes which are swept down by the applied field. The presence of the holes at the far end of the filament is detected as a change in current flowing to a second point contact which is biased in the reverse direction. Haynes has established that the current flowing to the collector points varies linearly with the concentration of holes in the filament. The time taken for the pulse to travel down the filament is measured from a trace of the collector current on a cathode ray oscilloscope. The drift velocity and mobility can thus be determined.

Figure 5 is a reproduction of a trace showing the leading edge of a typical pulse. The rounding off is due to diffusion, and increases with the time the holes spend in the filament. The derivative is very close to a gaussian curve from which the diffusion coefficient can be estimated. The values obtained in this way agrees with that obtained from the mobility and Einstein's relation, Eq. (2)*. The mean lifetime of the added holes can be obtained from the way the pulse height decreases with the time the holes spend in the filament.

Haynes has developed this method to be a precision method for measuring mobilities of electrons and holes. In order that the holes do not alter the conductivities and thus the field acting along the length of the filament, the injected current is kept small compared with the total filament current. The theory of interesting phenomena observed under transient conditions when large hole pulses are used has been discussed on a theoretical basis by Herring (1).

An experimental arrangement which has been widely used for investigating the steady state flow of injected holes in filaments and for measuring the influence of added holes or electrons on the characteristics of point contacts is shown in Fig. 6. As in Haynes's experiment, holes are injected from an emitter electrode near one end of the filament and are pulled down the filament by a field acting along the length. If diffusion is unimportant,

---

* An excellent direct check of Einstein's relation was made at the Bell Telephone Laboratories. Each of a group of some sixty persons, mostly university professors, attending an intensive course on physics of transistors, independently measured the ratio of the diffusion coefficient and mobility of conduction electrons in germanium. A pulse of holes was injected at an emitter point placed on a filament. The mobility is given by the drift velocity of the pulse in a known field and the diffusion coefficient by the spreading out of the pulse in time. The average results obtained gave for the ratio 0.027, in close agreement with the theoretical value of Einstein, kT/e, which is 0.026 ev at the temperature of the measurement. (*Phys. Rev.* **88,** 1368 (1952)).

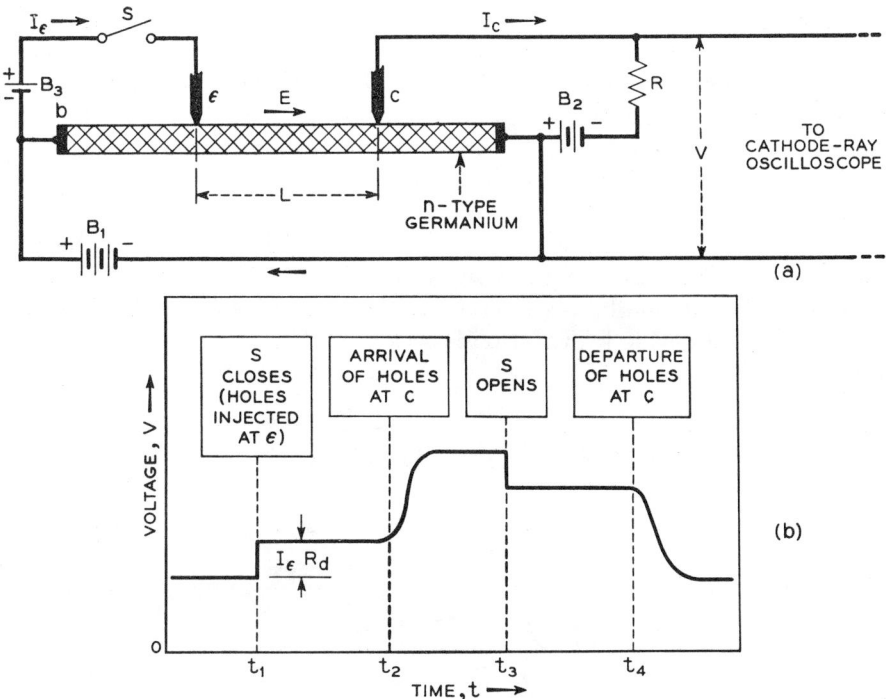

FIG. 4. Schematic diagram of method used by Haynes for measuring mobilities of holes in n type germanium filaments. Holes introduced in the form of a square current pulse at the emitter contact, E, flow down the filament under the influence of the field applied by the battery, B_1. The presence of the holes at the collector contact, C, is indicated by a change in current flowing through the resistance, R. The mobility is determined from the time taken for the holes to flow from E to C in a known field.

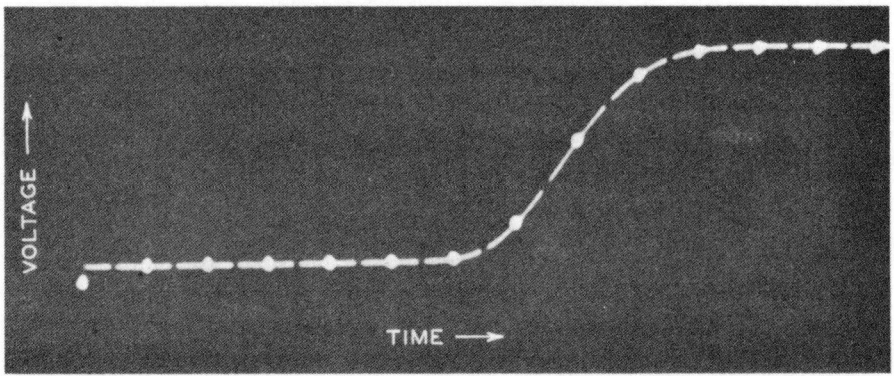

FIG. 5. Oscillograph trace showing leading edge of hole pulse in n type germanium filament (see Fig. 4). Vertical deflection is proportional to hole concentration in filament. The rounding off of the leading edge is due to diffusion. (From Haynes.)

the hole density in any portion of the filament can be determined from the resistance as measured by the voltage drop between potential probes. Electrical neutrality requires that the increase in electron concentration be equal to the added hole concentration. Because of recombination, the hole density decreases along the length of the filament. Knowing the concentrations, we may determine what part of the total current is carried by electrons and what part by holes.

In making these calculations, it is not always possible to neglect diffusion. Theoretical calculations of the flow which include the diffusion terms as well as recombination for a wide range of concentrations and filament currents have been made by van Roosbroeck (1). Diffusion is of importance

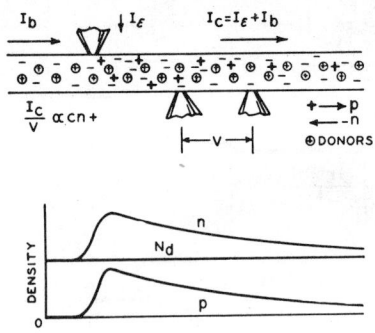

FIG. 6. Method of measuring hole and electron concentrations and currents. Schematic diagram of experimental arrangement used by Pearson and others for studying steady state flow of injected holes in n type germanium filaments. Holes injected at emitter contact are swept down by field of filament current, I_b, and alter the conductivity of the filament. The hole density, p, is determined from the conductivity as measured from the voltage drop, V, between potential probes. The hole density decreases with distance from the emitter contact by recombination of electrons and holes (2).

when the concentration of injected carriers is large compared with the normal concentration or when the filament current is small.

The characteristics of emitter contacts may be studied as follows. If recombination were negligible, the hole current flowing down the filament would be just equal to the hole current flowing from the emitter. The fraction of the emitter current carried by holes could then be obtained by dividing this hole current by the total emitter current. Recombination is important in actual filaments, but can be decreased by increasing the total filament current. The time the injected carriers spend in the filament is inversely proportional to their drift velocity and thus to the applied field. For example, the hole current flowing from the emitter to an n type filament can be estimated by measuring the hole current in the filament as a function of the total filament current and extrapolating to infinite filament current.

Filament may also be used to find out how the current voltage characteristics of contacts vary with concentration of injected carriers. The characteristics of a rectifying contact as a diode are measured at a place where the carrier concentration has been determined. In order to keep conditions in the filament from being disturbed by the measurement, the current flowing to the contact should be small compared with the filament current. Pearson (1) has shown that the low-voltage conductance of a point contact

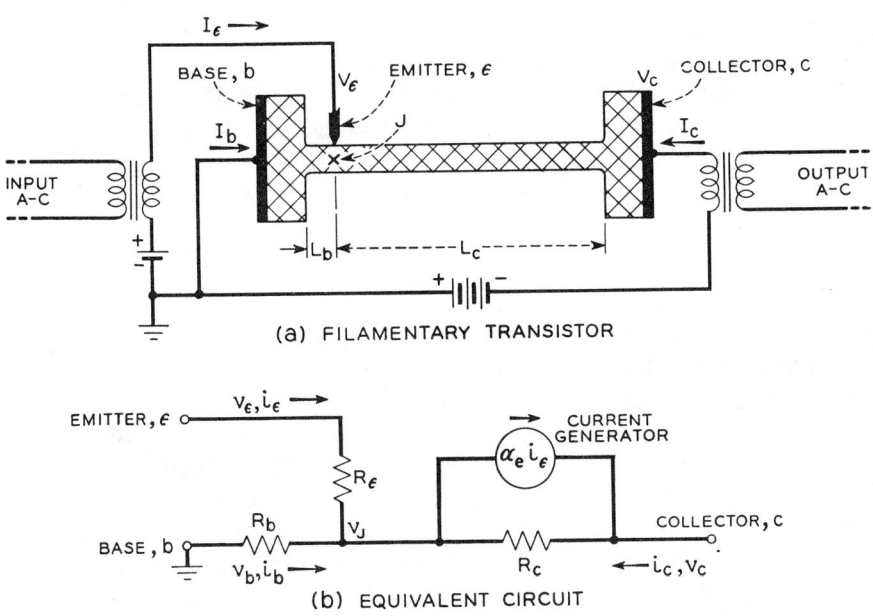

FIG. 7. Use of germanium filament as a transistor. The resistance of the filament can be altered in a controlled way by changing the current flowing from the emitter contact, E. The resistance is decreased by holes injected at the emitter and swept down the filament and by the added electrons which neutralize the space charge of the holes (2).

$$I = I_0(V) + I_p, \qquad I_p = -cp_a, \qquad I_0(V_f = cp_a \text{ for } I = 0)$$

to n-type germanium varies linearly with hole density. We shall discuss later the theory behind this relationship.

Suhl and Shockley (7) have shown how the paths of injected holes are bent by a magnetic field. By applying a large magnetic field along with a large sweeping current in the filament, the holes may be swept to one side of the filament. Changes in concentration of holes by the magnetic field are detected by measuring changes in conductance of a point contact.

Shockley, Pearson, and Haynes (1) have shown that a filament may be used as a transistor. A suitable arrangement is shown in Fig. 7. Holes injected at the emitter point are swept down the filament and lower its resistance. The resistance of the filament is made high compared to the resistance between emitter point and base when the emitter is operated in the forward direction. The hole current, introduced at low impedance, flows not to the base as would be the case for simple resistances, but to the collector electrode which may be connected to a high impedance load. This gives a voltage amplification of an input signal. There is also current amplification. The resistance of the filament is lowered not only by the added holes but also by the fact that for each added hole there is an additional electron to satisfy the requirements of electrical neutrality. Because of their higher mobility, the electrons have more effect than the holes. The maximum current amplification, achieved when all the emitter current consists of holes and when recombination is negligible, is equal to

$$\alpha = (\partial I_c/\partial I_e)\ V_c = \text{const} = (\mu_p + \mu_n)/\mu_p$$

which is about 3 for germanium. The theory of the filamentary transistor is relatively simple and agrees well with the experimental data.

CHARACTERISTICS OF RECTIFYING POINT CONTACTS

As another example, we discuss the influence of an added hole density on the current-voltage characteristics of a rectifying point contact to n-type germanium (8). The theory will be discussed with reference to Fig. 8. For simplicity we assume a metal contact in the form of a hemisphere. There is a space-charge layer in the germanium adjacent to the contact which gives a field in such a direction as to attract holes to the contact and repel electrons. In addition to the potential drop across the layer which exists under equilibrium conditions, there is an additional drop, V_e, which comes from a voltage applied between the contact and the germanium. We suppose that the non-equilibrium density, p_a, of added holes is small compared with the normal electron concentration, n_0.

We shall consider two limiting cases. If the applied voltage in the reverse direction is large, so that an appreciable current flows, the field in the germanium will be large and the conduction terms will predominate over the diffusion terms in the equations for flow. If the applied voltages and currents are very small, the added holes move mainly by diffusion.

If the field is sufficiently large so that diffusion can be neglected, the ratio of the hole and electron currents is equal to the ratio of the conductivities

$$I_p/I_n = \sigma_p/\sigma_n = \mu_p p_a/\mu_n(n_0 + p_a) \tag{15}$$

If $p_a \ll n_0$, this equation leads to a linear relation between the total current $I_n + I_p$, and p_a. Such a relation has been observed by Haynes (Fig. 9), and, as mentioned earlier, he has used the reverse current to a point contact as a measure of hole concentration in a filament. In the derivation of the correct relation, account must be taken of current multiplication in the contact (3, 4).

When the applied voltages and currents are very small, the added holes move to the contact by diffusion. The barrier layer acts as a sink for holes.

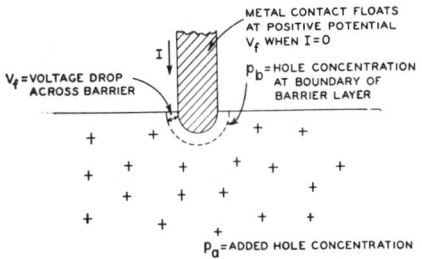

FIG. 8. Diagram used for discussion of influence of added hole density on current voltage characteristics of point contacts to n type germanium. Holes are indicated by (+); electrons are not shown. Holes which flow into the barrier layer at the contact are drawn to the metal contact by the field in the barrier region and are neutralized there by electrons from the metal. The contact floats at a positive potential, V_f, such that the current, I_p, from the added holes is compensated by a normal current, I_0 (V_f), flowing in the opposite direction.

Every hole which enters the space-charge region is drawn by the strong field existing there to the metal contact. The diffusion problem is easily solved, and the hole current flowing, I_p, is proportional to p_a when $p_a \ll n_0$. We may write

$$I_p = -K p_a \qquad (16)$$

where K involves* the diffusion coefficient, area of contact, etc. The con-

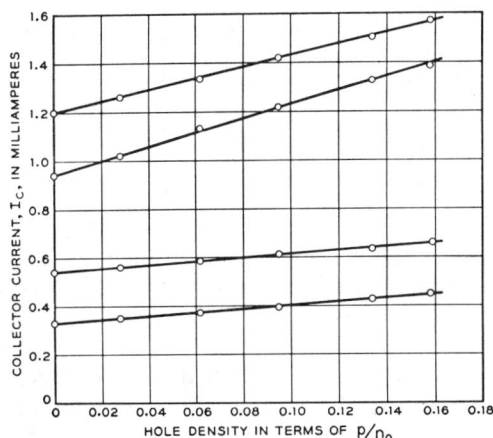

FIG. 9. Relationship between collector current and hole density in germanium filament for four different collector contacts. The hole density is expressed as a fraction of the normal electron density. The collectors are biased to 20 volts in the reverse direction (2).

---

* The expression for K is

$$K = 2\pi\, p_a e v_a r_b{}^2 / [4 + (e r_b v_a / k T \mu_p)]$$

where v_a is an average thermal velocity and r_b is the radius of the hemispherical contact area.

ventional direction of current flow is positive into the germanium, so that I_p is negative. In addition to the current, I_p, due directly to the added holes, there is the normal current flow, $I_0(V)$, which depends on the voltage drop, V, across the space-charge layer and which vanishes when $V_c = 0$. The total current is

$$I = I_0(V_c) - Kp_a \tag{17}$$

This equation has a number of important consequences and can be used for the analysis of experimental data.

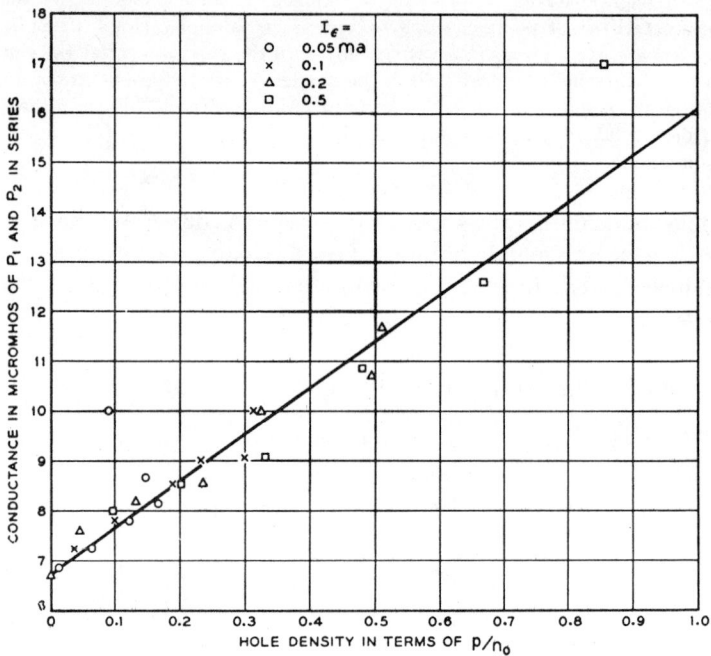

Fig. 10. Relationship between conductance of point contact for small currents and hole density in germanium filament. The hole density is expressed as a fraction of the normal electron density (2).

One consequence is that the contact will float at a positive voltage, V_{cf}, relative to the germanium. The value of V_{cf}, obtained by setting $I = 0$, is the solution of

$$I_0(V_{cf}) = Kp_a \tag{18}$$

The voltage V_{cf} is similar to a photovoltage and can be read on a voltmeter. The conductance of the contact near $I = 0$ is less than normal when added holes are present because it is that which corresponds to a forward voltage $V = V_{cf}$ rather than $V = 0$. Analysis based on using a semi-empirical expression of the form (5)

$$I_0(V_c) = A[\exp(\beta V_c) - 1] \tag{19}$$

for the forward characteristic indicates that the conductance should vary linearly with p_a.*

Some experimental data of Pearson (1) showing the relation between conductance and hole density in a germanium filament are shown in Fig. 10. Pearson used an arrangement similar to that of Fig. 6. The hole density was determined from the change in resistance of the part of the filament between the probes. The conductance at $I = 0$ was also measured. It was found that the conductance depends only on the added hole concentration and is independent of the filament current.

A measurement of conductance of a point contact has been a useful method for determining the concentration of added holes because the measurement does not change the distribution of carriers. As different contacts differ in their characteristics, it is necessary to calibrate each point that is used.

p-n Junctions and Junction Transistors

A p-n junction is a boundary between a region of a semiconductor in which there are excess acceptor impurities (p type) and one in which there are excess donor impurities (n type), as illustrated schematically in Fig. 11. It is a high resistance rectifying barrier (9).

The direction of easy flow (low resistance) is that for which the p side is positive and the n side negative, so that holes flow from the p side to the n side and conduction electrons from the n side to the p side. Because of their finite lifetime, holes may flow for some distance into the n side, increasing the conductivity, before recombination takes place. A p-n junction may thus act as an emitter contact in a transistor.

When the voltage is applied in the opposite, high-resistance direction, the current must consist of holes flowing from the n side to the p side and of conduction electrons flowing from the p side to the n side. These currents are small because there are in equilibrium few holes on the n side and few electrons on the p side.

There is a double layer at the boundary which adjusts the potential of one side relative to the other for equilibrium conditions. The p side is at a negative potential relative to the n side; the potential difference, V, is such that ratio of the concentrations on the two sides is given by the Boltzmann factor (see Eq. 14).

$$p_p/p_n = n_n/n_p = \exp(-eV/kT) \qquad (20)$$

As before, the subscripts in each case denote the conductivity type.

In the immediate vicinity of the junction, where the potential is changing rapidly, the concentrations of both types of carriers are small and there is

---

* The expression for the conductance, G, in terms of p_a is:
$$G = G_0 + K\beta p_a$$
where G_0 is the normal conductance with no added holes.

left the space charge of the fixed charges, negative on the p side and positive on the n side. This uncompensated space charge forms the double layer which in turn gives the potential change. The width of the layer is of the order of 10^{-4} to 10^{-3} cm.

When voltages are applied to the junction, most of the additional potential drop occurs across this space charge layer; the total potential drop is decreased in the forward direction and increased in the reverse direction. It is not true, however, that the resistance is to be associated only with this region. One might be tempted to calculate the junction resistance by

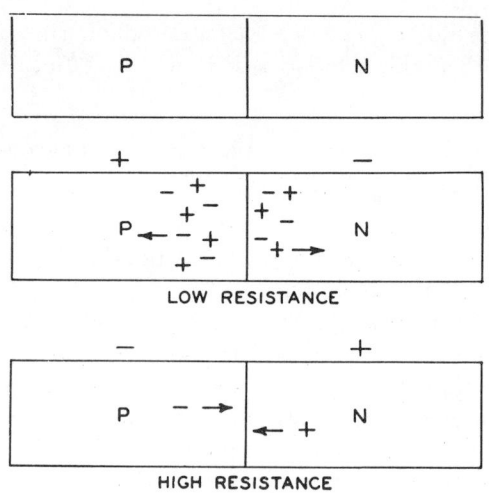

FIG. 11. Schematic diagram illustrating current flow in a p-n junction. In the direction of easy flow, holes flow from p-side to the n-side and electrons from the n-side to the p-side, increasing the concentration of carriers and thus the conductivity. In the direction of high resistance, the current consists of electrons thermally created on the p-side and of holes created on the n-side which diffuse to the junction. The large equilibrium concentrations of majority carriers, holes on the p-side and electrons on the n-side, are indicated on the diagram only by the letters "P" and "N."

dividing it into a large number of little elements in series, calculating the resistance of each element from the local concentration of electrons and holes, and then taking the sum of these series resistances. It is at first surprising that the actual junction resistance may be much larger than the resistance computed in this way. To get the correct result one must consider separately the current carried by holes, I_p, and the current carried by conduction electrons, I_n, and the corresponding resistances, R_p and R_n. If there were no recombination, each part of the current would be conserved and could be treated independently. One would calculate a resistance for flow of holes, R_p, and a resistance, R_n, for flow of conduction electrons, and the junction resistance would be equal to R_p and R_n in parallel. The main contribution to R_p is on the n side where the hole concentration is very small and correspondingly the main contribution to R_n is on the p side.

Actually, of course, recombination must be considered, but if the lifetime of the minority carriers is sufficiently large, they will move an appreciable net distance before recombination occurs. We shall see below that the relevant distance is the diffusion length,

$$L = (D\tau)^{1/2} \tag{21}$$

equal to the square root of the product of the diffusion coefficient and lifetime of the minority carrier. This is just the average net distance the carrier will move as a result of diffusion from point of origin to point of recombination. In calculating R_p, for example, one must include the resistance to hole flow for a distance L_p on the n side of the junction. This will be the main part of R_p, if, as is generally the case for germanium junctions, L_p is large compared with the width of the space-charge region of the junction.

When the applied voltages in either direction are not too large, the electric fields in the electrically neutral regions on either side of the junction are small and the minority carriers move mainly by diffusion. The basic equations which determine the flow of holes in the n region are the expression for the current density for vanishing electric field:

$$I_p = -eD_p(\partial p/\partial x) \tag{22}$$

and the equation expression conservation of holes:

$$\frac{\partial p}{\partial t} = \frac{p_n - p}{\tau_p} - \frac{i}{e}\frac{\partial I_p}{\partial x} \tag{23}$$

Here p is the actual hole concentration, p_n the equilibrium concentration, and τ_p the lifetime. The first term on the right gives the net rate of generation; it is the difference between the rate of thermal generation, p_n/τ_p and the rate of recombination, p/τ_p. The second term represents the rate of increase in hole concentration from flow. Equations (22) and (23) combine to give the diffusion equation:

$$\frac{\partial p}{\partial t} = \frac{p_n - p}{\tau_p} + D_p \frac{\partial^2 p}{\partial x^2} \tag{24}$$

The solution for steady-state conditions ($\partial p/\partial t = 0$) is of the form:

$$p = p_n + p_i e^{-x/L_p} \tag{25}$$

where L_p is the diffusion length, $(D_p\tau_p)^{1/2}$. This solution is chosen so that $p \to p_n$ when x is large. The value of p_1 is chosen to fit the boundary condition at $x = 0$, which is taken to be at the right-hand boundary of the space-charge region. When L_p is large compared with the width of the space-charge region, it follows as a consequence of Eq. (14) that to a close approximation:

$$p_1 = p_p [\exp(eV_a/kT) - 1] \tag{26}$$

where V_a is the applied voltage, taken positive when the p region is positive with respect to the n region. It should be noted that at $x = 0$, $p = p_n + p_1$ is small when V_a is large and negative (the reverse or high-resistance direction), and p is large when V_a is positive (the forward direction).

By introducing (25) into (22), we find that hole current across the junction is

$$I_p = (eL_p p_n/\tau_p) [\exp(eV_a/kT) - 1] \tag{27a}$$

Correspondingly,

$$I_n = eL_n n_p/\tau_n [\exp(eV_a/kT) - 1] \tag{27b}$$

The total current density is the sum of the two:

$$I = I_p + I_n = I_0 [\exp(eV_a/kT) - 1] \tag{28}$$

where

$$I_0 = \frac{eL_p p_n}{\tau_p} + \frac{eL_n n_p}{\tau_n} \tag{29}$$

This theoretical current-voltage characteristic has been checked in detail by measurements at the Bell Telephone Laboratories (10). To check the value of I_0 with theory, values of L_p and L_n were measured directly by a photoelectric method. From these and the known diffusion coefficients, the lifetimes can be obtained. The concentrations of the minority carriers are obtained from the measured conductivities. It was found that the observed value of I_0 is in reasonably good agreement with that deduced from Eq. (29).

When V_a is large and negative, I approaches at the limiting value $-I_0$. This saturation current consists of electrons and holes which are thermally generated in the vicinity of the junction. It is equal to the number of minority carriers generated within a diffusion length on each side of the junction, although this, of course, is an average distance; some diffuse from greater and some from smaller distances to the junctions.

Near $V_a = 0$, the resistance is given by

$$\frac{1}{R} = \frac{1}{R_p} + \frac{1}{R_n} \tag{30}$$

where

$$R_p = \frac{kT}{e^2} \frac{\tau_p}{L_p p_n} = \frac{L_p}{e\mu_p p_n}$$

$$R_n = \frac{kT}{e^2} \frac{\tau_n}{L_n n_p} = \frac{L_n}{e\mu_n n_p}$$

Here $e\mu_p p_n$ is the conductivity for holes and R_p is thus equal to the resistance for hole flow in the n region for a specimen of unit area and length equal to the diffusion length. The interpretation of R_n as a resistance to flow of conduction electrons in the p region is similar.

It is of interest to compare the magnitudes of the hole and electron currents. If, as is generally the case, L_p and τ_p are of the same order as L_n and τ_n, the relative magnitudes of I_p and I_n depend on ratio of the concentrations p_n and n_p of the minority carriers. These in turn depend inversely on the concentrations of the majority carriers, so that generally the current is carried mostly by carriers of the side which has highest conductivity. For example, if the conductivity of the p region is much higher than that of the n region, the current across the junction consists mostly of holes.

When the voltage in the reverse direction is larger than a critical value, it is found that the current increases rapidly above the saturation value.

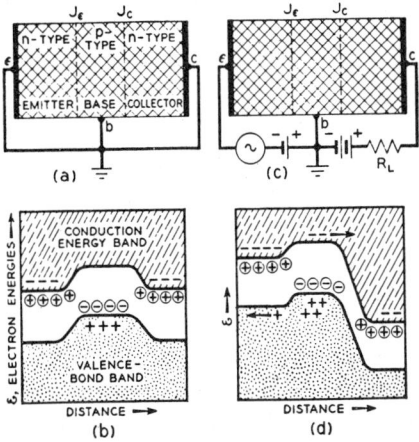

Fig. 12. The n-p-n structure, and the energy level scheme (a) and (b). Thermal equilibrium (c) and (d). Biased as an amplifier (12).

This added current results from electrons being pulled directly from the valence band to the conduction band by the electric field in the space-charge region. The possibility of such a current was discussed some years ago by Zener, but had not been observed directly prior to the experiments at the Bell Telephone Laboratories on p-n junctions in germanium (11).

Both the excellent rectification properties of p-n junctions and the Zener current promise to have important applications to electronics.

The junction transistor, invented by Shockley, consists in one form of two n type regions separated by a thin p type region in a single crystal of germanium (see Fig. 12). Large-area, low-resistance contacts are made to each of the n regions and a third electrode makes contact with the p layer (11). We shall not discuss here the remarkable properties of this device as a circuit element, very low power requirements, high gain per stage, low noise, etc., but shall confine ourselves to a few remarks concerning the nature of the current flow.

When operated as a transistor, one junction, say the left, is biased negatively relative to the p layer. This is the forward direction. The conductivity of the left n region is made large compared with that of the p layer so that most of the current flowing across the junction consists of electrons flowing from left to right rather than holes flowing from right to left. The second, right-hand junction is biased with a relatively large positive voltage relative to the p layer so that in the absence of current from the left-hand junction it would draw a saturation current in the reverse direction.

The left junction acts as an emitter, the right junction as a collector, and the thin p layer as the base region. Electrons introduced into the base layer at the emitter diffuse across the p layer to the right junction and thus contribute to the collector current. The magnitude of this current depends mainly on the voltage between emitter and base and is nearly independent of the voltage on the collector. If the diffusion length for electrons in the p layer is large compared with the thickness of the layer, few electrons will recombine in the layer, and relatively little current will flow from the base electrode. Practically, it has been found possible to design the structure so that less than 2% of the emitter current flows to the base.

The fraction of the emitter current carried by electrons is enhanced by making the p layer very narrow. As for the p-n junction, electric fields may be neglected if the applied voltages are not too large, so we need to consider only the diffusion terms. The excess concentration, n_1, of electrons at the left boundary of the p layer depends on the voltage difference between the left n region and the base and that at the right boundary depends on the voltage of the right n region. The rate of flow of electrons across the p layer is proportional to the concentration gradient and is thus inversely proportional to the width, W, of the layer. When the collector is at base potential, the concentration at the right boundary is the normal equilibrium value and the concentration gradient is n_1/W. For the case of an ordinary p-n junction with a wide p layer, the concentration gradient is determined by the diffusion length and is equal to n_1/L_n. The electron current is thus enhanced by the ratio L_n/W, which may be of the order of 10 in a practical design.

There are marked differences between point-contact and junction transistors. In the former, the collector current is large and holes from the emitter are attracted to the collector by the electric field in the crystal. In junction transistors, although the flow is determined almost entirely by diffusion, the geometry is such that almost all the emitter current reaches the collector. Relatively little current flows to the base, the control electrode, in close analogy with a vacuum tube in which the grid current is small. This is not true of a point-contact transistor in which the base current is large.

Acknowledgments

A number of the figures, namely Figs. 4, 6, 7, 9, and 10, are taken from the paper of W. Shockley, G. L. Pearson, and J. R. Haynes, on "Hole Injection in Germanium" in the *Bell System Tech. J.*, **28,** 344 (1949). These same figures are also used in Shockley's *Electrons and Holes in Semiconductors* (Chapter III), published by D. Van Nostrand Company, New York, 1950. Figure 12 is from W. Shockley, M. Sparks, and G. Teal, *Phys. Rev.*, **83,** 151 (1951). Thanks are due to the authors and publishers for permission to use the figures. The author also wishes to thank J. R. Haynes for use of Figs. 3 and 5.

References

1. Shockley, W., *Electrons and Holes in Semiconductors*. Van Nostrand, New York, 1950. This monograph gives a comprehensive treatment of this work. Two papers which deal particularly with the theory of flow have also appeared: W. van Roosbroeck, "Theory of Flow of Electrons and Holes in Germanium and other Semiconductors." *Bell System Tech. J.* **29,** 560 (1950); R. C. Prim, III, "Flow of Holes and Electrons in Semiconductors." *Ibid.* **30,** 1174 (1951). Important earlier papers are W. Shockley, G. L. Pearson, and J. R. Haynes, "Hole Injection in Germanium." *Ibid.* **28,** 344 (1949); C. Herring, "Transient Phenomena in the Transport of Holes in an Excess Semiconductor." *Ibid.* **28,** 401 (1949).
2. Shive, J. N., *Phys. Rev.* **75,** 689 (1949); Ryder, E. J., and Shockley, W., *ibid.* **75,** 310 (1949); Haynes, J. R., and Shockley, W., *ibid.* **75,** 691 (1949).
3. Bardeen, J., and Brattain, W. H., *Phys. Rev.* **75,** 1208 (1949).
4. Shockley, W., *Phys. Rev.* **78,** 294 (1950).
5. Torrey, H. C., and Whitmer, C. A., *Crystal Rectifiers*. McGraw-Hill, New York, 1948.
6. Haynes, J. R., and Shockley, W., *Phys. Rev.* **81,** 835 (1951).
7. Suhl, H., and Shockley, W., *Phys. Rev.* **75,** 1617 (1949); **76,** 180 (1949).
8. Bardeen, J., *Bell System Tech. J.* **29,** 469 (1950).
9. Shockley, W., "Theory of p-n Junctions in Semiconductors and p-n Junction Transistors." *Bell System Tech. J.* **28,** 435 (1949).
10. Goucher, F. S., Pearson, G. L., Sparks, M., Teal, G., and Shockley, W., *Phys. Rev.* **81,** 637 (1951).
11. McAfee, K. B. Jr., Shockley, W., and Sparks, M., *Phys. Rev.* **85,** 730 (1952).
12. Shockley, W., Sparks, M., and Teal, G., *Phys. Rev.* **83,** 151 (1951).

BARIUM TITANATE FERROELECTRICS*

A. VON HIPPEL

Massachusetts Institute of Technology, Cambridge, Massachusetts

INTRODUCTION

Ferromagnetics have been known since ancient times, and the utilization of ferromagnetic metals in the generators and converters of electric energy is one of the basic prerequisites of today's technical civilization. Ferroelectric materials have not been observed until recently, and their role in technical devices for the concentration of electric energy, for electromechanical transducers and in control mechanisms of all kinds is just beginning.

Ferromagnetism arises by a special kind of interaction between the electron systems of the atoms of a solid material. Each electron carries a permanent magnetic dipole moment, a magnetic spin. Normally these dipoles tend to compensate each other by forming antiparallel pairs, but under particular conditions certain electrons may align their magnetic moments spontaneously in parallel orientation.

The prefix "ferro" originally implied that a material showing this phenomenon contained iron (ferrum), later its meaning was extended to cover the group of iron-like elements (Fe, Ni, Co), and finally the term "ferromagnetism" became synonymous with the phenomenon of spontaneous magnetization whether or not one of these elements was present. The name "ferroelectricity" has been conceived in this metaphorical sense. It does not refer to iron but to the phenomenon of spontaneous alignment in parallel orientation and implies that electric dipoles are being aligned instead of magnetic ones.

That ferroelectrics might exist could have been anticipated after Debye (1) in 1912 postulated the existence of permanent electric dipoles. Such dipoles are bound to arise when atoms of different types form molecules since the partners differ in their binding strength for electrons and their interacting electron clouds are therefore displaced eccentrically toward the stronger binding atoms.† A closer analysis, however, shows that these permanent dipoles may not be the proper building stones for ferroelectric materials. This can be made clear by a kind of Socratic seesaw argument.

---

* Sponsored by the NDRC under Contract OEMsr-191 and later by the ONR, Signal Corps, and the Air Force under ONR Contract N5ori-07801.

† The separation of the nuclei in molecules is of the order of 10^{-8} cm, and the magnitude of the displaced charge is comparable to that of an electron ($e = 4.8 \times 10^{-10}$ electrostatic unit). The dipole moment $\vec{\mu}$, the product of dipole charge and dipole length, is therefore conveniently expressed in units of 1×10^{-18} esu, called 1 debye.

In an external field permanent dipoles experience a torque which tends to align them in field direction. However, as Langevin (2) first showed in the case of magnetic dipoles, nearly unattainable fields are required before some order can be enforced at normal temperatures against the randomizing action of the thermal agitation (Fig. 1).* This fact would rule out the existence of ferroelectrics as well as of ferromagnetics if we had to rely on the external field only. However, as Weiss (3) emphasized, the acting internal field may be much larger thanks to contributions from all the other dipoles. Thus the situation can arise as experienced by Munchausen's soldier who lifted himself out of the swamp by his own bootstraps. The field increases the orientation of the dipoles and the orientation of the dipoles, the acting field, until, at a critical "Curie temperature" T_c, the

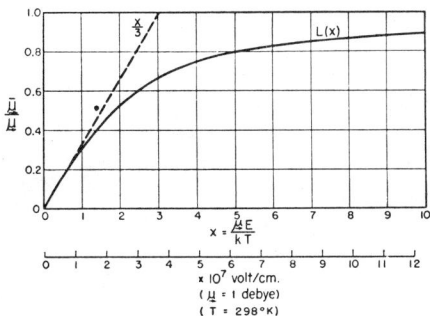

FIG. 1. Orientation of free dipoles by external field against randomizing temperature action.

disordering temperature motion can be overcome and alignment enforced. Mathematically this situation is described by the Curie-Weiss law,

$$P = \frac{3T_c}{T - T_c} \varepsilon_0 E$$

where the "polarization" P represents the dipole moment per unit volume, E the external field intensity, and ε_0 a conversion factor, the dielectric constant of vacuum.

There exists a tremendous number of permanent dipole types (4), hence, if this were the normal outcome of events, the world would be filled with ferroelectrics. However, man would not be here to tell the tale, since for water the catastrophe would occur near 1000°K, solidifying it at this high temperature and making life impossible on this earth. We are spared because the permanent dipoles in liquids and solids lose their freedom of

---

* The ratio of observed dipole moment to true dipole moment, $\overrightarrow{\mu} / \overrightarrow{\mu}$, is plotted as the ordinate. The abscissa x gives the ratio of orienting electric to disordering thermal energy, and, in the second scale, the field strength which would have to be applied at room temperature to a molecule of 1 debye to produce the average moment indicated.

action by association and steric hindrance (5). Thus their fields do not reinforce each other as predicted (6), and there is, in general, no space available in solids for the rotation of dipole molecules.

Thus, as in the case of the ferromagnetics, very special conditions must apparently exist to make ferroelectrics possible, and only three groups of materials are, as yet, known to belong to this class of dielectrics. Their representatives are rochelle salt, the tetrahydrate of potassium-sodium tartrate, recognized as a ferroelectric by Valasek (7) in 1921; dihydrogen potassium phosphate and arsenate, by Busch and Scherrer (8) in 1935; and finally barium titanate, noticed because of its unusual dielectric properties

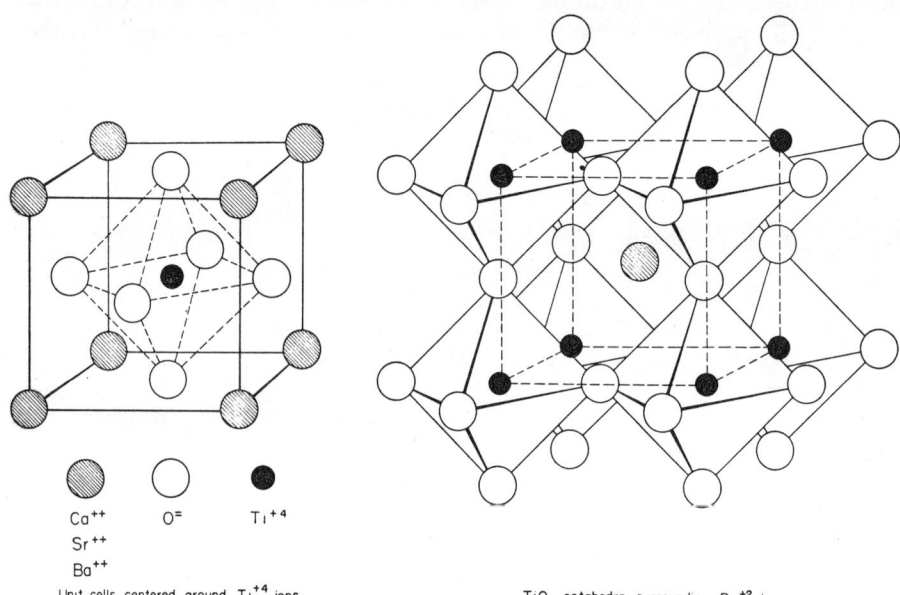

FIG. 2. Ideal perovskite structure.

by Wainer and Salomon (9) in 1942-43 and established as a new ferroelectric material in our laboratory at MIT (10) in 1943-44. The crystals of the first and second groups have relatively complicated structure; the ferroelectric range of the rochelle salt is narrow and that of the phosphates and arsenates limited to low temperatures. Barium titanate, in contrast, crystallizes in the simple perovskite structure (Fig. 2), is ferroelectric below 120°C, and may be employed as a single crystal or as a rugged ceramic material which can be formed into any shape desired. This substance therefore lends itself better than the earlier discovered groups to fundamental investigations and applications.

It is the purpose of this lecture to review what we have learned about $BaTiO_3$, and thus about ferroelectricity, in our investigations at MIT (11).

BARIUM TITANATE CERAMICS

When a multicrystalline sample of $BaTiO_3$ cools down through the Curie region, near 120°C, a number of its properties undergo rapid change. The relative dielectric constant, which measures how much more charge can be stored at the same voltage in a capacitor when $BaTiO_3$ instead of vacuum is used as a dielectric, traverses a sharp maximum that reaches about 7000 (Fig. 3). Simultaneously, the slope of the thermal expansion characteristic alters, and ferroelectric hysteresis loops appear (Fig. 4) manifesting that the

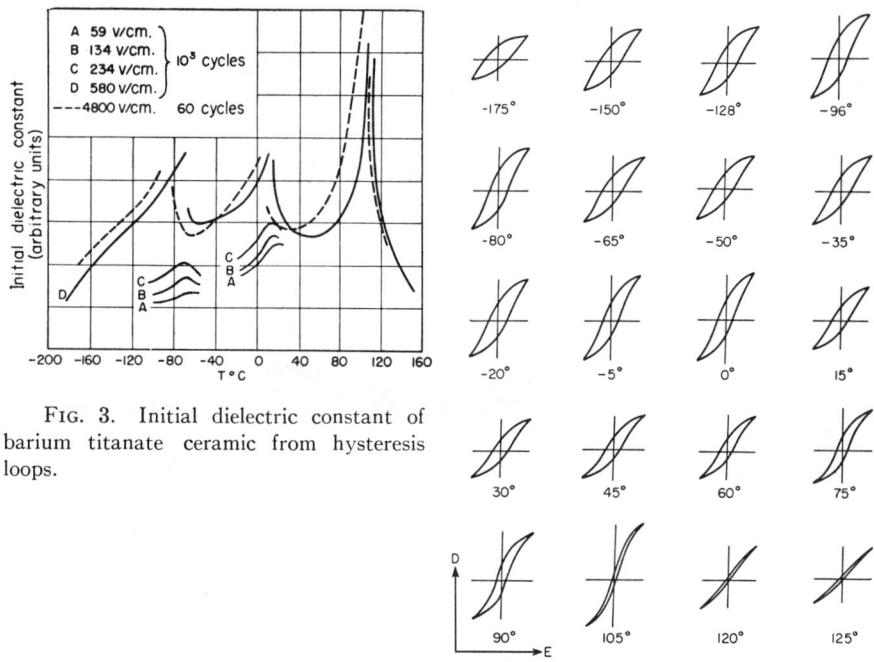

FIG. 3. Initial dielectric constant of barium titanate ceramic from hysteresis loops.

FIG. 4. Hysteresis loops of barium titanate ceramic.

dielectric constant, just like the permeability in the case of ferromagnetics, has become a function of the applied field strength. The x-ray diagram of the cubic structure undergoes at the same time a progressive transformation as the multiplicity of the Cu K_α-doublets of the back-reflection lines indicates (10). Accurate measurements of Megaw (12) in England identified this new phase as tetragonal (Fig. 5).

At lower temperatures two additional phase transitions appear near 0° and −70°C. They become more pronounced as the measuring voltage increases (cf. Fig. 3), but the material remains ferroelectric throughout, as the existence of the hysteresis loops certifies.

If one observes the temperature dependence of the polarization above the Curie point, one finds at low field strength the Curie-Weiss law fulfilled (Fig. 6), but at higher fields a strong dependence of the electric susceptibility $\chi = \underset{\rightarrow}{P} / \varepsilon_0 \underset{\rightarrow}{E}$ on the applied field (Fig. 7) (13). The useful range of this

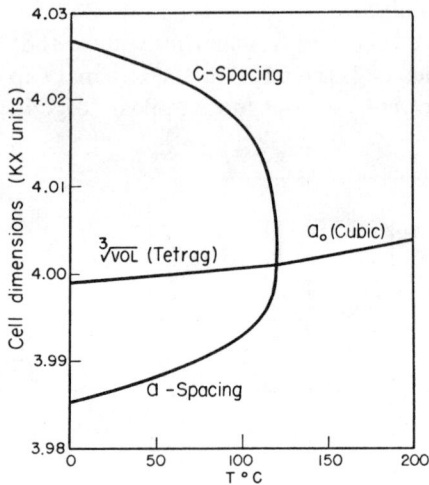

FIG. 5. Variations of cell dimensions of barium titanate with temperature (after Megaw).

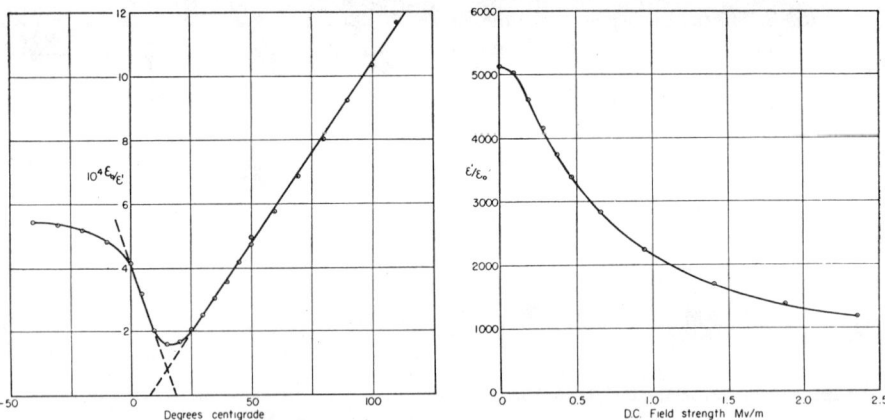

FIG. 6. Confirmation of the Curie-Weiss law in (Ba-Sr) TiO₃ (400 kc).

FIG. 7. Field strength dependence at 25°C of the dielectric constant of (Ba-Sr) TiO₃ (above Curie point).

nonlinearity of the polarization extends to about 40°C above the Curie point and is of special importance since the accompanying electric energy loss is much smaller than below the Curie point (Fig. 8). The constant of the Curie-Weiss law is not $3T_c$ or about 1200 as the simple theory predicts, but about 100 times as large.

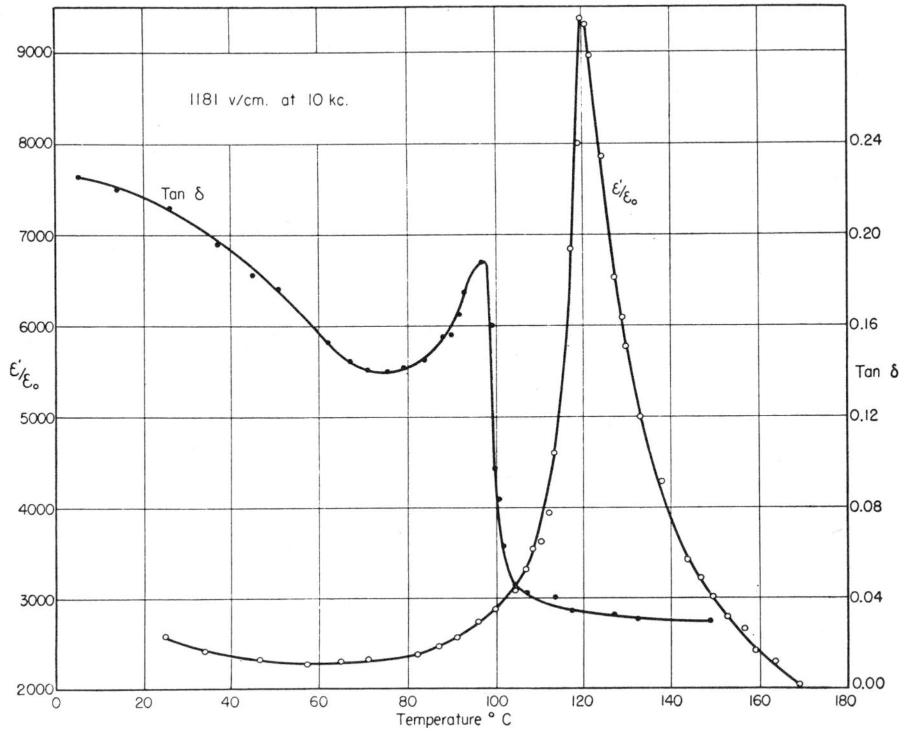

FIG. 8. Temperature dependence of dielectric constant and loss tangent of barium titanate ceramic under high field strength.

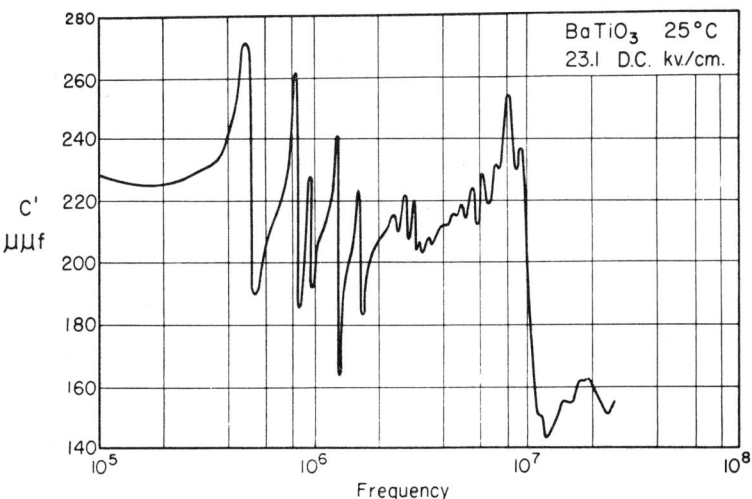

FIG. 9. Resonance frequencies of BaTiO$_3$ disk.

Since the field strength dependence of the dielectric constant makes the ceramics useful as nonlinear dielectrics in applications like amplifiers, modulators, and tuning elements of electric circuits, their behavior under the action of a biasing field was investigated in more detail. Extremely strong resonance effects occurred as the measuring frequency approached the megacycle range (Fig. 9) (13). These resonances remained even after the biasing voltage was disconnected, but could be wiped out by a counter-voltage. It became clear that the biasing field transformed the ceramic disk into a piezoelectric resonator which responds to electric fields by

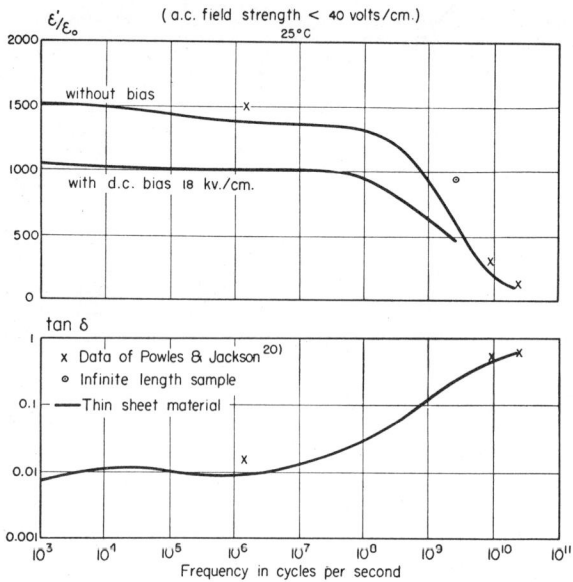

FIG. 10. Relaxation spectrum of the ferroelectric state of barium titanate.

mechanical deformations and vice versa; also without bias the piezoeffect persisted due to the remanence of the polarization.

This observation opened the way for the use of barium titanate ceramics as electromechanical transducers and memory devices. The development of ultrasonic generators, pickups, microphones, and numerous other devices has since been carried forward by a number of industrial laboratories.

In the field of ceramics our research continues with a twofold objective: We want to understand why the ferroelectric response weakens when the applied frequency exceeds 10^8 cps, and disappears in the range of 1-cm waves (3×10^{10} cps) (Fig. 10) (14). And we want to be able to control the size and shape of the ferroelectric hysteresis loops just as the metallurgists today are able to control the ferromagnetic hysteresis loops by alloying agents and special pretreatments. Only on the basis of such knowledge can these materials reach their full utility.

BARIUM TITANATE SINGLE CRYSTALS

We would not, however, expect to arrive at a real understanding of the ferroelectricity of BaTiO₃ by investigating only the polycrystalline ceramics. A study of single crystals was required, and such crystals were finally obtained from ternary melts according to the method of Blattner, Matthias, Merz, and Scherrer (15). Barium titanate crystals of hexagonal and of cubic symmetry may be grown and are generally obtained as flat plates up to several millimeters in length (Fig. 11).

A microscopic inspection of the pseudocubic type revealed that the crystals after cooling through the Curie region contained a variety of shaded areas (Fig. 12) (16). In an electric field these areas were seen to expand or

 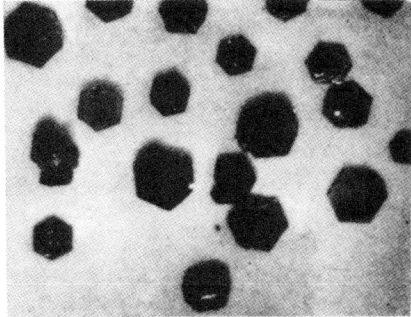

FIG. 11. Small crystals of the cubic and hexagonal modifications of BaTiO₃. 27×

contract, sections of new shading suddenly emerged, and disconnecting the voltage left a remanent state which required a countervoltage for its removal. Observed in an alternating field with stroboscopic illumination the whole crystal appeared in violent agitation and viewed in polarized light the flickering of the transmitted colors resembled Broadway at night.

The implications of this spectacular phenomenon become apparent by considering once more the ferromagnetics. If all the cooperating magnetic dipoles in a macroscopic bar adjust themselves parallel to each other, a permanent magnet results; a strong external magnetic field originates from the free north poles of the dipole chains at the one end and terminates in the free south poles of the other end. A part of this field returns directly through the bar and tends to turn the dipoles in the opposite direction, that is, to demagnetize the material. In soft iron this field suffices to break up the single macroscopic block of parallel dipoles into an array of microblocks or "domains" which eliminate the external field by mutual compensation as indicated schematically in Fig. 13. Hence soft iron and other "soft" ferromagnetics appear nonmagnetic until an external field enforces a preferential orientation of the domains in the field direction.

The actual domain structure of ferromagnetics can be derived only indirectly since they are opaque; in the case of the transparent ferroelectrics it lay in all its details before our eyes (16).

FIG. 12. Effect of electric field on domain structure of $BaTiO_3$ crystal.

A careful analysis of the pattern revealed their laws of formation in all ramifications (16, 17). As the cubic crystal cools through the Curie temperature, a polar axis develops and forces the crystal into tetragonal symmetry. Since any of the cube edge directions may become this polar c-axis, a twinning in the (110) planes tends to take place and laminae form with the c-axis alternating its direction by 90° (Fig. 14). Seen from the front these laminae produce a diagonal striation, while in the top and side views striae

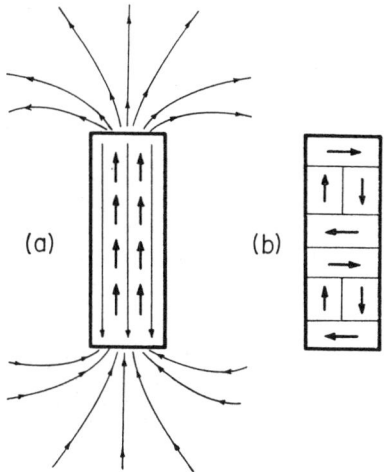

FIG. 13. (a) Single-domain crystal producing demagnetizing field. (b) Domain structure eliminating demagnetizing field.

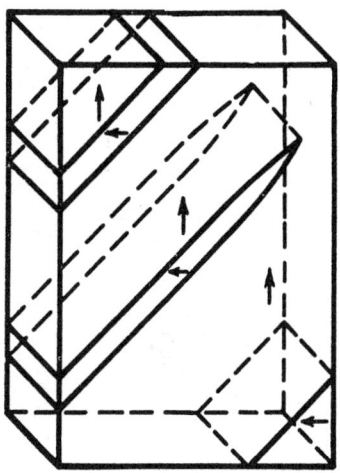

FIG. 14. Twinning in BaTiO$_3$ as determined by polar axes.

FIG. 15. BaTiO$_3$ crystal viewed in polarized light parallel to strong field.

parallel to the cube edges result; this is the origin of the shaded areas in Fig. 12. Since the optical index of refraction is smaller parallel to the polar axis than perpendicular to it, birefringence colors appear in polarized light which are an accurate measure of the position of the polar axis and the thickness of the domains. A strong electric field parallel to the direction of

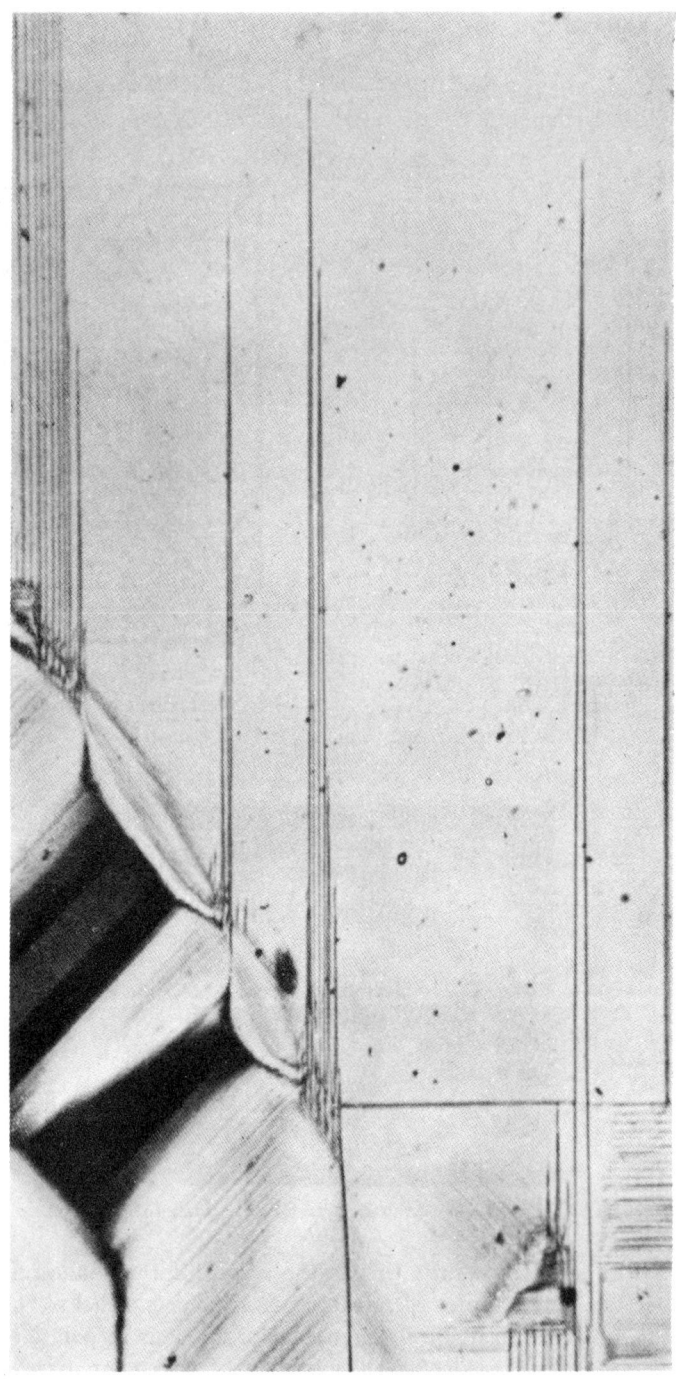

Fig. 16. Wedge-shaped laminar domains in BaTiO$_3$ single crystal.

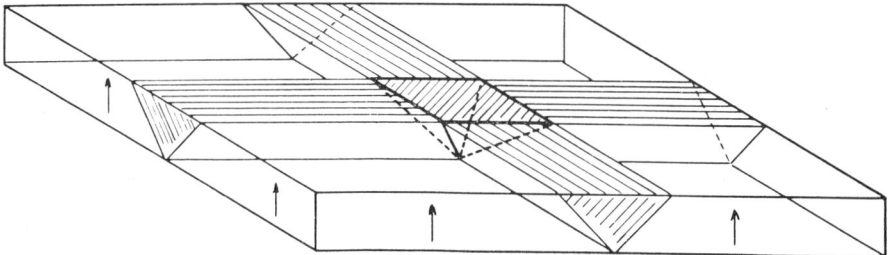

FIG. 17. Intersection of two perpendicular laminar groups.

observation forces all polar axes into the field direction and the well-known cross of a uniaxial crystal appears (Fig. 15). The crystallographer has thus to accept the unusual situation that the c-axis of the tetragonal $BaTiO_3$ crystal can be turned around at will by an external electric field.

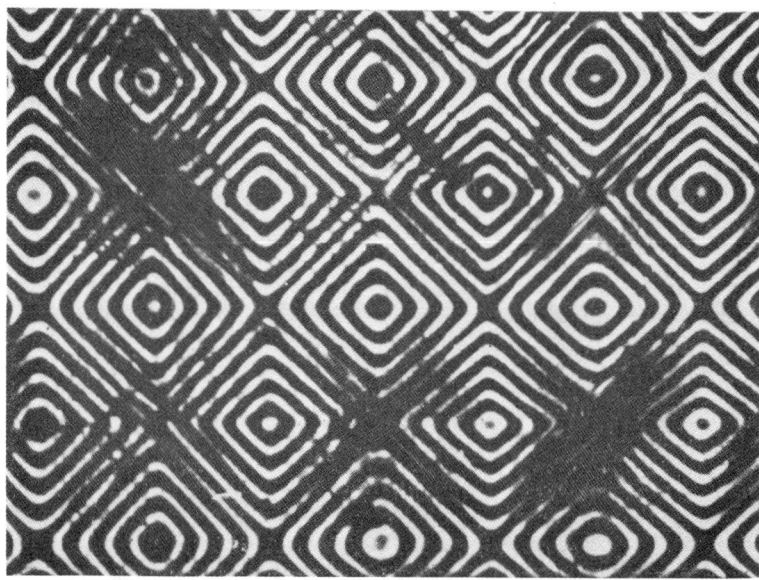

FIG. 18. Square-net domain pattern of $BaTiO_3$ single crystal.

The first step in the formation of domains is the penetration of the crystal by wedge-shaped laminae (Fig. 16). Electric fields or mechanical pressure may drive these wedges through the crystal or squeeze them out again; the final domain structure is therefore dependent on the strains originally contained in the material and on their modification by the unfolding domain pattern. By crossing the first set of wedges with a perpendicular one (Fig. 17), a system of intersecting lamellae can be developed which leads finally to the beautiful square-net pattern of Fig. 18. Its building blocks have been identified by birefringence measurements (Fig. 19) (17), and a detailed analysis has been obtained of the various other patterns which rival in beauty Persian carpet designs.

Above the Curie point and without field the cubic crystal appears dark between crossed Nicol prisms; as the temperature is lowered through the Curie region, the domain pattern develops as just described. Further cooling through the transition regions near 0°C and −70°C produces sudden changes in the domain structure: the polar axis, originally formed in the cube edge directions, jumps abruptly, first in the face diagonal, and near

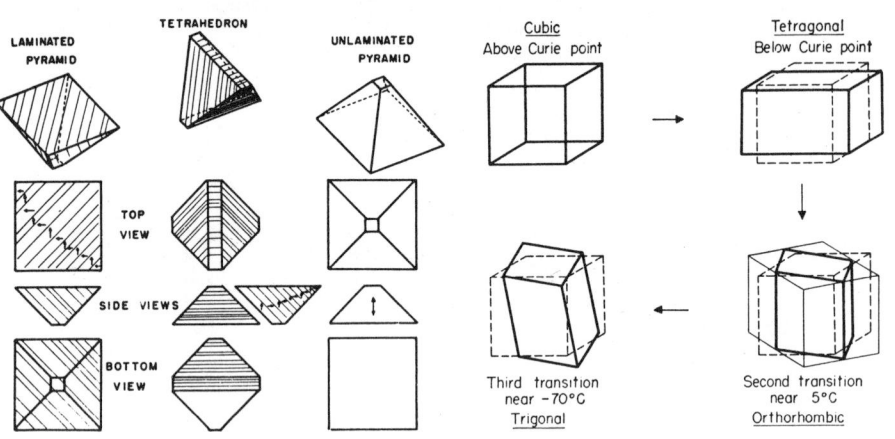

FIG. 19. Three types of building blocks composing square-net patterns.

FIG. 20. Phase transitions of barium titanate.

−70°C into the space diagonal position (Fig. 20). The cubic $BaTiO_3$, changing at the Curie point by a second order transition into a tetragonal crystal, is transformed by these two subsequent first order transitions into an orthorhombic and finally a trigonal modification (17).

It has become possible to grow crystals which give the optical appearance of a single domain; actually they consist of a system of antiparallel domains but a preconditioning treatment can force them into a parallel alignment. The properties of such single domain crystals are decisive for the theory of ferroelectricity and are therefore being investigated in our laboratory in much detail. Figure 21 shows the nearly rectangular hysteresis loop of such a crystal and Fig. 22 the temperature dependence of its dielectric constant (18). At the lower transition points the single domain pattern is destroyed by the reorientation of the polar axis, hence a temperature hysteresis appears, and the successive switching of the domains creates electric pulses in analogy to the Barkhausen noise of ferromagnetics.

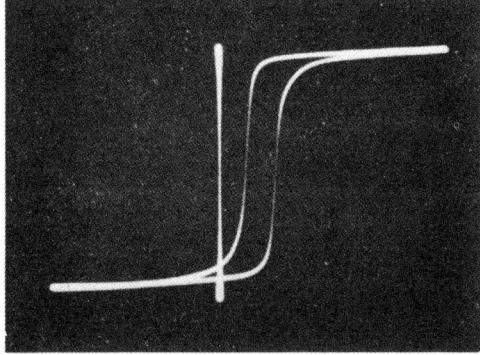

FIG. 21. Rectangular hysteresis loop of single-domain crystal.

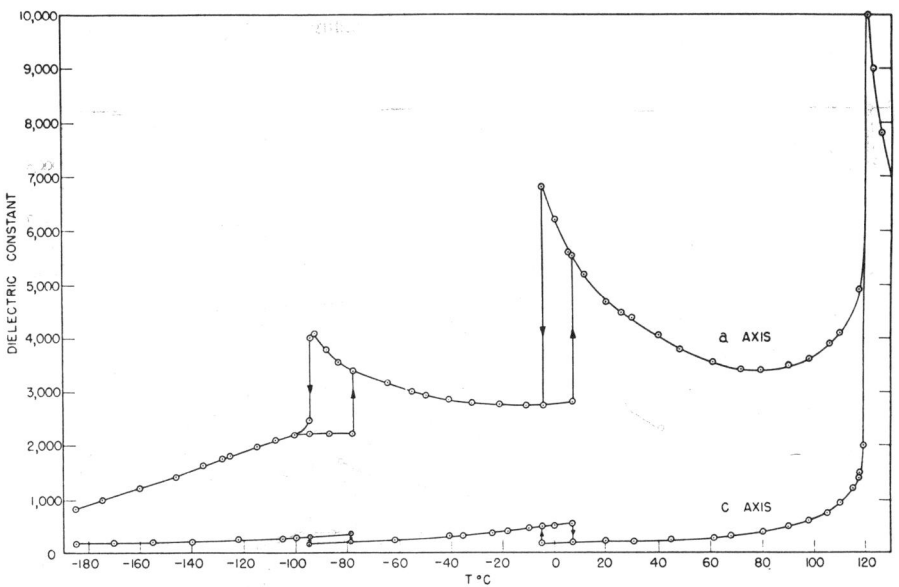

FIG. 22. Initial dielectric constant of single-domain crystal parallel to a- and c-axes.

Conclusions

It is obvious that the perovskite structure of the cubic $BaTiO_3$ (cf. Fig. 2) does not contain permanent dipole moments. Such dipoles might anyhow be of little use, as pointed out in the introduction, since they would probably be frozen into the structure and unable to align themselves freely in field direction. The ferroelectric principle lies in the titanium-oxygen octahedra of the material and in the proper space arrangement of these octahedra.

164 BARIUM TITANATE FERROELECTRICS

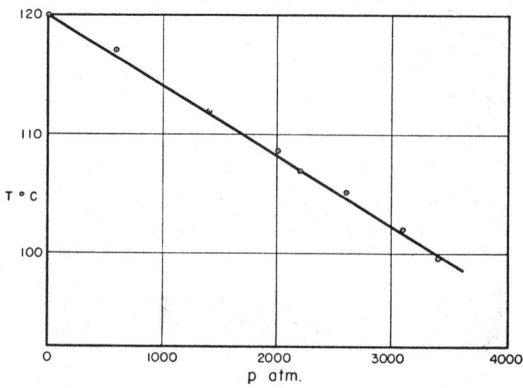

FIG. 23. Dependence of Curie temperature on hydrostatic pressure for $BaTiO_3$ single crystal.

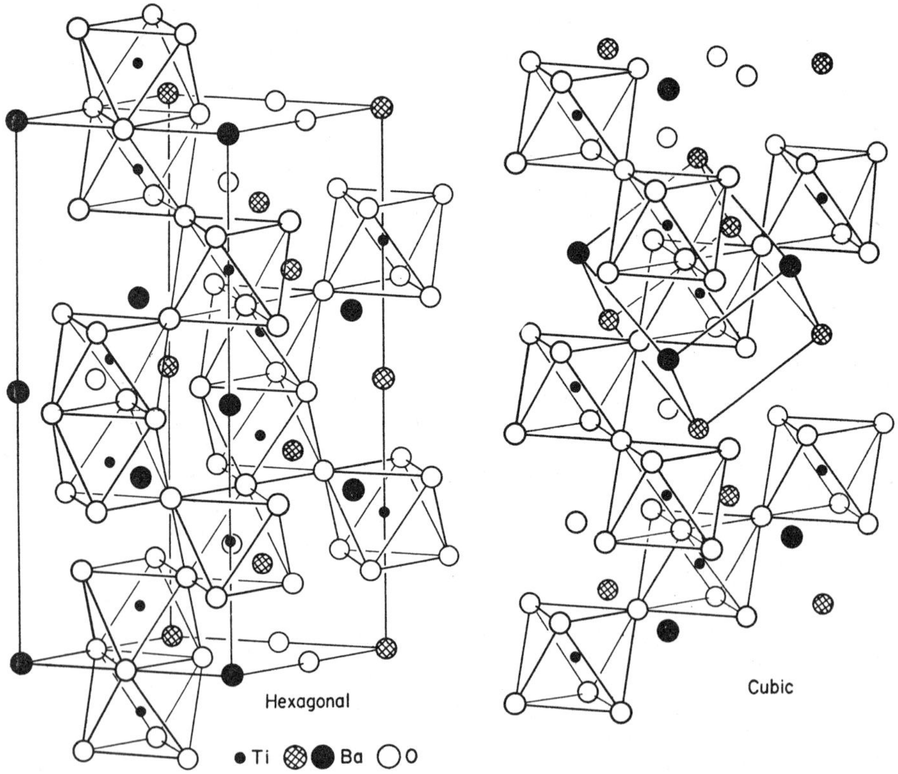

FIG. 24. Crystal structures of barium titanate.

The vibration of the Ti^{+4} ions against the O^{-2} ions in these $(TiO_6)^{-2}$ groups leads to very high induced dipole moments. Crystals containing these groups exhibit therefore high dielectric constants, but they are not ferroelectric until two additional conditions are fulfilled. The size of the octahedra must be just right and their arrangement must be such that the dipoles can mutually enforce their alignment by the bootstrap effect mentioned in the beginning. The first condition can be illustrated by the influence of hydrostatic pressure on the Curie point (Fig. 23) (19), the second becomes convincingly demonstrated by a comparison of the structures of the cubic and the hexagonal $BaTiO_3$ modifications (Fig. 24) (20). Both are built up by the octahedra, but in the cubic modification these share corners only, while in the hexagonal phase two-thirds of them form groups of two sharing one face. In this latter arrangement the two neighboring Ti^{+4} ions face each other relatively closely and unshielded. Their mutual repulsion prevents the concerted action needed for a spontaneous alignment; the hexagonal modification is therefore not ferroelectric.

Experimental evidence shows that in the cubic barium titanate, as the temperature is lowered into the Curie region, the titanium ion moves in a very shallow potential well. Any displacement of the ion from the center of the well results in large dipole moments. As soon as a number of adjacent moments are pointed in the same direction, whether under the influence of an external field or by thermal fluctuations, a local field develops which increases the polarization by mutual interaction and displaces the minimum of the potential well towards one of the oxygen ions. Thus a permanent moment results by a "displacive" transformation (10, 21), the structure becomes tetragonal, ferroelectric, and, since it loses its center of symmetry, piezoelectric. Work is in progress to transform this qualitative description into a more quantitative theory (22).

Acknowledgment

In presenting this report the author has acted as the spokesman for a group of co-workers who have investigated the ferroelectric behavior of barium titanate as physicists, chemists, electrical engineers, and ceramicists. Two extensive reports rendered during World War II were unfortunately not declassified before 1946, thus a number of our results were rediscovered in independent research in Russia, England, Holland, and Switzerland. This lecture does not intend to give a complete survey of the literature but a coherent account of our own investigations.

F. G. Chesley and later R. D. Burbank, and H. T. Evans, Jr., were responsible for the structure research; S. Roberts for the detailed electrical study of the ceramics, which led to the discovery of the piezoeffect; B. Matthias for the growing of single crystals from ternary melts and the initial investigation of the domain structure; P. W. Forsbergh, Jr., for its complete evaluation; W. J. Merz for the studies on single domain crystals; M. Caspari for measurements of the piezoelectric coefficients; and W. O. Statton for phase diagram studies. During World War II R. G. Breckenridge was in charge of the preparation of the ceramics and in general connected with the research; and L. Tisza advised on theoretical problems. In addition to these co-workers whose names appear as authors or co-authors of papers, the contributions of A. P. deBretteville, Jr., who established the lowest transition point by hysteresis measurements, of J. M. Brownlow, who prepared the ceramic samples and measured their expansion, and of W. B. Westphal, who determined the dielectric characteristics of many samples, should be gratefully acknowledged.

REFERENCES

1. Debye, P., *Physik. Z.* **13,** 97 (1912).
2. Langevin, P., *J. Phys.* **4,** 678 (1905).
3. Weiss, P., *J. Phys.* **6,** 667 (1907).
4. Cf. Wesson, L. G., *Tables of Electric Dipole Moments.* Laboratory for Insulation Research, MIT, Technology Press, 1948.
5. Fowler, R. H., *Proc. Roy. Soc. (London)* **149A,** 1 (1935); Debye, P., *Physik. Z.* **36,** 100, 193 (1935); Van Vleck, J. H., *Ann. N. Y. Acad. Sci.* **40,** 293 (1940).
6. Onsager, L., *J. Am. Chem. Soc.* **58,** 1486 (1936); Kirkwood, J. G., *J. Chem. Phys.* **7,** 911 (1939).
7. Valasek, J., *Phys. Rev.* **17,** 475 (1921).
8. Busch, G., and Scherrer, P., *Naturwissenschaften* **23,** 737 (1935).
9. Wainer, E., and Salomon, A. N., *Titanium Alloy Mfg. Co. Elec. Repts.* **8,** (1942); **9, 10** (1943).
10. von Hippel, A., and co-workers, NDRC Contract OEMsr-191 Rept. No. 300, Aug., 1944; No. 540, Oct., 1945; von Hippel, A., Breckenridge, R. G., Chesley, F. G., and Tisza, L., *Ind. Eng. Chem.* **38,** 1097 (1946).
11. Cf. for a more extensive report: von Hippel, A., *Revs. Mod. Phys.* **22,** 221 (1950).
12. Megaw, H. D., *Trans. Faraday Soc.* **42A,** 224 (1946); *Proc. Roy. Soc. (London)* **189A,** 261 (1947).
13. Roberts, S., *Phys. Rev.* **71,** 890 (1947).
14. von Hippel, A., and Westphal, W. B., National Research Council Conference on Electrical Insulation, Oct., 1948; Powles, J. G., and Jackson, W., *Proc. Inst. Elect. Eng.* **96, III,** 383 (1949).
15. Blattner, H., Matthias, B., Merz, W., and Scherrer, P., *Experientia* **3,** 4 (1947).
16. Matthias, B., and von Hippel, A., *Phys. Rev.* **73,** 1378 (1948).
17. Forsbergh, P. W., Jr., *Phys. Rev.* **76,** 1187 (1949).
18. Merz, W. J., *Phys. Rev.* **76,** 1221 (1949).
19. Merz, W. J., Ann. Meeting Am. Phys. Soc., 1950, R2.
20. Burbank, R. D., and Evans, H. T., Jr., *Acta Cryst.* **1,** 330 (1948).
21. Tisza, L., in *Phase Transformations in Solids.* John Wiley and Sons, New York, 1951. pp. 1-37.
22. Devonshire, A. F., *Phil. Mag.* **40,** 1040 (1949); **42,** 1065 (1951); Slater, J. C., *Phys. Rev.* **78,** 748 (1950); von Hippel, A., *Z. Physik* **133,** 158 (1952).

CHEMICAL PHYSICS

THE STRUCTURE OF POLYMERS

PETER J. W. DEBYE

Cornell University, Ithaca, New York

PART I

One of the first things a chemist wants to know about a substance is its molecular weight. For the determination of this fundamental quantity a number of methods are in current use which are or at least can be related to van't Hoff's law, with the help of the general principles of thermodynamics. This law states that the osmotic pressure of a solute in solution is proportional to the number of independent units per unit volume into which the original substance has been broken up. It further states that the proportionality constant in this relation is (at a given temperature) always the same, quite independent of such things as mass or size or structure of the individual particles. In the case of polymers we generally have to deal with units of high molecular weight. Direct osmotic pressure measurements can be made and are being used for the determination of this quantity. However, it is obvious that apart from the practical inconvenience that a semipermeable membrane is an essential part of the experimental arrangement, the osmotic pressure, which can be measured for a given concentration, will be smaller the larger the molecular weight of the substance happens to be. Under these circumstances it seems advisable to look around for some other method, which preferably should be especially appropriate for measuring high molecular weights. The measurement of the excess light scattered by a solution as compared to the solvent alone is such a method.

The principles involved can be explained in the following way. Suppose that in 1 cc of the solution n particles are distributed and suppose for the sake of the present argument we take each to occupy a certain very small volume v which has an index of refraction different from the surrounding liquid. If now a beam of light passes through the solution each particle will become the center of a scattered wave and this wave will have an amplitude proportional to the volume of the particle. The intensity of this scattered wave being proportional to the square of the amplitude will therefore be proportional to v^2 and the total intensity scattered by 1 cc, as far as it is due to the solute, will be proportional to nv^2.

Ordinarily not much attention is paid to this light scattered by so-called clear solutions, but it can be measured quite easily, for instance, with the help of a photocell and an amplifier. However, the determination of the scattered intensity by itself alone does not yet supply us with an evaluation of the number n, which determines directly the molecular weight. So we have to look for another experiment, which might give us another combination of n and v. The determination of the change in index of refraction going from the solvent to the solution is such an experiment. For small concentrations this change is small, usually of the order of one unit in the third decimal in a solution with a concentration of 1%. This difference is measured directly without difficulty and with a high enough precision in a differential refractometer. It is proportional to the product nv, so here we have what we were looking for. A combination of the results of the two experiments, the first determining the amount of scattered light and the second determining the change in index of refraction, obviously enables us to determine n.

Now although the preceding argument has been centered around the picture of a small homogeneous volume v, characterized by an index of refraction, this is by no means essential to the conclusions drawn from the argument. Any kind of particle will be influenced by the primary light in such a way that it acts as an antenna, with an electric moment proportional to the electric vector of the incoming wave. The proportionality constant is what is generally called the polarizability of the particle and it is this quantity which exactly replaces v in the general case. Therefore we conclude that the combination of the two measurements indeed leads to the evaluation of n independent of the structure of the particle.

In making this statement we have however to add two restrictions. The first concerns the size of the particle. Our statement is strictly true only so long as this size is very small compared with the wavelength of the light. For ordinary molecules this restriction is of no practical importance; their sizes are of the order of a few Angstrom units (10^{-8} cm), whereas the wavelength of visible light is of the order of 5000 Angstrom units. Polymer molecules of high molecular weight however may be so large that a consideration of the restriction becomes important. This will be a subject taken up in Part II.

The second restriction has to do with what is called the depolarization of the scattered light. In the argument it has been implicitly assumed that the direction of the electric moment induced in the particle by the electric vector of the primary wave is in the direction of this vector. If this is so the light scattered in a direction making an angle of 90° with the direction of the primary beam will be plane polarized light vibrating perpendicularly to the plane of primary and scattered beam. This will be so even though it is excited by natural light in which the electric vector has no preferential direction in the plane perpendicular to the direction of the beam. It is due to the well-known fact that an antenna does not emit any radiation in the direction of its vibrating moment.

In general, however, the directions of the electric vector of the primary wave and of the induced moment will *not* coincide, which is most easily seen thinking of a particle of ellipsoidal shape. In such a case part of the intensity of the 90° scattered light will be carried by a component with electrical vibrations in the plane of the incoming and the scattered beam, and this is the depolarization effect.

In itself the observation of the depolarization effect is interesting, because from it we can draw conclusions about the spherical symmetry or the lack of such symmetry of the particle. In measuring the scattered intensity in order to determine molecular weights such lack of symmetry also has an influence on the amount of energy scattered in a predetermined direction. This implies of course that in every case in which depolarization is established a correction has to be made to the molecular weight calculated from the experiments under the assumption that no depolarization effect existed. Fortunately this correction is so small, especially in the case of polymer solutions, that in most cases it can safely be neglected. If the depolarization should happen to be important, a separate measurement of its magnitude has to be made and the corresponding correction applied in the calculation of the molecular weight.

As a measure for the intensity of scattering the concept of "turbidity" is in current use. Turbidity is defined as the fraction of the primary intensity lost by scattering over unit distance traveled by the primary beam. If we say, for instance, a solution has a turbidity $\tau = 0.02$, we mean that over a length of 1 cm the primary beam loses 2% of its intensity by scattering. Expressed in terms of this turbidity τ our argument leads to the statement that for a substance of molecular weight M observed in a concentration c (in grams per cubic centimeter) we have for the excess turbidity

$$\tau = HcM \tag{1}$$

in which H is a constant which is determined by the second experiment in which the index of refraction is measured. If μ is the index of refraction of the solution and μ_0 that of the solvent H is to be calculated from the relation

$$H = \frac{32\pi^3}{3} \frac{\mu_0^2}{N\lambda^4} \left(\frac{\mu - \mu_0}{c}\right)^2 \tag{2}$$

in which N is Avogadro's number ($6 \cdot 10^{23}$) and λ the wavelength of the light in centimeters (measured in vacuum).

In all polymer solutions it is found that with a high degree of precision $\mu - \mu_0$ is proportional to the concentration which makes the quotient $(\mu - \mu_0)/c$ a constant for the special combination of polymer and solvent independent of the concentration and what is most important independent of the molecular weight. On the other hand, if the turbidity is followed with increasing concentration it turns out that only over a certain range of very small concentrations the excess turbidity is indeed proportional to the concentration, as relation (1) demands. With increasing concentration the relationship between τ and c can no longer be represented by a straight line as can be seen from Fig. 1. This is to be expected, for just as van't Hoff's law for the osmotic pressure is a limiting law holding strictly only in the limit for low concentrations, the same is true for relation (1). In higher concentrations we cannot avoid considering the effect of interaction of the particles, which of course is interesting too. As far as the molecular weight alone is concerned it is sufficient to consider only the tangent of the angle which the curve τ vs. c makes with the c-axis at its very beginning (in a representation in which τ is represented as an ordinate and c as an abscissa). Since it is awkward to measure the tangent of such an angle and with the theory of interaction in mind it now is customary to make a plot of the

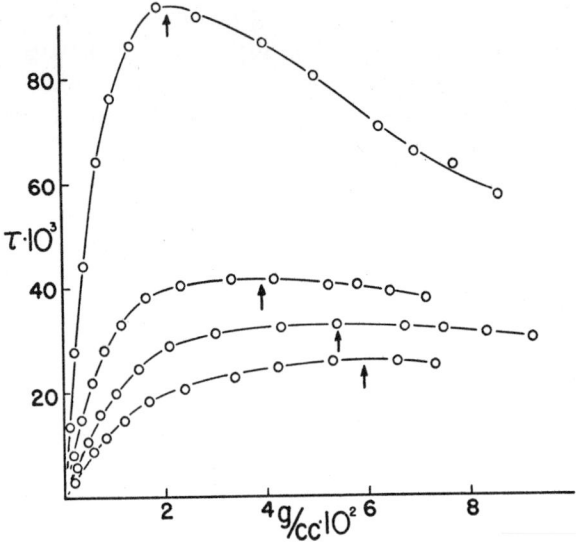

FIG. 1. Turbidity of different polystyrenes in methyl ethyl ketone.

combination Hc/τ as a function of c (Fig. 2). The intercept of this plot on the ordinate for $c = 0$ is equal to $1/M$. It is also found that now instead of on a curve the experimental points are situated with fair precision on a straight line over a considerable range of concentrations. The slope of this straight line is a measure for the curvature of the plot we considered originally in which τ itself was plotted vs. c. In the polymer case the slope of the straight line is large for good solvents and decreases with decreasing solvent power. It is this slope which gives information about the mutual interaction of the particles and their interaction with the solvent, which is of so much importance in the theories of Flory and of Huggins.

In the preceding part an analogy has been mentioned between scattering and osmotic pressure regarding their dependence on concentration. There is more than just a superficial analogy. Actually the two effects are related very intimately to each other. Why and how this is can be seen if light scattering is considered from a point of view which at a first glance may seem quite different from the reasoning we used before.

At the time when Smoluchowski and Einstein were concerned with the explanation of the light actually scattered by pure liquids for reason of their molecular structure, Einstein hit upon a device of calculation in which the liquid was considered in a first approximation as perfectly homogeneous and in which the effect of the actual molecular structure was represented by spontaneous density fluctuations of this homogeneous liquid. Since on the one hand a change in density corresponds to a change in index of refraction and a medium in which we have fluctuations of the refractive index will scatter light it was possible to connect the amount of light scattered with the intensity of such fluctuations, applying Maxwell's electromagnetic

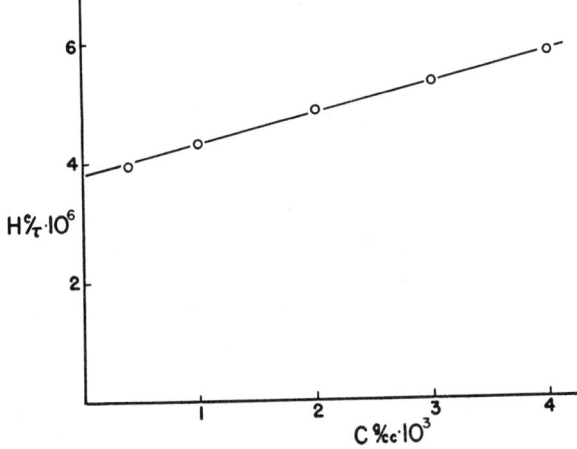

FIG. 2. Variation of reciprocal specific turbidity of polystyrene in butanone with concentration (for diluted solutions).

theory to the problem. On the other hand Einstein could show how by confronting statistical theory and thermodynamics the intensity of the density fluctuations could be predicted from the thermal energy and the compressibility. In fact it is easy to understand that at the same temperature a medium with a high compressibility will show stronger fluctuations than a medium with low compressibility, the decisive quantity being the work necessary to accomplish compression as compared with the universal thermal energy per degree of freedom.

If now we dissolve in the liquid another substance, we will add to the density fluctuations another type of fluctuations which is due to the fact that the distribution of the solute molecules will never be absolutely homogeneous. We will have fluctuations in concentration and since with every change in concentration we will generally also observe a change in refractive index, we have a second reason for light scattering. How intense the concentration fluctuations are will depend on how much work we have to do to change a given concentration. One can think of this change as being accomplished by the use of a semipermeable membrane, doing work against the osmotic pressure. So we see that the second type of scattering, due to fluctuations in concentration is determined by the osmotic pressure. In this way it turns out that what we have called here the excess turbidity of a solution is directly proportional to the reciprocal of the concentration gradient of the osmotic pressure. If for a given change in concentration the increase in osmotic pressure is large, the natural fluctuations in concentration and of course also the excess turbidity will be small.

It is because of this reciprocal relationship that it is advisable, as we did before, not to consider the specific turbidity τ/c itself but its reciprocal c/τ. This reciprocal will be directly proportional to the concentration gradient of the osmotic pressure. We can therefore say that after all, measuring turbidity is nothing but another, of course rather peculiar, method of determining the osmotic pressure of a solution in its dependence on concentration. Considering things this way makes it clear all at once that the straight line representing Hc/τ vs. c is again only an approximation and that in going to higher concentrations terms with higher powers of the concentration will have to be taken into account. The theoretical calculation of the coefficients of the first and possibly higher terms in such a series from the structure of the polymer molecule and its interaction with the solvent has recently been undertaken independently by Zimm and by Flory.

The theory states that provided we use the appropriate refraction constand H in every case, the same polymer measured in different solvents should give the same intercept in a Hc/τ vs. c diagram regardless of the slope of the line, since this intercept represents $1/M$. This has been checked and confirmed for instance by Doty except in some special cases. Of these special cases one type has to do with association phenomena, and the explanation prevailing here is obvious.

Another type is more interesting. At the occasion of light scattering measurements of McCartney and of Ewart and Roe on solutions of polystyrene in mixtures of benzene and methanol, in which the percentage of methanol was gradually increased in order to obtain a solvent with decreasing solvent power, it was observed that the intercept changes with the composition of the solvent mixture. The larger the percentage of alcohol in the mixture, the smaller was the intercept, as if the polymer knew that with still more alcohol in the mixture it was going to precipitate and was preparing for it. Now this of course is impossible, and Ewart hit upon the correct explanation. If a polymer like polystyrene is added to a mixture of say benzene and alcohol, we are not just adding polymer molecules to a solvent mixture which remains the same. A preferential adsorption effect will exist, which will deprive the bulk of the mixture of an amount of benzene which will be concentrated within the range of each polymer molecule, from where at the same time alcohol will be driven out. These changes, which are proportional to the polymer concentration, are accompanied with simultaneous changes of the refractive index of the actual mixture which surrounds the polymer particles. Such changes have not been anticipated in relation (1), and therefore if we apply this relation to the calculation of a molecular weight, without appropriate correction, we will make a mistake. Once this is realized and the correction is established, a comparison of the apparent value of M calculated according to relation (1) with the correct value (as for instance derived from measurements in a single solvent) will be a measure of the degree of preferential adsorption. Since then the general

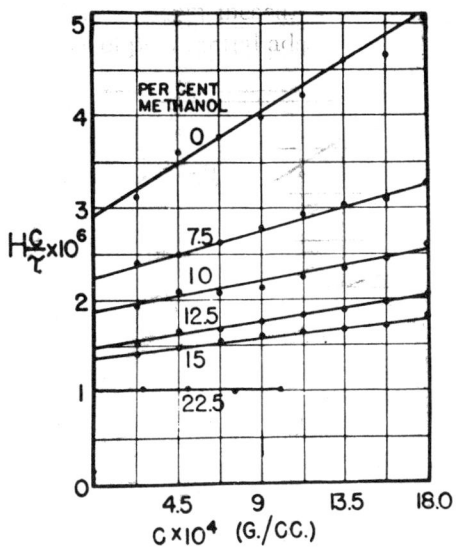

FIG. 3. Reciprocal specific turbidity of polystyrene in benzene-methanol mixtures.

theory of light scattering in mixtures containing more than two components has been worked out by Hermans and Brinkman, by Kirkwood, and by Stockmayer. Their results agree with each other, and the effect is being used in experiments by Badger as a measure for preferential adsorption.

Usually the question is being asked, what the smallest molecular weight is, which can practically be measured by the light scattering method. There obviously is no limit in the opposite direction, provided the correction to be discussed in Part II is applied. Small molecular weights however are difficult to measure, not so much because the instruments are not sensitive enough to measure small turbidities, but more because with decreasing concentration and decreasing molecular weight, the scattering effect of small amounts of accidental colloidal material (dust) in the solutions becomes increasingly important. Nevertheless McCartney could show some years ago that a fair value for the molecular weight of sucrose dissolved in water could be obtained by light scattering. Since then Halwer has investigated sucrose solutions again and in going to higher concentrations than McCartney did, he established that the observed turbidity dependence on concentration really is in accordance with the concentration gradient of the osmotic pressure derived directly from osmotic pressure measurements.

A final question, which came up very early was: What do we measure if the solute molecules are not all the same but have a distribution of molecular weights? The answer is: What we measure is the *weight average* molecular weight. Osmotic pressure measurements can be shown to give the *number average* molecular weight. So if both methods have been

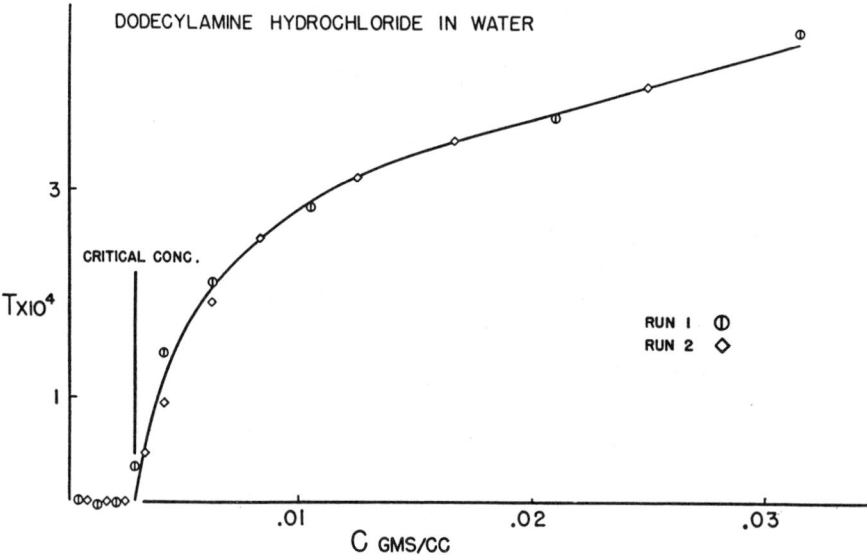

Fig. 4. Turbidity of a solution of dodecylamine hydrochloride (DAH) in water, showing the critical concentration.

applied, we can by comparing the two averages draw conclusions about the width of the distribution curve.

Soaps and many detergents are of a type of substances for which we expect, according to McBain, Harkins, Hartley, and others, that under certain conditions in solution conglomerates of ions, called micelles, will appear. It seemed interesting to us to investigate whether the light scattering method could be applied to this case and what we could learn about the micelles.

If we make an alkali salt of a higher fatty acid or if we replace in an NH_4Cl molecule one hydrogen atom with a hydrocarbon radical of sufficiently long chain length, we obtain a soap. Anacker in his experiments on light scattering worked for instance with dodecylamine hydrochloride ($NH_3(C_{12}H_{25})Cl$), abbreviated to DAH in the text which follows. It has long been known that solutions of such substances in water change their properties (as for instance their electrical conductivity) rather strikingly at a definite concentration called the critical concentration, which for the DAH solution is at 1.3×10^{-2} molar. The picture is that below this critical concentration the substance is in solution in the form of single molecules and (or) ions as a product of their dissociation. As soon, however, as the critical concentration is exceeded, single ions containing the long hydrocarbon chain (in the DAH case the $NH_3(C_{12}H_{25})$ ions) agglomerate to micelles. If the law of mass action is applied to such a reaction, it is seen that although the transition never can be absolutely sharp, it is very nearly so as soon as the number of single ions in a micelle is taken to be large. We approach then the ideal case in which addition of solute below the critical concentration merely increases the number of single molecules and (or) ions, whereas from the critical point on further addition produces only micelles and leaves the monomolecular contents of the solution unchanged. The experimental curve representing the turbidity as a function of concentration is in accordance with this expectation. Up to the critical concentration the turbidity of the solution is barely distinguishable from the turbidity of the solvent. As soon as the critical concentration is reached, the turbidity curve starts up with a sharp bend and with increasing concentration assumes the form familiar in solutions of ordinary polymers. Only a quantitative difference exists, inasmuch as the observed excess turbidity is lower than for common polymer solutions containing polymers of higher molecular weight. Refraction measurements reveal that the refraction difference between solution and solvent is as usual and of the usual order of magnitude. Therefore the molecular weight of the micelles cannot be very large at least under ordinary circumstances.

Since the micelles practically do not appear before the concentration reaches the critical point, it is easy to see how relation (1) has to be changed in order to be applied to the problem in hand. If we plot not Hc/τ but $H(c - c_0)/\tau$ vs. c, with c_0 standing for the critical concentration, we should obtain a straight line and the intercept at $c - c_0$ should be the reciprocal of the molecular weight of the micelle. This is how the experimental curves obtained by Anacker could be interpreted. In the case of the DAH in water Anacker obtained, for instance, a molecular weight of 12,000, which corresponds to 55 molecules per micelle.

If different soaps of the same type, with different lengths of the hydrocarbon tail, are compared, it is seen that the critical concentration decreases with increasing length. Every additional CH_2 group reduces the critical concentration roughly by a factor $\frac{1}{2}$. At the same time it is revealed by light scattering measurements that the molecular weight of the micelle increases rapidly. Whereas a micelle of n-alkyl trimethyl ammonium bromide with a C_{10} chain contains 36 monomeric ions per micelle, this number increases to 75 for the soap with a C_{14} chain, according to Anacker's measurements.

The same effect, namely a decrease of the critical concentration and a simultaneous increase of the molecular weight can be induced by adding salts like NaCl and BaCl$_2$. (See Fig. 5.) Addition of NaCl, enough to make the solution 0.46 molar changes the critical concentration from 1.31×10^{-2} M to 0.72×10^{-2} M and the molecular weight from 12,000 to 31,000 in the case of aqueous DAH solutions. The most outstanding feature observed in such experiments is that what is important is not the ionic strength of the salt solution but merely the concentration of the ions with a charge of opposite sign to that of the micelle (in our case the Cl$^-$ ions).

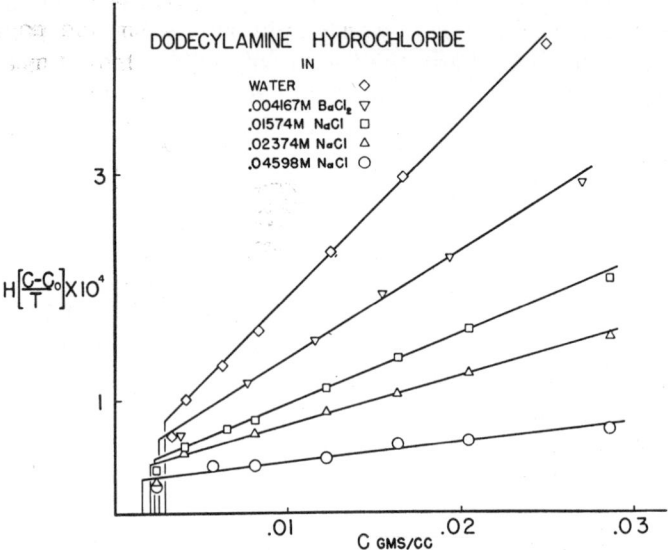

Fig. 5. Straight-line representation of reciprocal specific turbidity, showing effect of salt additions.

The question arises whether a model for the micelle can be found with the help of which this great variety of observations can be adequately understood. When monochromatic x-rays are scattered by a soap solution and the scattering effect is observed in the small angle region (of the order of a few degrees) some intensity maxima are found, which appear on the photographic plate as lines or better bands. From a discussion of the atomic distances corresponding to the angular position of these bands and the manner in which their intensity changes with concentration, Harkins and his co-workers come to the conclusion that the micelle might be something resembling a sandwich or a package of two flat brushes touching each other with the bristles. This last is a picture already previously used by McBain; the bristles stand for the hydrocarbon tails and the ionic heads which remain after the small ions have dissociated off are pictured to be arranged in two planes represented in the brush model by the plates in which the bristles are fastened. It is remarkable that if a picture of this kind is taken more seriously and the energy relations, which are a consequence of it, are investigated, the result is most satisfactory. In such calculations the ionic heads are supposed to repel each other with electrostatic Coulomb forces, which fall off slowly with increasing distance. As a result of this long-range character the work expended in building up a micelle due to these forces increases more rapidly than the number of monomers in the micelle. If this number is N the work is proportional to $N^{3/2}$. On the other hand work is gained in the building process because hydrocarbon tails are transferred from their water surroundings to surroundings which are mostly hydro-

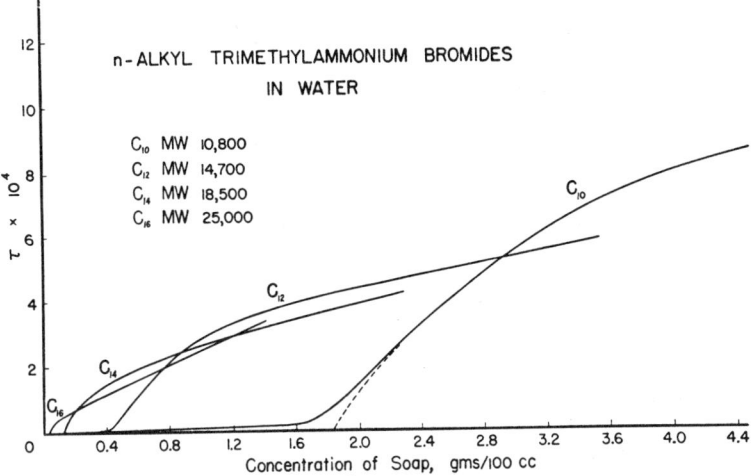

FIG. 6. Turbidity of several n-alkyl soaps, showing effect of chain length.

carbons. In this case, since only ordinary short-range molecular forces are involved, the gain is proportional to N. Considering the total work, this will have a smallest value for a definite number $N = N_0$ of monomers. If this number N_0 is reached, either a further decrease or increase of N involves expended work, so this number should characterize the equilibrium of the micelle.

It turns out that the amounts of the two different kinds of work, which can be derived from the actually observed molecular weight and the critical concentration, can be predicted with a fair degree of approximation from the electrical repulsion between two electronic charges on the one hand and the heat of vaporization of hydrocarbons on the other. It can also be understood in a qualitative way why added electrolytes have effects as described and why the principle of ionic strength is not applicable in the case of highly charged micelles.

Recently giant micelles, obtained by using derivatives containing a long

Alkyl chain dissolved in more concentrated salt solutions, have been investigated. They have the shape of a rod, the interior of which probably contains the lyophobic hydrocarbon tails, whereas the lyophillic, ionic heads occupy the cylindrical surface.

Part II

In Part I some complications were anticipated in the case of particles with sizes comparable with the wavelength of the light. We know indeed that if a light wave travels through a particle of larger size appreciable phase differences exist along the path of the primary ray and therefore the individual electric moments induced at different points in the particle will also have phase differences. From each of these points a spherical wave emerges, and after a choice of a point of observation has been made, the

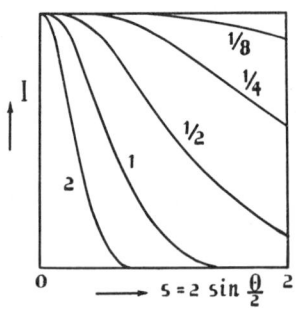

Fig. 7. Angular dependence of intensity scattered by spherical particles. Numbers indicate quotient diameter to wavelength.

scattered waves arriving at this point will have traveled over different distances and therefore will have added new phase differences to those already existing. The total observable amplitude in the point of observation which is the result of the overlapping of all the scattered waves will depend essentially on those phase differences; we may get a large or a small or even zero total amplitude, depending on whether all the amplitudes are positive or partly positive and partly negative. This is the well-known interference principle, illustrated by the familiar effects observed with a grating and ordinary light or with x rays in the Laue diagram of a crystal.

In the cases just mentioned the observed interferences are due to a highly regular arrangement of scattering centers fixed in space. Suppose now we destroy this arrangement. We take the crystal which gave the Laue diagram, crush it to a fine powder, and this time we illuminate the powder, not as we were compelled to do in the Laue case with x rays having a continuous spectrum but with monochromatic x rays. As we know we still will have interferences, high intensities on definite cones surrounding the primary ray. Suppose then that we go farther and make the powder finer and finer.

We will still observe interference maxima and minima, but our narrow lines which had been printed say on a photographic plate, where the cones cut through the plane of the plate, will begin to broaden, the more so the finer the powder.

We realize that we still are working with crystals, although very small, in which the atoms are arranged in a highly regular lattice. We can now ask whether such a lattice arrangement really is necessary in order to produce observable interference maxima and minima. The answer is no. For we know that if we illuminate a gas, consisting of molecules which are oriented at random and are changing their orientation continuously, with monochromatic x rays or with a beam of electrons of definite speed, we can observe interference effects. Obviously it is only necessary that the single molecule, in which the different atoms act as scattering centers, has a definite architecture. Knowing the wavelength of the radiation we can deduce from the angular position of the interference maxima the mutual distances of the atoms in the molecule.

Should we have to deal with proteins, which are particles with a rigid form, we would conclude at once that an interference experiment is feasible. The same will be true, for instance, for a rubber latex, which is a suspension of spherical particles. However, many of the polymers we are interested in like rubber or polystyrene or polymethylmethacrylate or the silicones are of another type. In solution they are long flexible chains, which continually change their form. So the definite architecture of ordinary molecules or colloidal particles is lost. One feature, however, is left. Since such molecules are chains which do not come apart, each molecule can in the average occupy not more than a space (with spherical symmetry) of finite dimensions, and of a size which must depend on the degree of polymerization, or in other words on the extended length of the chain. We can well ask whether we cannot devise an interference experiment in which the size of this space is revealed by its interference effects. This now is exactly what we do if we observe the light scattered by a small volume of a polymer solution and compare its intensities for different directions of observation. However in all such cases it is not enough merely to see the theoretical possibility of interference, we must of course also make sure that it can be observed in practice. In all interference problems the angle between primary and secondary ray for which a peculiar behavior of the intensity distribution can be expected is of the order of magnitude λ/d, if we use radiation of the wavelength λ and if d is the dimension we want to measure. So the question comes up what are the expected dimensions of the space occupied by a coiling polymer molecule in solution.

Since the time Staudinger established that chain molecules are held together by the same chemical bonds as small organic molecules, the statistical behavior of such chains in which the bonds can rotate freely around each other has been considered by many authors, W. Kuhn being one of the outstanding and very first in this field. The fundamental problem can be presented in the following way. Suppose we have a chain of particles (monomers) linked together by rigid bonds but in such a way that every bond is permitted to take up any direction in space, independent of the direction of the preceding or following or of any other bond. Suppose further that we measure the actual distance D from beginning to end of the chain in any of the very large number of configurations such a chain can assume. We can then ask what the probability is that this distance D will have a value between D and $D + dD$. This is a well-known problem in mathematics; the answer is that the probability function approaches a gaussian distribution when the number of links becomes large. However, for our immediate purpose we need not go into these details. It will be enough if we know what the average value of D^2 will be. The whole problem is known in mathematics as that of the random walk in which a man is supposed to take consecutive steps, not relating at all the direction of his next step to the direction of the preceding and in which we want to know how far this individual will get from his starting point. The most probable event will be that he will end up near his starting point, but the probability of this most probable occurrence may be very small. Indeed if we consider all possible occurrences, the average square of all the possible final distances between starting point and end point will have a definite finite value. This is just the same situation as that of the velocity distribution of molecules in a gas. The average of the square of the distance traveled, however, will not go up as fast as the square of the number of steps taken; in fact it is only proportional to the first power of this number.

For our polymer chain the result is that the average of the squares of the distances D, an average which we will call R^2, is for a chain with N links

$$R^2 = Na^2$$

in which a is the length of a single bond. In the chain, we have been considering so far, every bond could take up any direction whatsoever, without any discrimination. In a polymer which is built up around a backbone of a C—C— chain we will wish to acknowledge the fact that the valence angle between consecutive C—C— bonds will have to be conserved. This introduces a stiffening and makes R^2 larger, so we will have in this case

$$R^2 = fNa^2 \tag{3}$$

with a numerical factor f, which assuming the customary valence angle turns out to be $f = 2$.

By taking as an example a polystyrene molecule of a molecular weight 1,000,000, the distance R becomes 300 A. The wavelength of the light, measured in the solution, can be made about 3000 A. So according to this calculation we do not seem to have much chance of observing any pronounced interference effect in scattering experiments even with polymers of rather high molecular weight. At this point we decided to try it anyway, and it turned out that interference effects were observed and were much more pronounced than could be expected from the calculation.

Experiments of this kind, in which not the absolute intensity is measured, but in which merely the relative change of scattered intensity is measured with increasing angle between scattered and primary beam, have been carried out by Doty, A. Bueche, Ewart and Roe, LaMer, and others.

As an example let us consider the experimental results obtained by A. Bueche for different fractions of polystyrene running in molecular weights from 100,000 to 1,000,000 and dissolved in benzene. He finds that the quotient of the scattered intensity measured at an angle of 45° to that

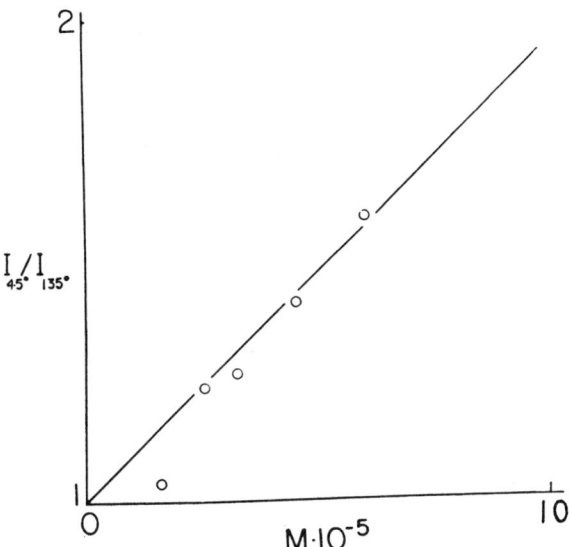

FIG. 8. Dissymmetry in its dependence on molecular weight for polystyrene in benzene.

measured at 135° increases linearly from its value 1 for small molecular weights to approximately 2 for a molecular weight of 1,000,000. So for high enough molecular weights there is a steady decrease of scattering going from the forward to the backward directions; no interference maxima or minima are observed. This agrees with a theoretical calculation of the interference effects to be expected from a coiling chain, which goes through all possible configurations, if it is realized that only the average effect can be observed. However, in order to explain the actually observed angular dissymmetry, we have to assume for R a much larger value than that expected for the free coiling chain. Instead of $R = 300$ A, the experiment demands $R = 1100$ A for the 1,000,000 molecular weight sample. This means that the chain is much stiffer than expected from the principle of absolutely free rotation around bonds. Expressed in terms of the factor f in relation (3) it means that this factor is approximately 24 instead of 2.

Similar effects have been observed for other polymers. Cellulose and cellulose derivatives seem to have an unusually stiff chain according to measurements of Doty, recently confirmed by Badger.

At this point we can easily understand why a correction has to be applied to the molecular weight calculated from relations (1) and (2) if the molecule becomes so large that it shows interference effects. Usually the turbidity is not measured like ordinary absorption because it is so small. Instead the scattered light is measured directly, and in most cases for an angle of 90°. The numerical coefficient $32\pi^3/3$ in relation (2) which serves to calculate the refraction constant H, which in turn defines the total turbidity (that is the total scattered radiation integrated over all angles), has been computed under the assumption that interference effects do not occur. It is obvious that, since these effects show up as a gradual decrease in intensity with increasing angle, the numerical coefficient will be in error. Instead of correcting this coefficient, tables have been worked out which show the factor with which the apparent molecular weight following from relations (1) and (2) has to be multiplied in order to arrive at the correct value, as determined for instance by the quotient of the scattered intensities observed at 45° and 135° or at 60° and 120°.

If the total turbidity should be observed like ordinary absorption, relations (1) and (2) predict that τ should change strongly with the wavelength, being proportional to the reciprocal of the fourth power of the wavelength (Rayleigh's law). But this also is correct only for small particles. As soon as interference becomes important, another factor depending on the wavelength has to be added. Sometimes practical use can be made of this fact by measuring the dependence of absorption, solely due to scattering, with a spectrophotometer over a range of wavelengths. If deviations from Rayleigh's law are found, the particles must be so large that interference effects become observable, and those deviations can be used in order to determine particle size. Heller was one of the first to propose this method.

It is interesting to speculate how the stiffening of the chain is accomplished in more structural detail. A model can be made in which it is assumed that when we define a plane by two consecutive bonds, the following third bond cannot go around freely but prefers positions near this plane, making the whole structure resemble more the usual plane zigzag picture of a —C—C—C— chain at least over the range of a few consecutive bonds. This device leads of course to a stiffer chain, and the observed R values can be obtained by restricting the rotation to a range of sufficiently small angles with respect to the plane just mentioned. However, this is but one of many more possible models which represent stiffening. We can ask whether or not additional experiments of another kind can be performed which might be used to restrict this ambiguity.

In the course of 1949, F. Bueche began experiments on the dielectric constant of polymer solutions with this in mind. In such a solution the solute contributes to the electrical polarization and therefore to the dielectric constant, partly because it is polarized under the influence of the electrical field. A second, sometimes very important part, is due to permanent dipoles in the molecule, which to some extent are being oriented in the field. Suppose now that in a chain molecule we make electric dipoles out of the monomers without breaking the chain. In polystyrene, for instance, this can be done by substituting a chlorine atom for the H atom say in the para-position of the benzene ring. It is not important for our discussion that, if we do this, we have in the polystyrene case not made a dipole out of every bead of the chain but only out of every second bead. Let us assume for the sake of simplicity of the argument that every bead is a dipole. The contribution of such a chain to the polarization due to its dipole character will depend essentially on the directions of the monomer dipoles with respect to each other. If the chain was rigid and all dipoles were rigidly connected to this chain all pointing in the same direction, the average contribution to the observed polarization per chain element would be proportional to Nm^2, where N is number of dipoles and m is the electrical moment of a single dipole. This is so because the total contribution of the whole rigid chain is proportional to the square of its total moment, which under the circumstances is Nm. If on the other hand we would assume that we are dealing with a —C—C— chain and have free rotation, a simple calculation leads to the result that for a geometrical arrangement like that of the parachloropolystyrene chain the contribution per chain element is only $0.92m^2$.

Other arrangements can be imagined in which the dipoles mutually cancel their effect on the polarization completely or nearly so. In this way our expectation can be any value from 0 passing $0.92m^2$ up to the huge contribution Nm^2. Under these circumstances it was interesting to see how much the actual contribution per chain element is. F. Bueche measured the parachloropolystyrene polymer and found it to be $0.54m^2$. In some other cases investigated up to now contributions always smaller than those calculated from free rotation were found. In the literature values of the dielectric constant of solutions of starch are reported by Dumanskii and Kurilenko,* which would mean that in this case the contribution is much larger than calculated for free rotation. Coming back to the parachloropolystyrene the question now had to be answered whether the special model of hindered rotation mentioned before can account for both the observed size of the coil *and* for the observed dielectric constant. The answer is no. So a model with another arrangement had to be found. It may be sufficient to say at this place that it can be done. The main feature of such models is that instead of a kind of oscillatory freedom around a plane zigzag arrangement of consecutive bonds, their freedom of oscillation now centers around a twisted chain arrangement. In this way it is possible to introduce

---

* *Doklady Akad. Nauk. S.S.S.R.* **60**, 1197-99 (1948).

enough mutual compensation of dipoles to account for the small contribution to polarization, without making the chain too stiff and thereby exceeding the experimentally determined size of the coil.

So far we have centered our attention on a type of experimental evidence which still is a newcomer in the field. From the very beginning, following work by Staudinger, a lot of attention has been given to the viscosity of polymer solutions as a means of estimating molecular weights and possibly other properties of the polymer. If a polymer of the coiling type is dissolved in a liquid, it is observed that the viscosity goes up very fast, much faster than for ordinary molecules, (added so as to obtain the same concentration in grams per cubic centimeter). This property is measured by the so-called intrinsic viscosity, which is the relative increase in viscosity per concentration unit in the limit for small concentrations (i.e., the increase in viscosity divided by the viscosity of the solvent, this divided by the concentration and calculated for the very first additions of solute to solvent). As is apparent from the definition the dimension of the intrinsic viscosity is that of a specific volume. In a well-known paper by Einstein the intrinsic viscosity of a suspension of rigid spheres is calculated. He finds that it is equal to 2.5 times the specific volume of the material of the spherical particles. This is the factor which is correct if grams and cubic centimeters are used as units. The current practice differs from this inasmuch as not 1 cc but 100 cc is used as the unit of volume in stating the concentration. Intrinsic viscosities are most of the time reported based on this concentration unit. If so the Einstein factor is 0.025. Many polymer solutions may show intrinsic viscosities of say 2 or 3 (in practical units). So if their molecules really were homogeneous spherical particles, the material of such particles would have to consist of a substance with a specific volume of from 80 to 120 cc/g. Moreover, if we would believe that increasing the molecular weight would mean merely increasing the volume of these hypothetical spherical particles, we would have to expect an intrinsic viscosity which is independent of the molecular weight. Staudinger has instead proposed the rule that the intrinsic viscosity of polymer solutions is proportional to the molecular weight and has used this relation to estimate molecular weights. It certainly is true that with increasing molecular weight the intrinsic viscosity increases also, but careful measurements, for instance, by Flory have shown that in most cases the intrinsic viscosity increases more slowly than the molecular weight. (See Fig. 9.)

We can now ask the question how we expect the intrinsic viscosity of chain molecules to behave. Kuhn, Huggins, Kramers, and others have represented chain molecules by a model consisting of small balls, one ball for every monomer, on which the surrounding liquid can apply frictional forces and which are connected to each other by bonds with the same freedom of rotation around each other, as C—C bonds have. These bonds are supposed to move freely through the surrounding liquid without suffering frictional forces. Their calculations show that for such models the intrinsic viscosity is proportional to the average of the squares of the distances between the ends of the polymer chain, the quantity which we formerly called R^2. Since on the other hand R^2 is proportional to the number N of monomers in the chain, the final result was that Staudinger's rule should hold. We know that in general it does not. In looking through the calculations it becomes apparent that they are built upon the implicit assumption that no interaction between the balls of the model exists, i.e., that the frictional force on one of them is independent of the simultaneous motion of all the others. Obviously this cannot be so. Attempts to take this effect into account have been made by Kirkwood and by myself. The

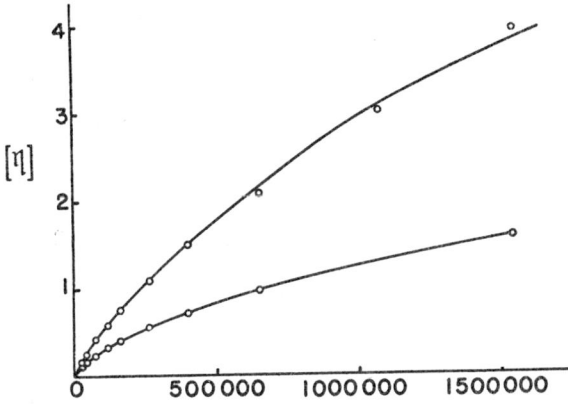

FIG. 9. Variation of intrinsic viscosity of polystyrene with increasing molecular weight in benzene (upper curve) and butanone (lower curve).

result is that proportionality between intrinsic viscosity and degree of polymerization, although still maintained for low molecular weights, gradually changes over with increasing molecular weight to another relationship according to which the intrinsic viscosity rises more slowly than the molecular weight, and this agrees with the experimental facts. Similar results for diffusion and sedimentation have been obtained by A. Bueche.

Since R^2 is a quantity of primary importance in the explanation of viscosity, experimental values for the intrinsic viscosity can be used to estimate the average size of the space occupied by the polymer molecule. The agreement with values derived from the angular dissymmetry of light scattering is fair. We can therefore say that the picture of the chain molecule with more or less free rotation around bonds is satisfactory in every respect.

Recently the theory of light scattering has been considered from a more general point of view. So far we have been focusing our attention on liquids in which particles are distributed, which are responsible for the excess scattering. We can however choose a more remote starting point and build the theory up on the picture of distributed inhomogeneities, say of the refractive index, without pretending to know from the beginning what these inhomogeneities are. The question then comes up how large the average square of the fluctuations of the refractive index is and how much of a correlation exists between the fluctuations in two points of the medium some distance apart. Such questions can again be answered by measuring the scattered intensity absolutely and by paying attention to its angular distribution. This more general formulation has shown decided advantages in the interpretation of scattering measurements performed by A. Bueche on solids like lucite or glass and is well adapted to the understanding of the critical opalescence, which has recently been investigated by Zimm from a similar point of view.

I should like to make use of this occasion to acknowledge gratefully that most of the work reported here has been continuously supported by the Reconstruction Finance Corporation and has been performed by a team in the Chemistry Department of Cornell University. I also want to emphasize that where I have devoted so much space to the discussion of our own work and so little to that of others, this should be no means be interpreted as a lack of appreciation of the work in other places and by other men. The reason simply is that I have tried to present those things that I know best.

IMPLICATIONS OF THE CHEMICAL KINETICS OF SOME BIOLOGICAL SYSTEMS*

Rufus Lumry† and Henry Eyring

University of Utah, Salt Lake City, Utah

A conceptual basis for the understanding of the living organism is probably complete at the present time. Increasing success in nuclear investigation will produce new fundamental bases for natural law requiring modification in the present physical and chemical formulations of quantum mechanics. The advances in nuclear theory may be expected to have no more effect on the quantum theory of molecules than quantum mechanics itself had on Newtonian mechanics for systems with large quantum numbers. It is safe to predict that processes in living systems will be explained without recourse to new force fields, special variations in thermodynamic law, and the like. No physical or chemical process in any living system that has been analyzed shows any divergence from the laws of chemistry applicable to inanimate systems. The first and second laws of thermodynamics seem to hold unquestioned sway. The essential problem is one of complexity rather than fundamental understanding.

The origin of life, though probably unstudyable because of probability considerations, has a possible explanation in chemical terms more acceptable at the present time than vital force theories. Oparin (1), Horowitz (2), and numerous other authors have presented reasonable pictures of life initiation in chemical language. In a sterile world there was continual synthesis of all sorts of complicated organic compounds, part of which accumulated in swampy places decomposing slowly, if at all. Finally, a molecular configuration arose capable of self-duplication, and before the reservoir of food was exhausted, some relatively simple molecular configurations acquired the photosynthetic power and so became independent of external, organic food. This "individual's" reproduction outstripped all competitors and starved them out. Thus it became the ancestor of all living things. Arguing (3) for a common origin for living things is the overwhelming preponderance of the L-amino acids. This is understandable if all such molecules trace their ancestors back to a single self-duplicating template. As time went on, Walden inversions occurred in organisms, but these are in effect a fatal malady, since an enzyme of the wrong configuration is extraordinarily nonreactive so that organisms with such afflictions lose out and disappear. The

---

† Merck Fellow in the Natural Sciences.

* This work was supported in part by the U. S. Atomic Energy Commission and was carried out by the Department of Chemistry and the Institute for Rate Processes.

result is the survival of only those individuals with molecules configurationally related to the original templates. Occasional exceptions such as the existence of D-pinenes in conifers in this hemisphere with L-pinenes in the eastern hemisphere are understandable, if we suppose the pinenes are among the final products of metabolism and are therefore inconsequential for the life of the organism. The origin of the mutant could then come from a simple Walden inversion in the templates synthesizing pinene followed by a later independent mutation anywhere in the plant with survival value.

The original living organism must have had the unique ability selectively to accelerate the trend toward equilibrium of only those parts of its environment which would favor production of its own species. Further spontaneous mutations in the catalytic system lead to successful elaboration of these systems in some individuals either by addition of new catalysts or through a reorganization of existing catalysts. Eventually, it became not only possible but necessary for any given chain of cooperating organisms to compete with other chains for the free energy available. Improvements in efficiency become necessary for survival. In the local sense, efficiency denotes superior ability to make use of the free energy of the environment with the ultimate result of out-reproducing competing systems. In a general sense efficiency is a measure of the approach to reversibility and hence has an overall single direction determined by the second law of thermodynamics (4). Progress along this direction of increased efficiency we call evolution. Organisms are extraordinarily improbable collections of chemicals and hence are characterized by relatively low entropies. In a system in thermodynamic equilibrium, they would represent the extremes of spontaneous fluctuations. Fortunately, the earth exists in the center of a system at a steady state so geared as to pump entropy away from the earth even more rapidly than it is increased by the normal tendency toward equilibrium. Whereas a very small fraction of the energy absorbed by the earth is stored, the fact that in a year vast amounts of energy, E, are first taken up as sunlight with the high characteristic temperature of the sun and later lost as low-temperature black-body radiation into space results in a continuous yearly removal of entropy to the value

$$\Delta S \approx E \left(\frac{1}{T_{\text{earth}}} - \frac{1}{T_{\text{sun}}} \right) \qquad (1)$$

Chains of organisms must be able to degrade the energy, E, successfully within the limits of entropy change set by the above expression. Clearly, those with the greatest number of chemical processes most closely approaching reversibility will be able to utilize a given amount of free energy most efficiently providing only that the processes are also rapid. Time itself is an important factor in competition. Since the heat content of the individual is not much greater than that of a random collection of the substances from which it is formed, it is the utilization of free energy to decrease entropy which provides the fundamental *primus mobile* of life. Indeed viewed as intermediates in a stationary rather than equilibrium situation, the existence of living things no longer appears so remarkably improbable.

Reversibility must be approached by harnessing chemical chains so that minimum free-energy changes take place in individual reactions, a condition which can be achieved by carrying out an overall process in a number of steps, each step involving but a small part of the total free-energy change. The spontaneous generation of biological catalysts (enzymes) made possible the satisfaction of requirements for rapidity and step multiplicity by making available a very great number of clear-cut chemical reactions which can proceed at world temperatures. Enzymes, in lowering the level of activation free energies for living systems, act as chemical lubricants to reduce non-useful heat production. In so doing, they not only make useful rates of reaction possible at physiological temperatures, but also provide the means for their control. Control of small catalyst concentrations suffices to control the entire system.

Some Observations on Protein Structure

Most, though probably not all, enzymes are protein molecules. Nucleic acids, large polysaccharides and other large biological molecules may contribute similar functions or they may serve to support the production of proteins. The catalytic function of proteins is not yet well understood. Though all proteins are composed of polypeptide chains, the number order and types of functional groups (R groups) attached to the polypeptide skeleton determine the stable configuration of the protein in its natural form and its functional activity.

Thermal data recently secured by Sturtevant and colleagues (5) suggests that the polypeptide itself is only slightly unstable with respect to hydrolysis of the peptide bonds. On the other hand, the natural configuration is unstable, being readily destroyed by application of heat and chemical agents. Mirsky and Pauling (6), Huggins (7), Astbury (8), and others suggested that hydrogen bonding between carboxyl and amino nitrogen of different peptide bonds provided the stabilizing interaction for folding. In the last year, Pauling and Corey (9) have emphasized certain spiral and planar structures in which all but terminal peptide bonds are thus linked to neighbors. Kauzmann and co-workers (10) have shown that such bonds are weakly stable with respect to similar bonds with water molecules. Salt linkages and the types of secondary interaction known as van der Waals forces make less important interactions with the solvent and may provide stabilizing energies in secondary interaction comparable with hydrogen-bonding energies. There are a variety of types of proteins, and it is probable that the importance of any one type of secondary interaction varies from type to type. The cooperative hydrogen bonding, always possible with peptide chains, probably makes such structures the predominant form.

In fibrous proteins such as wool or silk, cross links between peptide chains is provided by primary bonds, apparently limited to the disulfide groups of cystine. These bonds provide the ultimate strength of the fibers. The secondary polar interactions between peptide bonds of the parallel polypeptide chains determine the elastic properties (11). A class of proteins characterized by pepsin depends for folded stability on salt bonds (12). Serum albumen does not fall into any single category. Some proteinoid structural materials may not require unique configurations of secondary bonds, but in general, it now appears that the functional activity of the polypeptide chains depends on specific orientation under the influence of some combination of hydrogen, van der Waals, or other weak bonds. To form a useful protein the configuration must not only have an appreciable lifetime but must also have some functional ability. It is not certain that the natural configuration is actually that of lowest free energy. Perhaps the protein is synthesized in its natural configuration and lacks activation energy at normal temperatures to assume less useful, but more stable forms. Logically, the stepwise synthesis in natural configuration of a molecule possessing hundreds of amino acid residues of many different types in a definite order is very unsatisfying. If the ordering is a function of the protein as a whole, a different catalyst would be required for each peptide bond.

If proteins are not continuously structures of primary bonds but rather aggregates of smaller primary bonded units connected by secondary bonds, the problem becomes less extreme. All proteins then might consist of the same few building blocks. On the other hand, the mechanism of template synthesis better fulfills most of the requirements for protein synthesis (13). In such theories the amino acids are assembled and connected in unfolded form on some templates, perhaps the essential genetic structures. At some size the protein is able spontaneously to take on the natural configuration and so pass into solution. This mechanism would require that the natural form be the most stable in comparison with folded and unfolded structures, if not with respect to hydrolysis which is clearly prevented by the protection which the semi-rigid structure affords for the peptide links. Other evidence also suggests that the natural form is the most stable. There is, for instance, the fact that many proteins undergo a reversible change (denaturation) from the natural form at elevated temperatures (14). Typical data for such a change are provided by Kunitz for the reversible denaturation of the crystalline trypsin inhibitor of soybeans (T.I.) (15):

$$\text{TI (active)} \rightleftarrows \text{TI (inactive)}$$

Forward: $\Delta H^\ddagger = 55{,}000$ cal/mol; $\Delta S^\ddagger = 95$ eu/mol
Backward: $\Delta H^\ddagger = -1{,}900$ cal/mol; $\Delta S^\ddagger = -84$ eu/mol
Equilibrium: $\Delta H = 57{,}000$ cal/mol; $\Delta S = 180$ eu/mol
Irreversible process (Forward): $\Delta H^\ddagger = 80{,}000$ cal/mol

Similar quantities have been observed for chymotrypsin (16), luciferase (17) and other proteins.

Two types of explanations have been advanced for the very large heat and entropy values. Mirsky (16), Pauling, Stearn, and Eyring (21), suggested that the reversible denaturation process was an unfolding from a stable folded configuration to a very much less folded configuration stable at the higher temperature. Eisenberg and Schwert (16) and others (18) have proposed that the values represented the unfreezing of water molecules from ionized groups on the protein which, because of their positive heats of ionization, became unionized at the higher temperatures. The matter has recently been considered in more detail by Johnson, Eyring, and Polissar (14). Their conclusion is that both types of processes are involved. To achieve the activated state, nineteen completely frozen water molecules, each absorbing 5 entropy units, would have to be liberated from the protein and from the hydroxyl ions which accept the dissociated proton. The subsequent entropy change to the high-temperature form probably involves an unfolding process requiring very little uptake of heat energy but a considerable increase in entropy. Entropy changes of the order of 15 eu per neutralizing pair are not unusual for small highly hydrated ions (14). Contributions of this order per neutralizing pair are probably necessary to explain the observed entropy values. It is somewhat surprising that such high values should be encountered for the deionization of amino groups. However, Schellman and Lumry (19, 20) have considered the question of effective dielectric constant at the surface of a charged protein molecule and they find it to be, in most instances, considerably smaller than the value for the bulk solvent. We may, therefore, expect hydration of amino and carboxyl groups in their ionized forms to be considerably greater at the protein surface than that of similar groups attached to small molecules. Such an explanation also applies to ion binding by proteins. Chloride ions for instance are bound tightly to the positive ions of proteins.

Thus far, in addition to the hydrolyzed polypeptide we have distinguished two forms of the protein connected in reversible reaction (21). At lower temperatures the stable form is characterized by low energy and low entropy. At higher temperatures both quantities have much higher values. The free energy difference between the two is, however, small ($\Delta F = 3000$ cal/mol for TI). As pH, ionic strength and solvent composition are altered, continuous changes can be expected to take place in both configurations. At some extreme states of the medium or at selected temperatures, the free energy difference becomes zero. Beyond these conditions only the unfolded form is stable.

There is still a more common configuration of the protein connected to one or both of the others by an irreversible process. It is well known that proteins are readily and irreversibly precipitated by heating as witness the poaching of an egg. Reversibly denatured forms, especially, show heat sensitivity (16). If approximately half of the entropy change existing between the reversibly connected forms is due to unfolding and unfolding introduces three rotational positions for each one previously existing, and here we may neglect losses in vibrational degrees of freedom since these are largely compensated by interactions with the solvent, at least 45 bonds must be released to rotation. In a protein, the size of trypsin, which has about 200 amino acid residues, this change must represent a very considerable unfolding. Primary bonds previously protected by the semi-rigid configuration are now exposed to the random brownian movements of one part of the protein against other parts and may be expected to be readily broken. The process of unfolding itself probably releases ionic groups of opposite sign which can neutralize each other and thus decrease the affinity for the solvent. Non-polar side chains previously buried in the protein are now exposed to the solvent again favoring precipitation. These interactions must be quite efficient since the net volume change is negative.

It does not seem possible to explain irreversible denaturation strictly as a dehydration process. A case in point is the thermal denaturation of pepsin studied by Steinhardt (12). At pH 5.7, five protons must be dissociated from the pepsin molecules to produce the activated complex. These processes require 45,000 cal/mol of a total activation energy of 67,500 cal/mol, much too large a fraction to be explained by a destruction of hydrogen bonding or other low-energy bonds. It seems probable, therefore, that a few salt bonds of about 9000 cal in key positions determine the folded configuration. The corresponding entropy change of 127 entropy units of a total entropy of activation of 135 eu (21) is too large to be explained as the simple dissociation of water from five neutralization processes. The rupture of each salt linkage must trigger the unfolding of an extensive volume of the pepsin molecule.

La Mer (22) has criticized the high activation parameters calculated for thermal inactivation of pepsin on the grounds that since these quantities vary with pH, a standard state of constant acid dissociation is preferable to that of constant pH on which the original calculations were based. Use of the standard state, he suggests, would result in substracting, from activation values, the very large contributions made by proton dissociation. The important process is the transition from folded to unfolded form at constant pH, so that the standard state of constant pH is certainly the more useful. Indeed, it is probably the only meaningful state since reduction in the values of the activation parameters at increased pH may be interpreted to mean that less extensively bonded or less tightly folded configurations of pepsin are being unfolded in the denaturation process. Further increases in pH should produce further decreases in the activation parameters as the molecule assumes a more and more unfolded configuration. At some value of pH there will be no difference between folded and unfolded forms.

Further evidence for the denaturation picture we have presented is forthcoming from studies on the mechanisms of biological luminescence undertaken by Johnson, Eyring, and co-workers (14, 17). The work is discussed in detail in another part of this volume. We will confine ourselves to some of its implications.

The process of oxidation of luciferin in the presence of luciferase in luminescent bacteria is sensitive to temperature and pressure. When bacteria are heated to 23° for a short period of time, an enzyme involved in the process is reversibly converted to an inactive form. Evidence cited by Johnson *et al.* and the recent confirmation of activation-parameter values by Chase (23, 24), using a purified system, indicate that this enzyme is luciferase itself. Pressure studies show an average volume increase of 64 cc per mole of enzymic center at 25°. This value, correlated with a large entropy change of 184 eu per mole of catalytic site, suggests an unfolding process similar to that discussed for trypsin inhibitor. The volume increase may be due to an exposure of hydrophobic groups which repel H_2O. Incubation at higher temperatures produces a further increase in volume (to a total increase of 70 cc per mol at 32°C) which is followed immediately by irreversible inactivation. The activation energy for the irreversible process is but 25,000 cal greater than that for the reversible process (55,300 cal). Primary bonds are, in all probability, ruptured in the irreversible reaction but 55,300 cal of activation energy must be expended against secondary (and coulombic) forces before the primary bonds become vulnerable.

Because soluble proteins exist in water solutions, the stable form will generally be that in which a major fraction of the ionizable groups of the protein are exposed to solution. Hydrophobic groups will be enclosed in the resulting hydrophilic surface. Increases in volume of proteins on unfolding probably results from the exposure of hydrophobic groups of water. Consequently the magnitudes of such increases serve to measure the extent to which hydrophobic groups have been isolated in the interior of the folded molecule. Detergents and organic solvents by adsorption on hydrophobic groups will stabilize unfolded forms of proteins (18) thus tending to decrease the magnitude or reverse the sign of the free-energy difference existing between folded and unfolded protein forms. The efficiency with which a given detergent denatures a series of proteins should therefore increase in proportion to the magnitude of volume increases in unfolding for the series of proteins.

Very high pressures (15,000 psi and greater) will cause denaturation (25). It must not be concluded therefore that all unfolded forms of protein molecules have greater volume than the biologically active, folded forms. Activated complexes and reversibly unfolded molecules generally appear to have the same or larger volumes than the folded forms. However, structural requirements for these forms will prevent optimum interaction of hydrophilic groups with similar partners or with the aqueous solvent. Similarily, hydrophobic groups are not allowed to interact with each other maximally. At extreme unfolding, probably following in most cases rupture of primary bonds, there will be no strong restrictions on orientation. Interactions of like groups will predominate and the net volume of the system can be reduced.

Mechanics of Enzymic Catalysis

Enzymes as a class are the most diverse and efficient group of catalysts known. There is no reason for believing that they all function in the same way, but it seems probable that all have the ability to act as sources or sinks for electrons. This requirement seems to be the unifying feature of catalysts in general. Lu Valle and Goddard (26) have investigated the application of this concept to enzymic catalysis. Recent studies of Chance (27) and of George (28) suggest that the ability of the iron atom of peroxidase in combination with heme group and protein to achieve an unusual plus four valence may be the principal function of this enzyme. However, most enzymes must fulfill an additional function of selecting, from among many potential substrates, all with correct electronic configuration at the point of reaction, only those with a certain definite arrangement of specific side chains. An example is the selection by some peptidases of substrates containing only L-amino acids. To achieve such specificity, multiple linkages must occur between enzyme and substrate.

In the main, the linkage points of the protein must be provided by the amino acid side chains or by prosthetic groups such as the metalloporphyrins or perhaps metal ions (29). Enzymes like leucine amino peptidase are able to distinguish between two and three CH_2 groups in a substrate side chain (30), an accomplishment shared by other proteins and indicative of rigid orientation of the bonding groups in space. Hence many, if not all, enzymes classify themselves closely with the heterogeneous catalysts of the laboratory. It is probable that the rigid orientation provided by the folded peptide provides the most important reason for the large size of the protein, though size is unquestionably important in keeping the protein in the cell and in proper position. The previously discussed role of the protein as a whole in determining the effective dielectric constant at reaction sites (19, 20) may be important for the catalytic function. Proteins because of their large mass with relatively low "polarizability," again attributable to extensive secondary cross linking, produce a lowering of the effective dielectric constant at their surfaces. In a water medium the result is greater coulombic attraction and repulsion between charges or charges and dipoles. Interactions between substrate and enzyme may consequently be considerably stronger than previously recognized. Furthermore, since it is probable that most enzymic reactions involve ionic intermediates, the behavior of the enzymic site may be strongly dependent on the charged state of neighboring ionizable groups. In fact in many instances there must exist a spectrum of catalytic forms consisting of all possible statistical arrangements of charged and uncharged ionizable groups in the neighborhood of the site.

Dispersion forces provide an important class of interactions as witnessed by the dependence of carboxypeptidase action on the hydrocarbon R groups of substrates (31) of the typical form shown here. There is no

$$\begin{array}{c} \text{O} \qquad \text{COO}^- \\ \| \qquad | \\ \text{R}'\text{—C—N—C} \\ |\quad | \\ \text{H} \quad \text{CH}_2 \\ | \\ \text{CH}_2 \\ | \\ \text{R} \end{array}$$

electronic coupling of the R group and the labile peptide bond as witnessed by the independence of the pK's for phenylalanine on substitution in the phenyl ring (74). Larger size and greater polarizability of the R group increase the rate constant for initial binding of substrate and enzyme. A similar story exists for leucine amino peptidase (30) with substrates of the form.

$$\begin{array}{c} \text{NH}_2 \\ | \\ \text{R—C—NH}_2 \\ \| \\ \text{O} \end{array}$$

Pauling (32) has pointed out that the analogy between many surface-catalyzed reactions and those enzymically catalyzed may be very close. In the initial interaction of reacting partners strain may be introduced at the expense of the negative free energy change of this interaction to labilize the susceptible bond. Such a situation is indicated for carboxypeptidase (31). In this case increased size of the R group not only increases the rate constant for initial compound formation but also increases the rate constant for the subsequent catalytic step leading to bond rupture. The very old lock and key mechanism to explain enzyme specificity must be extended (33, 34). Another old idea in enzyme work is that the protein is drawn into a strained configuration which on regaining the normal configuration tears the substrate apart. The entropies of activation for the various component slow steps in the enzymic process do not support this idea. Large negative entropies of activation usually accompany the reactions

Enzyme · Substrate → Enzyme + Product

whereas the same quantities for the reactions

Enzyme + Substrate → Enzyme · Substrate

are usually small or zero (35). Present evidence from a large variety of sources suggests that, aside from the factors of rigidity and dielectric constant, the participation of the protein involves only that part contiguous with the "catalytic site."

In addition to whatever strain is produced in the substrate it is probably necessary that the enzyme provide a group or groups acting as generalized acids or bases to facilitate the necessary rearrangement of charge according to well-known mechanisms of acidic and basic catalysis. It may be, as Laidler (36) has suggested, that enzymes provide both an electron-donating group and an electron-receiving group, the two perhaps being internally connected. Swain has found that a variety of reactions can be very greatly accelerated by the provision of both pushing and pulling groups. Such a mechanism shown below for the hydrolytic action of carboxypeptidase (38) is consistent with this picture and with the experimental data. Changes in binding of carboxyl and R group present additional complications not pictured here.

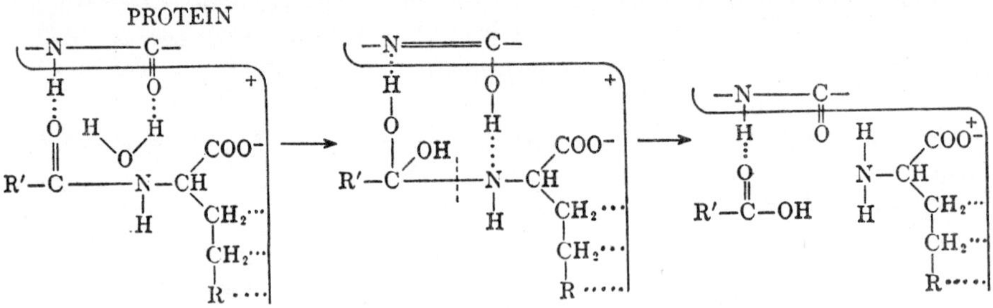

For enzymes which catalyze oxidation-reduction reactions, the picture is likely to be slightly different. As in the case of peroxidase, the protein and its prosthetic group may serve primarily for the easy production of a stable one-electron intermediate, reacting more readily with the cosubstrates of the reaction than they can react with each other. Peptide bonds display a considerable amount of chemical versatility and Geissman (39) has suggested that electron transfers from donor to receptor molecules can occur through hydrogen atom migrations of the following form:

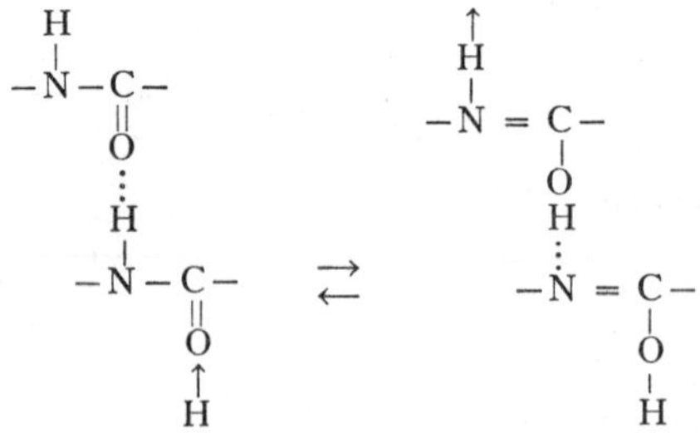

Whether or not such elaborate mechanisms are necessary, we do not yet know. For the present it is probably sufficient to assume that proteins acting as catalysts do nothing that less refined surface catalysts cannot do, but simply do it better.

Neural Phenomena and the Role of Cellular Organization

Nature not only manufactures better catalysts than man, but also makes better use of the trick of organization than has man. By suitable geometric arrangement, concentration, potential or osmotic gradients are set up within cells. Products of a preceding reaction moving down these gradients become available at the site of the next reaction at higher concentrations than would be possible if the substances were able to diffuse at random into the cytoplasm. Thus, part of the entropy increase which would result from dilution is avoided. These gradients also serve to direct reactions along necessary lines. The cytochromes are a case in point. These substances form a chain of reducible intermediates which provide a means for degrading the free-energy difference existing between carbohydrate metabolites and their oxidation products. Each cytochrome in the series is constrained by geometry to reduce only its next neighbors so that the degradation occurs in small steps which can be harnessed to drive a number of reactions in which substances necessary to the body are synthesized (40). Free dispersion of the cytochromes would allow reactions between extreme neighbors with the consequent loss of free energy.

Another process which depends for the integrity of its functioning on geometrical organization is that of nerve. Kinetic studies of neural phenomena have in the main taken the form of studying the changes in current and impedance across the nerve wall at rest and during functioning (41, 42). Though the electrical signs of nervous activity undoubtedly provide the best practical experimental approach to the problem, interpretation of the results in terms of electrical theory may be misleading. In this case, as in all biological cases, electrical gradients and their changes are manifestations of ionic processes. A preliminary attempt to translate electrical theories of nerve action into chemical terms has been made by Eyring, Lumry, and Woodbury (43). The approach has been extended by Johnson, Eyring, and Polissar (14). We will briefly summarize the ideas involved. While it must be acknowledged that the theory as given here does not give exactly the equations recently derived by Hodgkin and Huxley (44) to fit a wide variety of neural phenomena, the equations for the nerve current are surprisingly alike, demonstrating similar dependencies on potential and concentration. Since electrical theories do not lead to the chemical mechanism which must be the ultimate goal of investigation, it seems advisable to proceed in the manner outlined below, even though the preliminary attempts be incomplete.

In nerve cells (neurones), the basic cellular property of irritability has been developed to such a high degree that it provides a method for rapidly transmitting information from one part of the body to another. Neurones can be functionally pictured as long bags of highly oriented membrane (plasma membrane) filled with cytoplasm and surrounded by extracellular fluid (Fig. 1). The membrane itself is pseudo-crystalline as shown by Curtis and Cole (45), who demonstrated that a very low kinetic temperature pertains. This conclusion is in agreement with the optical studies of Schmitt and co-workers (46) and the activation parameters for non-ionic diffusion calculated by Zwolinski, Eyring, and Reese (47). The ordering of the membrane structure which exists in the resting nerve is in part owing to the high electrical potential gradient of 25,000–100,000 volts per centimeter existing across the membrane (50–100 millivolts across a membrane thickness of 100 to 200 A (48)). A reduction in this gradient greatly increases the ease of diffusion of ions through the membrane as shown by a drop in membrane resistance from 1000 to 25 ohms per square centimeters as calculated by Curtis and Cole for the giant (49) squid axion. *The most reasonable explanation for the potential gradient is that material formed within the cells adds to one or more ions, especially sodium, selectively to speed up their exodus from the cell.* Since the cations can, as a result, diffuse more rapidly than

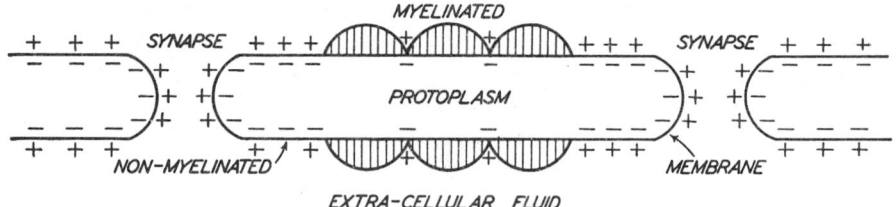

FIG. 1. Functional diagrams of a nerve cell.

anions, an inner excess of anions and hence a membrane potential will be built up until potential and concentration differences are adequate to produce steady-state conditions such that anions and cations leave the cell in equivalent amounts. This process may be described mathematically in the following sketchy manner. Considering for simplicity that the membrane provides but one important free-energy barrier (Fig. 2) to the diffusion of ions (the general treatment is given by Zwolinski, Eyring, and Reese (47), the net outward diffusion flux, q, across this barrier in the absence of an electric field can be represented thus

$$q = c\lambda k - c'\lambda' k \tag{2}$$

where c is the concentration of the species under consideration in the region preceding the potential maximum; λ is the distance traveled in passing over the following metastable position; k is a specific velocity constant measuring the number of jumps per second across the barrier. The unprimed quantities refer to conditions near the inside of the membrane; the primed quantities to conditions near the outside and hence to the inward diffusion process. When a potential, V, exists across the free-energy barrier, the net flux is given by

$$q = c\lambda k e^{-(ZV/2RT)23,060} - c'\lambda' k' e^{(ZV/2RT)23,060} \tag{3}$$

where Z is the valence of ionic species under consideration; R is the gas constant; Ions not formed or destroyed within the cell yield $q = 0$ and

$$\frac{ck\lambda}{c'k'\lambda'} = e^{(ZV/RT)23,060} \tag{4}$$

as required by thermodynamics. This is just the Nernst equation for a concentration cell. Steady-state ions, however, have $q \neq 0$ and, simplifying the flux expression by the use of the relations $\lambda' = \lambda'$, $k = k'$, $c = c'$:*

$$V = -\frac{2RT}{23,060 Z} \sinh^{-1} \frac{q}{2c\lambda k} \tag{5}$$

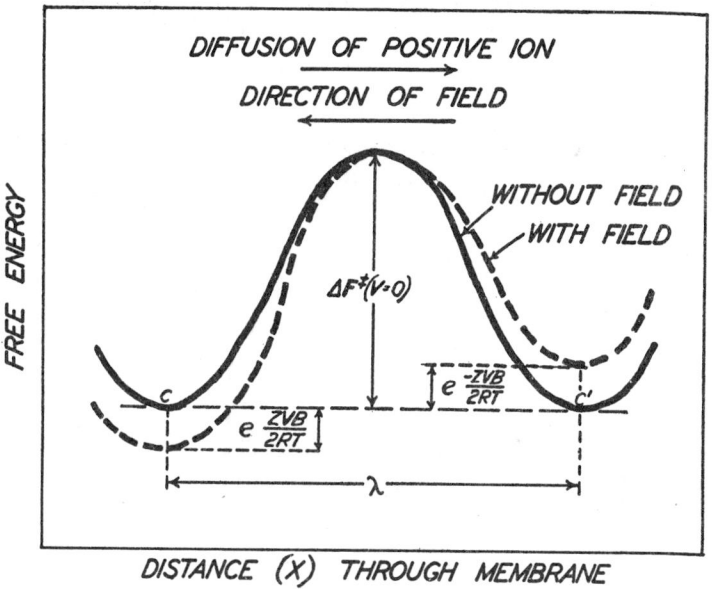

Fig. 2. Free-energy barrier for diffusion in ideal systems with and without an applied electrical field.

Similar relations apply to all non-equilibrium ions with

$$V_1 = V_2 = \cdots = V_i = \cdots = V_n$$

These considerations for all important ions describe the resting state of the nerve.

According to the best modern evidence the actual process of nerve transmission involves a local physical depolarization followed by a chemical repolarization (41). The mechanics of this process can be best described by means of Fig. 3 in which a longitudinal section of nerve membrane is pic-

---

* Justification for the first two relations is apparent. The third relation is probably not justified but for purposes of this exposition it is adequate.

tured as a series of small electric condensers, with leakage properties dependent on potential. The internal and external resistances R_I and R_E are small since they measure the difficulty of ionic diffusion through essentially water solution. At rest R_M, the resistance to ionic movements in the membrane is about 10^3 ohms/cm² (49). In nerve action when ions are allowed to diffuse away from a local site, the potential value across the membrane drops. When a critical lower potential value characteristic of

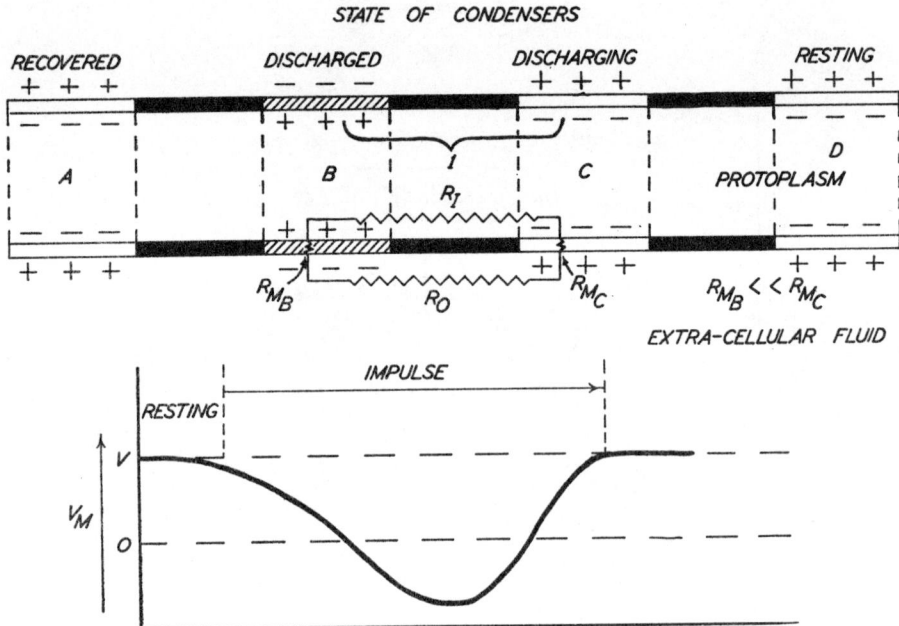

Fig. 3. The transmission of a nerve impulse along a short section of nerve cell. The nerve cell is represented as a series of cylindrical electric condensers in the upper drawing. The electrical potential across the membrane at each longitudinal distance is shown in the lower curves.

the nerve and its physiological condition is reached, the resistance of the membrane falls very rapidly to a small fraction of its resting magnitude (one-fortieth for the giant squid axon as determined by Curtis and Cole (49)). The rate constants k and k' are therefore nearly discontinuous functions of potential. For present purposes the expression for R_M may be obtained in simplified form from the summation of flux expressions and the expression for membrane potential (Eqs. 2 and 3).

$$R_M = \frac{-2RT}{\rho_M \cdot 23{,}060 \Sigma_j Z_j^2 \lambda_j e \, [c_j k_j (V) - c_j' k_j'(V)]} \tag{6}$$

ρ_M is an area factor, determinable from the geometry of the nerve.

As the impulse, decay, and growth of membrane potential, proceeds down the nerve, condenser A is discharged, R_{MA} drops to a low value permitting ions to leak away from the surfaces of condenser B. At the critical value of potential, R_{MB} drops to a low value thus permitting condenser C to discharge. In this way the potential decays progressively down the nerve. Discharged condensers may be recharged by means of increased metabolic activity. For instance, the relaxed and hence highly permeable membrane may present new catalytic sites to speed up the production of metabolites which selectively act to pump ions out of the cell by complexing with them. Residual differences in ease of diffusion of cations and anions will cause the rebuilding of a small potential which in turn decreases the permeability of the membrane and further emphasizes the difference between anions and cations. Such a process would proceed autocatalytically until the steady state of the resting potential is reached. The velocity, v, of the overall process for a single impulse is given by an expression which follows immediately from flux and resistance expressions (3) and (6).

$$\frac{-l}{v} = \tau_1 + \tau_2 \equiv \left[\frac{RTC(\ln V_M - \ln V_C)}{23{,}060\,e} \left(2\rho_M \sum_j Z_j^2 \lambda_j [C_{ij}k_{ij} - C_{oj}k_{oj}] \right)^{-1} \right.$$
$$\left. + \left(\rho_i \sum_j Z_j^2 \lambda_{ij}'' k_{ij}'' \right)^{-1} + \left(\rho_o \sum_j Z_j^2 \lambda_{oj}'' k_{oj}'' \right)^{-1} + \frac{h}{kT} e^{\Delta F_M^\ddagger/RT} \right] \quad (7)$$

where c = capacity of condenser,
  i = inside membrane,
  o = outside membrane,
  $\Delta F_M^\ddagger$ = free energy of activation for change in permeability properties of the membrane;
  double primes refer to processes taking place on one side of the membrane only.

The first sum of terms in the bracket is the time for discharge, τ_1, of the condenser from the resting to the critical potential. The last term measures the succeeding period, τ_2, for transition of the membrane to the permeable state.

A possible mechanism for the sodium pump which would explain many of the features follows. Some metabolite formed by enzymes inside the cell (conceivably acetyl choline) solvates sodium ions preferentially. The complexes formed diffuse readily through the membrane to the outside of the cell where the metabolite is destroyed. If acetyl choline is the metabolite, destruction is by deesterification. The sodium ion in the absence of metabolite has no easy way of return to the cell's interior. The resting potential persists so long as the metabolite gradient maintains the necessary associated cation gradient. Part of the enzymes maintaining this metabolite gradient presumably are buried in the membrane and only come into action when the membrane through depolarization becomes permeable.

This more rapid metabolism then restores the polarization to the resting potential. Any imposed change tending to lower the metabolite gradient by lowering the action potential tends to excite the nerve. If this gradient decrease is carried far enough, the resting state becomes unstable, leading to convulsions. The potential gradient can of course be decreased in a wide variety of ways without changing the metabolite gradient. Diisopropyl fluorophosphate destroys choline esterase, thus causing the concentration of acetyl choline outside nerve cells to rise and so causing convulsions. Atropine by tying up the acetyl choline prevents convulsions. However, experiments which would definitely establish the nature of the important metabolites are lacking.

The behavior of the membrane is best described as a phase change with low activation energy for the change to permeable (H) state.

$$L \text{ (low)} \rightleftarrows H \text{ (high)}$$

Under conditions of rest the chemical potential change, $\Delta\mu$, in this reaction is positive. Reduction of potential will change the sign of $\Delta\mu$ allowing the reaction to take place. A few molecules of some chemical such as are inhaled from a smelled substance by changing the activity of enzymes at the end plate promote chemical reactions which reverse the sign of $\Delta\mu$ of the nerve ending. Pressure and temperature could have similar effects on $\Delta\mu$. The process of sight is extremely sensitive, one quantum of radiation per rod for several rods being sufficient under optimum conditions to produce an impression on the brain (50). One excited molecule itself acting on a nerve ending is certainly inadequate to produce nerve function. It may therefore be expected that the metastable molecule excited by absorbed radiation will be found to act as an enzyme inhibitor or activator to change the rate of production of a number of other molecules which then act on the nerve ending.

Photosynthesis and the Hill Reaction

Photosynthesis is another process completely dependent on cellular organization. The overall reaction is readily studied only in intact plant cells, a condition restricting investigation to involved chemical kinetics. However, fragments of the highly organized grana of the chloroplast may be isolated, washed free of small cytoplasmic molecules, and studied as the apparatus by which oxygen is produced from water on illumination. Hill (51) first observed the reaction in cell-free systems when a suitable oxidant replaced the carbon dioxide system as an electron sink. Improvements in technique are now such that both velocity and yield are as good as those for photosynthesis. By converting the half-cell potential (determined by the ratio of oxidized and reduced forms of the added oxidant) to potential time data, good accuracy in velocity determination can be obtained even with $10^{-6}M$ oxidant concentrations. Applying potentiometry to chloroplast fragments from higher plants, Spikes, Lumry, and their co-workers have undertaken quantitative investigation of the Hill reaction (52). Examples of their potential time and oxygen time results appear in Fig. 4. In all known respects the Hill reaction and the light reaction of photosynthesis are identical. The former can be more precisely studied and hence forms the better subject for quantum requirement experiments. Correct knowledge of the requirement is essential to mechanism development.

Velocity-light intensity (I) data for higher plants obey, with good precision, the rectangular hyperbolic rate law provided by a system of reactions requiring a minimum of three conserved intermediates thus (53)

$$\begin{array}{ll} \text{Chl} + h\nu \xrightarrow{k_1} \text{Chl}^* & \text{Chl} + \text{Chl}^* = a \\ \text{Chl}^* \xrightarrow{k_{-1}} \text{Chl} + \text{heat} & \\ \text{Chl}^* + B \xrightarrow{k_2} \text{Chl} + \{B\} & B + \{B\} = b \quad (8)\\ \{B\} + C \xrightarrow{k_3, H_2O\,?} B + \{C\} & \\ \{C\} \xrightarrow{k_4} C \to \tfrac{1}{4}O_2 & C + \{C\}^* = c \end{array}$$

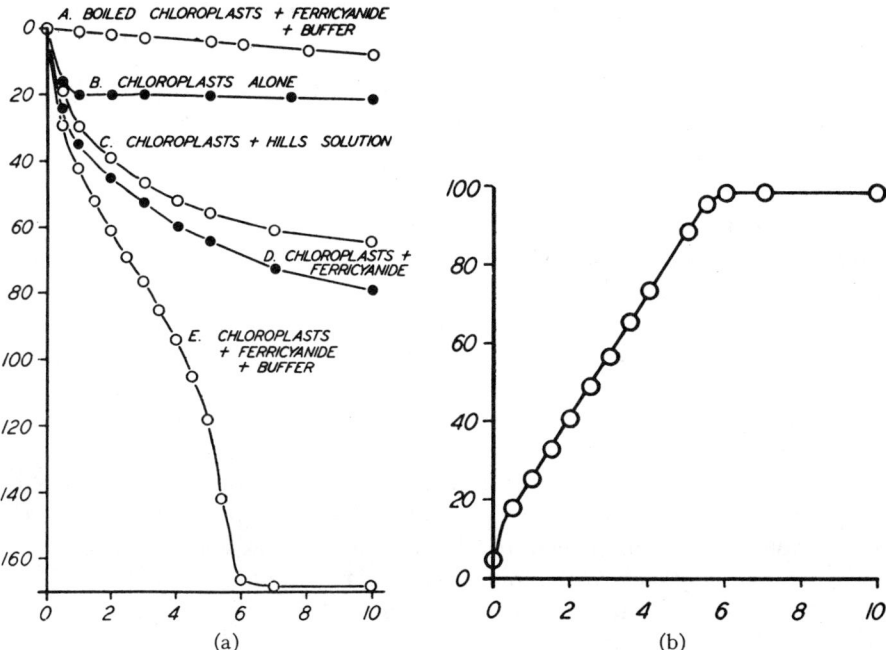

FIG. 4. (a) Curves showing the changes in potential upon illumination of suspensions of washed spinach chloroplasts under different conditions as indicated. Chloroplast concentration equal to 150 mg of chlorophyll per liter. Potassium ferricyanide concentration $0.001 M$ where used. Potassium phosphate buffer concentration $0.1 M$, pH 6.7, where used (44). Ordinate: illumination in minutes. Abscissa: change in potential (millivolts). (b) Curve showing per cent reduction of ferricyanide to ferrocyanide with time by washed spinach chloroplasts upon illumination. This curve is derived from the data of curve E of (a) by applying Eq. (8). Ordinate: illumination in minutes. Abscissa: per cent ferricyanide reduced to ferrocyanide.

B and C are probably enzymes at least one of which shows the characteristic sharp occurrence of irreversible denaturation at 35°C (54). The rate expression partially simplified to fit some of the available observations has two useful forms

$$v = \frac{dO_2}{dt} = \frac{4k_1k_2k_3k_4\ abcI}{k_1k_2k_3\ ab\ I + k_1k_2\ a\ I + k_1k_3k_4\ I + k_{-1}k_3k_4C + k_2k_3k_4C} \quad (9)$$

$$\frac{I}{v} = \frac{k_2k_3ab + k_2k_4a + k_3k_4C}{4k_2k_3k_4abc} + \frac{k_{-1} + k_2b}{4k_1k_2ab} \quad (9')$$

The latter form coupled with the total rate for quanta absorption,

$$\frac{dQ}{dt} = \alpha aI \quad (10)$$

where Q = number of quanta absorbed,
α = rate constant for absorption,
allows extrapolation of quantum requirement, $1/\varphi$, to its minimum value at zero intensity according to

$$\frac{1}{\varphi} = \frac{dQ/dt}{dO_2/dt} = \frac{\alpha\ (k_2k_3ab + k_2k_4a + k_3k_4c)\ I}{4k_2k_3k_4bc} + \frac{\alpha\ (k_{-1} + k_2b)}{4k_1k_2b} \quad (11)$$

The constant term above is the minimum requirement and is equal to the number of useful primary processes if, as the available evidence suggests, $k_2\ b > k_{-1}$. By such extrapolation a requirement of 8.2 quanta per oxygen molecule was observed using washed sugar-beet chloroplast fragments (55). This value is somewhat lower than the best previous determinations in photosynthesis as reported by most American observers (56). It is twice as large as the value consistently measured by Warburg et al. (57). Chard values were found to be about 10 to 11, in good agreement with quantum requirement studies of the Hill reaction by other investigators (58, 59). Pre-illumination or preheating of inadequately washed chloroplast fragments will produce values less than 8 due, it appears, to a strong reducing agent coupled to the photochemical system. The Warburg value of 4 which depends on special pretreatment of Chlorella might be explained in this way.

The minimum theoretical requirement of 4 quanta is very difficult to reconcile with the number and type of chemical intermediates which must be involved (60). A more readily acceptable mechanism would consist of 8 primary processes followed by 4 chemical dismutations though the observed requirement of 8 makes it impossible to include "slippage" (non-useful degradation of quantum energy into heat) in such a scheme. Although this mechanism is less than fifty per cent efficient in energy conversion, it is very nearly one hundred per cent in the use of quanta.

There is very little evidence bearing on the means by which the chemical energy produced in eight individual primary processes may be collected and dismuted to evolve a single oxygen molecule and four electrons. Recent observations demonstrate that energy absorbed by those pigments of the grana which have absorption and fluorescent spectral peaks to the blue side of the red peak of chlorophyll a not only excite the fluorescence of the latter but also produce efficient photosynthesis (61, 62, 63). These observations lead to the belief that electronic quanta move freely among the pigment molecules (64, 69) (but see Frank and Livingston (65)). The mechanism of this transfer is supposed to depend on strong dipole coupling of neighboring molecules and close balancing of available excitation levels in interacting pairs (66). The facts are adequately treated by the "exciton" theory of Fraenkel (67) or the resonance-fluorescence theory of Forster (66) and others. If such movements are rapid, it might be possible to secure the collection, from a number of absorbing dye molecules, of the required number of quanta at a single trapping center and thus avoid the necessity for a collection process depending on the diffusion of small, high-energy chemical intermediates (69) (but see Franck (76)). For example, the trapping center could consist of a single molecule of enzyme B associated with chlorophyll a in such a way as to speed up its internal conversion (68).

It is now possible to explain the small spectral shifts which occur when chlorophyll is incorporated in grana not as the result of combination with protein but rather as a result of the aggregation of dye molecules (70, 71). Most chlorophyll molecules are probably not connected to protein or lipid in such a way as to modify their electronic structure and may as a result be unsatisfactory places for internal conversion. Emerson and Arnold (72) observed that there were about 2300 chlorophyll molecules (1800 of chlorophyll a) per molecule of the enzyme (probably B above) which limits photosynthesis at high light intensity. Their "photosynthetic unit" could be the collection unit of which we speak. A rough calculation indicates a slight favorable probability for such a mechanism. The minimum lifetime of fluorescence is 10^{-8} sec for chlorophyll a, and the fluorescence yield is 1% leading to an excitation lifetime of 10^{-10} sec for an electronic quantum in a homogeneous matrix of chlorophyll a molecules among which are randomly distributed trapping centers. Assuming an eight-quantum requirement for each oxygen molecule, the "photosynthetic unit" of Emerson and Lewis reduces for the single-quantum process to about 200 chlorophyll a molecules per trapping center since one-fourth of the chlorophyll molecules are of type b which have been shown to transfer their energy irreversibly to the a type (75).

If a customary time for internal conversion of 10^{-13} sec is attributed to the trapping center and is assumed to be the lower limit for a jump or transfer time, 10^3 jumps are on the average possible. In a random trip through a three-dimensional lattice of pigment molecules, the quantum may be trapped or lost as heat, thus:

$$\text{Chl } a + h\nu \xrightarrow{k_1} \text{Chl } a^*; \quad \text{Excitation} \tag{12}$$

$$\text{Chl } a^* \xrightarrow{k_2} \text{Chl } a + \text{Heat}; \quad \text{Loss}; k_2 \sim 10^{10} \tag{13}$$

$$\text{Chl } a^* \xrightarrow{k_3} \text{Chl } a + \text{Useful energy}; \quad \text{Trapping} \tag{14}$$

$k_3 \sim 10^{13}$ if

$$n = \frac{\text{Trapping centers}}{\text{Chl a molecules}} \sim 200$$

$$\frac{d \text{ Chl a*}}{dt} = k_1 \text{ Chl a } I - k_2 \text{ Chl a*} - k_3 \text{ Chl a*} = 0$$

at the steady state, I being the light intensity.

$$\frac{d \text{ (Useful energy)}}{dt} = k_3 \text{ Chl a*}$$

$$\text{Trapping efficiency} = \frac{k_3 \text{ Chl a*}}{k_1 \text{ Chl a } I} = \frac{k_3}{k_2 + k_3} \sim 0.84$$

A similar calculation based on the theory of Förster yields 0.80 (75). More careful consideration based on various one-dimensional, ordered arrays using the "game of the gambler's ruin" yields slightly lower efficiencies. More reliable values must await better understanding of the transfer process. Several investigations in this direction are underway. The problem is typical of most modern problems in photosynthesis in that kinetic studies provide the best available means of attack. Though structural and analytical chemical evidence would be most satisfactory, the complexity of the system and the low concentrations of intermediates which must be involved make chemical studies very difficult. It is probable therefore that, as in the past, we must turn for the bulk of knowledge about this reaction to the implications of chemical kinetics.

Thus in a cursory and almost aristotelian manner we have considered a variety of important biochemical and physiological systems with which, because of their complexity, investigational techniques have been largely restricted to chemical kinetics. The various component discussions bespeak the authors' biases and inadequate knowledge and they are not meant as a guide to the workers in the various fields, to most of whom we owe apologies for inadequate coverage of their work.

Progress in the application of conventional chemical techniques and theories to problems of living organisms since World War II has been remarkable. Within the next few years we may expect to have correct chemical mechanisms for enzymic catalysis, muscle function and other problems for which the knowledge of participating molecules is well advanced. It is the authors' belief that progress in many less well analyzed systems can be accelerated by serious attempts to integrate existing facts in the light of established chemical-kinetic principles.

Acknowledgment

The authors are indebted for helpful discussion to Drs. E. L. Smith, J. D. Spikes, J. Schellman, K. J. Laidler, and W. J. Woodbury.

References

1. Oparin, A. I., *The Origin of Life*. Macmillan, New York, 1938.
2. Horowitz, N. H., *Proc. Natl. Acad. Sci. U. S.* **31**, 153 (1945).

3. Eyring, H., Johnson, F. H., and Gensler, R. L., *J. Phys. Chem.* **50**, 453 (1946).
4. Blum, H. F., *Time's Arrow and Evolution*. Princeton University Press, Princeton, New Jersey, 1951.
5. Dobry, A., Fruton, J. S., and Sturtevant, J. M., *J. Biol. Chem.* **195**, 149 (1952). See also, Breitenbach, J. W., Derkosch, J., and Wessely, F., *Nature* **169**, 922 (1952).
6. Mirsky, A. E., and Pauling, L., *Proc. Natl. Acad. Sci. U. S.* **22**, 439 (1936).
7. Huggins, M. L., *Chem. Revs.* **32**, 195 (1943).
8. Astbury, W. T., *Advances in Enzymol.* **3**, 63 (1943).
9. Pauling, L., Corey, R. B., and Branson, H. R., *Proc. Natl. Acad. Sci. U. S.* **37**, 205-11, 235-85 (1951).
10. Kauzmann, W., Private communication, "Kinetics of Protein Denaturation."
11. Reese, C. E., and Eyring, H., Mechanical Properties and Structure of Hair. Technical Report XI, Navy Research Project NR-032-168 Salt Lake City, Utah, 1949.
12. Steinhardt, J. V., *Kgl. Danske Videnskab. Selskab Mat.-fys. Medd.* **14**, No. 11 (1937).
13. Reviewed in Haurowitz, F., *Chemistry and Biology of Proteins*, Chap. 17. Academic Press, New York, 1950.
14. Johnson, F. H., Eyring, H., and Polissar, M., *The Kinetic Basis of Molecular Biology*. John Wiley and Sons, New York, 1953.
15. Kunitz, M. J., *J. Gen. Physiol.* **32**, 241 (1948).
16. Eisenberg, M. A., and Schwert, G. W., *J. Gen. Physiol.* **34**, 583 (1950).
17. Johnson, F. H., and Eyring, H., *Ann. N. Y. Acad. Sci.* **49**, 376 (1948).
18. Neurath, H., Greenstein, J. P., Putnam, F. W., and Erickson, J. O., *Chem. Revs.* **34**, 157 (1944).
19. Schellman, J., Dielectric Participation of the Protein in Ion Binding. To be submitted to *J. Phys. Chem.*
20. Lumry, R., and Schellman, J., To be submitted to *J. Am. Chem. Soc.*
21. Stearn, A. E., and Eyring H., *Chem. Revs.* **29**, 509 (1941).
22. La Mer, V. K., *Science* **86**, 614 (1937).
23. Chase, A. M., and Lorentz, P. B., *J. Cellular Comp. Physiol.* **25**, 53 (1945).
24. Chase, A. M., *J. Cellular Comp. Physiol.* **27**, 1 (1946).
25. Bridgman, P. W., In *Colloid Chemistry*, J. Alexander, Editor **5**, 327 (1944). (Reinhold Publishing Corporation, New York.)
26. Lu Valle, J. E., and Goddard, D. R., *Quart. Rev. Biol.* **23**, 197 (1948).
27. Chance, B., *Advances in Enzymol.* **12**, 153 (1951).
28. George, P., *Nature* **169**, 612 (1952).
29. Smith, E. L., *Proc. Natl. Acad. Sci. U. S.* **35**, 80 (1949).
30. Smith, E. L., and Lumry, R., *Cold Spring Harbor Symposia Quant. Biol.* **14**, 168 (1950).
31. Smith, E. L., Lumry, R., Poleglase, J., Armstrong, M., and Glantz, R., Kinetics of Carboypeptidase III. Effects of Substitution in the Phenyl Ring of Substrates and Inhibitors. To be submitted to the *J. Am. Chem. Soc.*
32. Pauling, L., *Chem. Engr. News* **24**, 1375 (1946).
33. Haldane, J. B. S., *Enzymes*. Longmans-Green, London, 1930.
34. Smith, E. L., Personal communication.
35. Laidler, K. J., The Molecular Kinetics of Enzyme Catalyzed Reactions. Paper presented in Symposium on Biochemical Kinetics at the Diamond Jubilee Meeting of the American Chemical Society, September, 1951.
36. Barnard, M. L., and Laidler, K. J., Solvent Effects in the Chymotrypsin Hydrocinnamic Ester System. Abstract of thesis submitted by M. L. Barnard to Catholic University, 1952.
37. Swain, C. G., and Brown, J. F., Jr., *J. Am. Chem. Soc.* **74**, 2538 (1952).
38. Lumry, R., Smith, B. L., and Stockell, A., Kinetics of Carboxypeptidase IV. Medium Effects. To be submitted to *J. Am. Chem. Soc.*, 1952.
39. Geissman, T. A., *Quart. Rev. Biol.* **24**, 309 (1949).

40. Chance, B., *Nature* **169**, 215 (1952).
41. Erlanger, J., and Gasser, H. S., Electrical Signs of Nervous Activity. University of Pennsylvania Press, Philadelphia, 1937.
42. Katz, B., *Electric Excitation of Nerve.* Oxford University Press, London, 1939.
43. Eyring, H., Lumry, R., and Woodbury, J. W., *Record Chem. Progress (Kresge-Hooker Sci. Lib.)* 100, (Summer, 1949).
44. Hodgkin, A. L., and Huxley, A. F., *J. Physiol.* **117**, 500 (1952).
45. Cole, K. S., Four Lectures on Biophysics, Instituto de Biofisica, University of Brazil, Rio de Janerio, 1947.
46. Schmitt, F. O., and Bear, R. S., *Biol. Revs. Cambridge Phil. Soc.* **14**, 27 (1939).
47. Zwolinski, B. J., Eyring, J., and Reese, C. E., *J. Phys. and Colloid Chem.* **53**, 1426 (1949).
48. Cole, K. S., and Curtis, H. J., *Cold Spring Harbor Symposia Quant. Biol.* **4**, (1936).
49. Cole, K. S., and Curtis, H. J., *J. Gen. Physiol.* **22**, 37 (1938); **23**, 649 (1939).
50. Hecht, S., Shlaer, S., and Pirenne, M. H., *J. Gen Physiol.* **25**, 819 (1942).
51. Hill, R., and Scarisbrick, R., *Proc. Roy. Soc. (London)* **B129**, 238 (1940).
52. Spikes, J., Lumry, R., Eyring, H., and Waynynen, R., *Proc. Natl. Acad. Sci. U. S.* **36**, 455 (1950); *Arch. Biochem.* **28**, 48 (1950).
53. Lumry, R., Rieske, J., and Waynynen, R., Kinetics of the Hill Reaction II. To be submitted, 1953.
54. Bishop, N., Thesis submitted in partial fulfillment of the requirements for M. S. degree, University of Utah, 1952.
55. Waynynen, R., Thesis submitted in partial fulfillment of the requirements for Ph.D. degree, University of Utah, 1952.
56. Reviewed in Rabinowitch, E. I., *Photosynthesis and Related Processes*, Vol. II, Pt. 1, Chap. 29. Interscience Publishers, New York, 1951.
57. Warburg, O., and Burk, O., *Naturwissenschaften* **37**, 560 (1951); *Z. Naturforsch.* **6b**, 12 (1951); *Arch. Biochem.* **25**, 410 (1950).
58. French, C. S., and Rabideau, G. S., *J. Gen. Physiol.* **28**, 329 (1945).
59. Ehrmantraut, H. C., and Rabinowitch, E. I., *Arch. Biochem. and Biophys.* **38**, 67 (1952).
60. Rabinowitch, E. I., *Photosynthesis and Related Processes*, Vol. I. Interscience Publishers, New York, 1945.
61. Emerson, R., and Lewis, C., *Am. J. Botany* **30**, 165 (1943).
62. Reviewed in Rabinowitch, E. I., *Photosynthesis and Related Processes*, Vol. II, Pt. 1, Chap. 30. Interscience Publishers, New York, 1951.
63. Dutton, H. J., and Manning, W. M., *J. Phys. Chem.* **49**, 380 (1941).
64. Duysens, L. N. M., *Nature* **168**, 548 (1951).
65. Franck, J., and Livingston, R., *Revs. Mod. Phys.* **21**, 505 (1949).
66. Förster, T., *Ann. Physik.* **2**, 55 (1948).
67. Fraenkel, J., *Phys. Rev.* **37**, 17, 1276 (1931).
68. Arnold, W., and Oppenheimer, J. R., *J. Gen. Physiol.* **33**, 423 (1950).
69. Förster, T., *Z. Naturforsch.* **2b**, 174 (1947).
70. Strain, H. H., *Science* **116**, 174 (1952).
71. Jacobs, E. E., and Holt, A. S., *J. Chem. Phys.* **20**, 1326 (1952).
72. Emerson, R., and Arnold, W., *J. Gen. Physiol.* **15**, 391 (1932); **16**, 191 (1932).
73. Rabinowitch, E. I., *Photosynthesis and Related Processes*, Vol. II, Pt. 1, Chapt. 28. Interscience Publishers, New York, 1951.
74. Nachmansohn, D., and Wilson, I. B., *Advances in Enzymol.* **12**, 259 (1951).
75. Duysens, L. N. M., Transfer of Excitation Energy in Photosynthesis. Thesis, University of Utrecht, 1952.
76. Franck, J., Carbon-Dioxide Fixation and Photosynthesis, *Society of Experimental Biology Symposium* **5**, p. 160. Academic Press, New York, 1951.

BIOPHYSICS

SOME PHYSICAL AND CHEMICAL PROPERTIES OF AXONS RELATED TO CONDUCTION OF NERVE IMPULSES

Frank Brink, Jr.

The Johns Hopkins University, Baltimore, Maryland

One way to characterize a field of science is to indicate the origin of the problems dealt with and the area of significance of the anticipated results. Thus, the problems of biophysics originate as problems in biology and the final results of such investigations become part of modern biology rather than a part of physics. The most penetrating biophysical investigations are concerned with the analysis in physical terms of the processes and properties of living cells. For example, this is a characteristic of the outstanding biophysical work of our chairman, Dr. Cole. His studies of the impedance of cell membranes have been made upon living cells, at rest and during activity. The results of such investigations, having a direct bearing on living phenomena, are a part of modern biology.

All problems in cellular physiology have chemical as well as physical aspects. For example, the transformation, in animal cells, of the internal energy of chemical compounds into electrical or mechanical work, or into radiation, is intrinsically such a problem. In general, the adjectives biophysical and biochemical describe aspects of biological investigations.

It is often convenient to choose the single cell as the frame of reference in a biological investigation. In the study of multicellular organisms two general kinds of questions immediately arise: (1) What is the primary role of a cell in the organism and which of its properties are of immediate importance to this role? (2) What physical and chemical processes underlie these specialized properties of the cell? These two aspects of biological studies are prominently exhibited in experimental neurophysiology. There are, on the one hand, the problems of expressing the properties of the integrated nervous system in terms of the properties of constituent neurons. On the other hand, the properties of the single neurons require explanations in physical and chemical terms.

Historically the phenomena of initiation and conduction of the nerve impulse were classified as electrobiology or electrophysiology. This simply indicated that the cellular phenomena under study were initiated or detected by physical instruments designed to produce or measure electrical currents. At present the study of the electrical properties of nerve cells is considered to be part of biophysics. In the following discussion there is no particular emphasis on the electrical nature of the phenomena or on the kind of instruments employed in the research. Rather an effort will be made to formulate some problems that are of biological significance in terms suitable for experimental and theoretical analysis by methods of modern physical science. In this discussion we will focus our attention upon some of the coordinated processes that together create the properties of an axon that are essential to its physiological function within the nervous system.

Physiological Function of the Axon

The axon is the communication line of the nervous system. It is an extended part of a nerve cell that connects with other nerve cells, glands, or muscles. Its physiological function is to conduct impulses initiated at one end to the other end of the cell. The axon can conduct an impulse in either direction but the direction of travel of an impulse is imposed normally by the site of excitation being at only one end of the cell. In a sensory nerve the signals are developed in the nerve by action of the receptor organ that is acted upon by some physical agent in the environment, such as a sound wave, or a hot object on the skin. The signals enter the central nervous system after transmission along the axon of the sensory nerve. In a motor nerve, changes produced by activity in contiguous endings of other nerve cells lead to excitation and conduction of impulses along the motor axon to its ending on a distant muscle. The axon of a nerve has the following properties that are of immediate significance to this physiological function in the organism: (1) it transmits signals at a definite velocity; (2) it normally remains quiescent unless excited through action at its receptive terminals; (3) it recovers excitability after each signal in a short time so that trains of impulses can be transmitted within a wide range of frequencies.

The physiologically significant properties of an axon have been localized in its surface regions and are closely connected with the electrical properties of these molecular structures. This has given rise to a definition of nervous action in biophysical terms. On the other hand, the maintenance of the capacity for transmitting impulses is closely related to the rate of oxidation of each part of the axon. The energy for continued transmission and recovery therefrom is developed from chemical reactions at each local area of the axon. Thus, this conducting mechanism is not a passive one; on the contrary local sources of energy are involved in the propagation of the electrochemical disturbance that constitutes the signal within each cell. Thus it is clear why an adequate discussion of the specialized physiological properties of these living structures requires biochemical and biophysical information.

Physical Nature of a Nerve Impulse

Our present concepts of the physical character of the nerve impulse are derived mainly from measurement of (1) the membrane potential difference and (2) the electrical impedance of the membrane. The prominent aspects of these properties of the membrane can be illustrated by two experiments that are especially clear-cut and direct. It is possible to insert a small electrode inside the giant axon of a Squid, which may be as much as 700 microns in diameter. Thus the potential difference between the inside and outside of this nerve fiber can be measured directly (Curtis and Cole, 1940; Hodgkin and Huxley, 1942). When a nerve impulse is propagated along this axon there is observed a rapid decrease and reversal in the potential difference between the inside and outside electrode, Fig. 1. This marked change in the spatial distribution of charged particles occurs in less than a millisecond and is evidence of a change in the structure of the axon surface.*

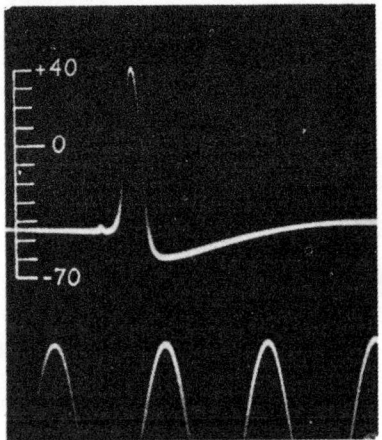

FIG. 1. Potential difference between an electrode inside a living axon and an external electrode. The inside was 45 mv negative to the outside electrode with the nerve at rest. During the passage of an impulse this potential difference was reversed during the transient change in the membrane. Time in milliseconds. Scale in millivolts. (Hodgkin and Huxley (22), reproduced with permission of the publisher, Cambridge University Press.)

The decrease of membrane potential in a region of the axon is followed by a decrease in impedance across the surface of the cell. This can be measured by the transient imbalance of an impedance bridge that is recorded simultaneously with the action potential, Fig. 2. The analysis of this impedance change as a function of frequency indicates a large decrease in the radial component of resistance across the axon surface (Curtis and Cole, 1939).

Thus the nerve impulse is a change of cell structure that spreads along the surface of an axon. The changes in membrane potential difference and in radial resistance characterize, in physical terms, the response of an axon to a stimulus. These structural changes permit current to flow from adjacent

* Since this paper was written in 1949 there has been much progress in such experimental investigations. There is evidence that the sudden change in structure is a change in permeability to sodium ions. The inward transport of Na^+, decreasing the membrane potential, is followed by an outward surge of K^+, restoring the original potential difference. (cf. A. L. Hodgkin, *Biol. Revs., Cambridge Phil. Soc.* **26**, 339, 1951; and A. L. Hodgkin and A. F. Huxley, *J. Physiol.* **116**, 449, 1952.

areas of the membrane through a conduction path including the external solution bathing the axon. The action currents are sufficient to initiate a response in the adjacent areas and the structural change is propagated, as a nerve impulse (Hodgkin, 1937). The action currents flowing externally provide a physical sign of the passage of an impulse since the longitudinally distributed potential difference in the external solution can be measured as the action potential (Bernstein, 1866). Clearly, the action potential is an essential accompaniment of a nerve impulse.

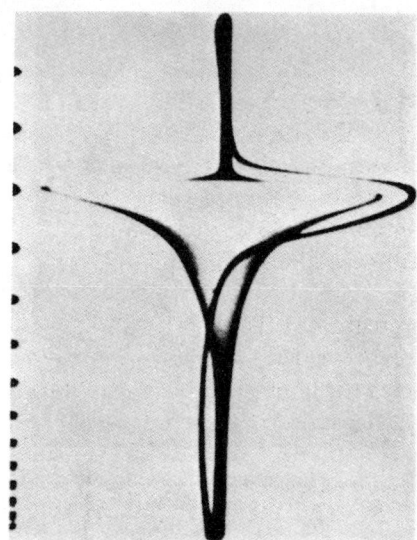

FIG. 2. During the action potential the impedance of the membrane decreases. The earlier deflection from the baseline is the action potential. The other trace measures the imbalance in an impedance bridge. Time in milliseconds. (Curtis and Cole (8), reproduced with permission.

The familiar action current and the demarcation current of a resting nerve measured between two external electrodes had revealed previously the major role of electrical phenomena in nervous activity (cf. Erlanger and Gasser, 1937). Indeed, relative values of the membrane action and resting potential, like that shown in Fig. 1, can be readily measured by external electrodes (Shanes, 1949). The recent experiments mentioned above, on a special kind of axon, have afforded more satisfactory conditions for quantitative study of the absolute value of the membrane potential of a cell at rest and during activity.

It is not necessary for the membrane potential change shown in Fig. 1 to be reversed in sign, for propagation of the impulse can occur in this kind of nerve even when the change is less than the initial membrane potential difference (Hodgkin and Katz, 1949). In other words, propagation can occur by means of action currents developed by partial depolarization of a nerve (cf. Lorente de Nó, 1947 (a), page 247).

The visible and otherwise optically defined structures of the axon are not identical with the structures identified by these electrical measurements. For example, during activity there are no changes in structure detectable by polarized light (Schmitt and Schmitt, 1940). However, a change in opacity of some nerves has been noted during activity (Hill and Keynes, 1949). The accumulated evidence suggests that the optically revealed structures are a stable molecular framework upon which is superimposed the more labile spatial arrays of molecules associated with excitation, response, and recovery.

Method of Signalling

The signals conducted along axons are stereotyped electrochemical changes detected as a series of pulselike action potentials. The signals transmitted to the central nervous system from receptor organs have been detected as action potentials in axons from visual sensory cells (Hartline and Graham, 1932), from stretch receptors of muscles (Matthews, 1933), from taste receptors (Pumphrey, 1935), from pressure receptors (Bronk and Stella, 1932) and from pain fibers (Adrian, 1932). The signals coming from the central nervous system have been similarly examined in axons leading to muscles (Adrian and Bronk, 1928). Two principles of neuronal action of interest here have been revealed by this work: (1) an increase in intensity of a stimulus to a receptor organ is reflected in an increase in the average frequency of action potentials transmitted along the axon, as well as in a transient change in temporal pattern; (2) the intensity of mechanical response of a muscle increases with the frequency of impulses in a motor nerve, within limits. Thus the frequency of impulses measures the intensity of a stimulus from outside the organism and, in turn, determines the intensity of stimulus delivered to a muscle, or, in general, to the next nerve cell in a chain of neurons in the central nervous system. Consequently in the study of the performance of axons, viewed as the communication lines of the nervous system, their properties during activity at various frequencies are of importance.

Diagram of Some Relations

Because of the incomplete and provisional character of our knowledge of axonal properties it is convenient to consider a unit length of axon as composed of four interacting systems. These are (1) a metabolizing system consisting of an organized enzyme structure together with all the generated intermediates between primary metabolites and products; (2) a polarizable structure located in the surface regions that is defined by its electrical impedance properties; (3) a battery with an electromotive force generated by metabolic processes that maintains a potential difference across the polarizable structure; and (4) a metastable structure in the interface that can undergo a cycle of change in response to a stimulus. This latter structure is very labile and is responsible for the property of excitability. It is closely integrated with the polarizable structure. These are the operationally defined aspects of neuronal structure. As mentioned before, one can imagine the axonal processes to be structural changes in labile molecular arrangements that take place within the framework of a relatively slowly changing cellular structure such as that revealed thus far by microscopic and other optical methods. Establishing connections between these various defined systems in an axon is an important type of investigation. In Fig. 3 is a diagram of these ideas.

The evidence for independent existence and for interaction between postulated structures can be summarized in relation to this diagram. It has been demonstrated repeatedly that withdrawal of oxygen is followed by loss of membrane potential, failure of conduction, and changes in impedance.

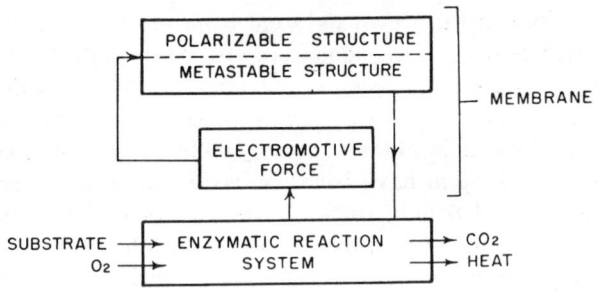

Fig. 3. Block diagram of some relations between operationally defined structures postulated for convenience in discussing relationships among properties of axons.

These phenomena can occur without visible changes in axons and are readily and rapidly reversed when oxygen is supplied again. This proves that continuation of the oxidative reactions is essential to the maintenance of the electromotive force and of the metastable structure in the surface. An experimental result reported recently by Lorente de Nó (1947 (b), page 147) may be used to reveal the relation between these postulated systems in frog axons. A region of nerve depolarized by anoxia and non-functional can be restored by anodal polarization imposed from an external battery. Thus, a sufficient potential difference created and maintained across the polarizable structure leads to a restoration of the labile structures upon which excitation and response depend. Therefore we can say that the most likely sequence of connections is that the flux of metabolites through the oxidative chain of reactions leads to the development of the electromotive force that keeps the polarizable structures charged to a level such that the excitable mechanism is operative. Of course, on a longer time scale, the polarizable structures and the optically revealed structures are created and maintained similarly by oxidative metabolism. One difference is that the labile structures deteriorate faster during anoxia and can be restored by re-establishing the oxidative reactions.

STEADY STATES OF ACTIVITY AND OF METABOLISM

When an axon conducts a train of impulses the rate of oxygen consumption increases. This extra metabolism of activity has been studied by Gerard (1927, 1932), by Fenn (1927), and by Schmitt (1936) in relation to amount of activity. Thus we may say that successive waves of electrochemical change sweeping over an axon cause an increased rate of reaction between some substance and oxygen.

In order to study the kinetics of the reaction system thus set into action by nerve impulses at low, as well as high, frequencies of impulses, it was necessary to have a respirometer with greater sensitivity and stability than was readily achieved with the gasometric type. Furthermore, it was frequently necessary to change the chemical content of the solution bathing the nerve in order to study the effect of inhibiting enzymes in the metabolic chain of reactions in the nerve. Such considerations led to the development of the constant flow respirometer shown in Fig. 4. A solution with known oxygen concentration is drawn past the nerve in a 1-mm. capillary tube. The concentration of oxygen is measured by an oxygen cathode (Davies and

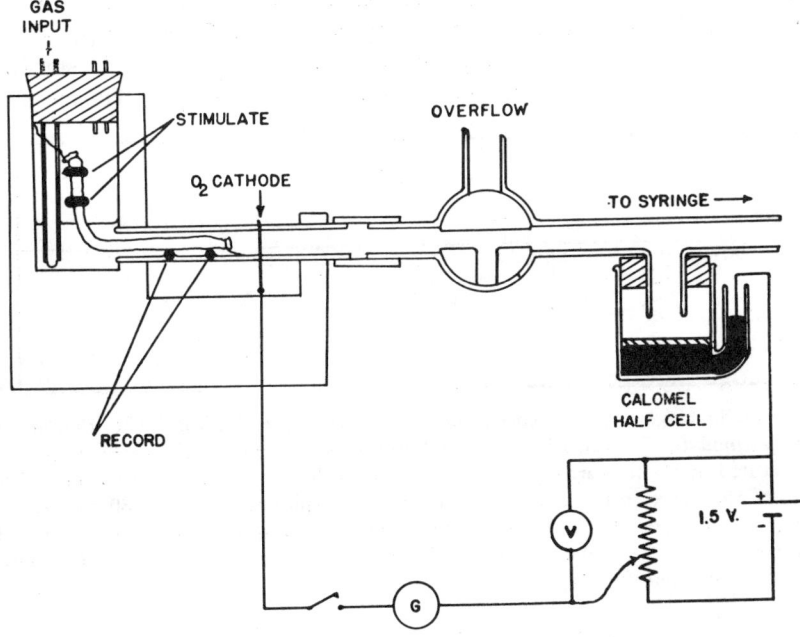

FIG. 4. Flow diagram of a respirometer. The solution in the well is equilibrated with a gas of known oxygen content. This solution is pulled past the nerve in the capillary at a known rate by a motor-driven syringe. The oxygen concentration is measured by the calibrated oxygen cathode. The nerve can be stimulated and the action potential recorded within the capillary tube. The nerve fits into the tube more snugly than is indicated in the diagram.

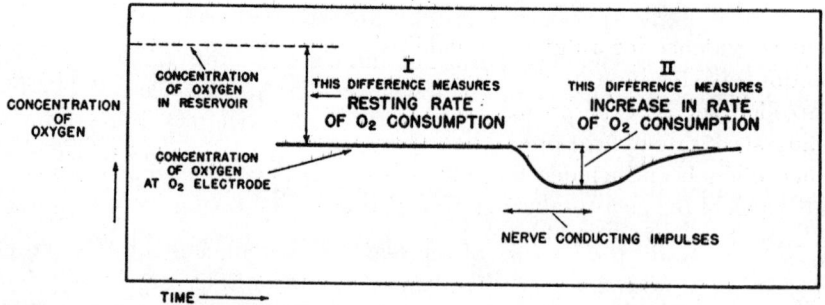

Fig. 5. Diagram of a record from typical experiment. The concentration of oxygen is measured with the nerve at rest and during conduction of impulses. The area under the transient curve produced by activity measures the total amount of extra oxygen used as a result of the activity.

Brink, 1942), that is, in the flowing stream below the nerve, Fig. 5. From the volume flow rate and the measured change in concentration, the amount of oxygen removed by the nerve can be calculated (Carlson, Brink, and Bronk, 1950).

The steady-state rate of oxygen uptake of a sciatic nerve of a frog has been measured at several frequencies of response from 5 volleys* per second to 200 volleys per second. An approximately steady-state rate of respiration is reached after 25 to 30 minutes of continued activity, Fig. 6. The measure-

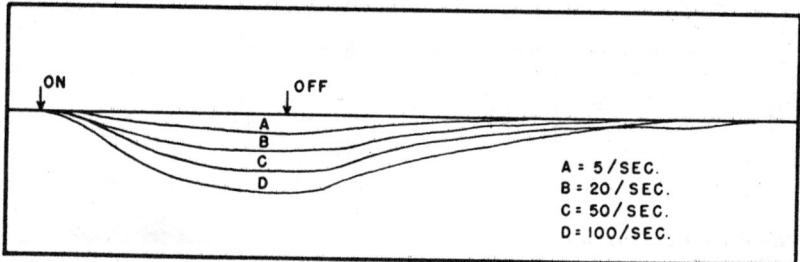

Fig. 6. Tracings of actual experiments on a sciatic nerve of a frog. The stimulus was chosen to stimulate all of the α-fibers of the A group. The period of stimulation begins at a time indicated by the first arrow. The initial delay is the transit time of the solution from the end of the nerve to the oxygen cathode in the respirometer. After 30 minutes the stimulation was stopped (second arrow). The steady-state rate at each frequency is readily determined by use of a calibration curve of current vs. oxygen concentration, the known volume flow rate, and the weight of the nerve. Thus

$$R_{ss} = \frac{f(\alpha I_1 - \alpha I_2)}{w} = \frac{f(c_1 - c_2)}{w}$$

where

f is in liters/hr, c in mols/liter, and w in grams wet weight.

---

* The word volley refers to a synchronized group of impulses made up of one impulse in each of many axons. This activity is recorded as the compound action potential developed by the synchronization of all the action currents from many fibers.

ment is evidence for a flux of an oxidizable material through a sequence of reactions at steady state. In Fig. 7 the averaged results from a carefully controlled set of experiments on ten different nerves are plotted. The steady-state rate of heat production (Feng and Hill, 1933) for a similar kind of nerve has been included for comparison. It is obvious that the steady-state rates of oxygen uptake and of heat production do not increase linearly with the frequency. At higher frequencies some property of the reaction

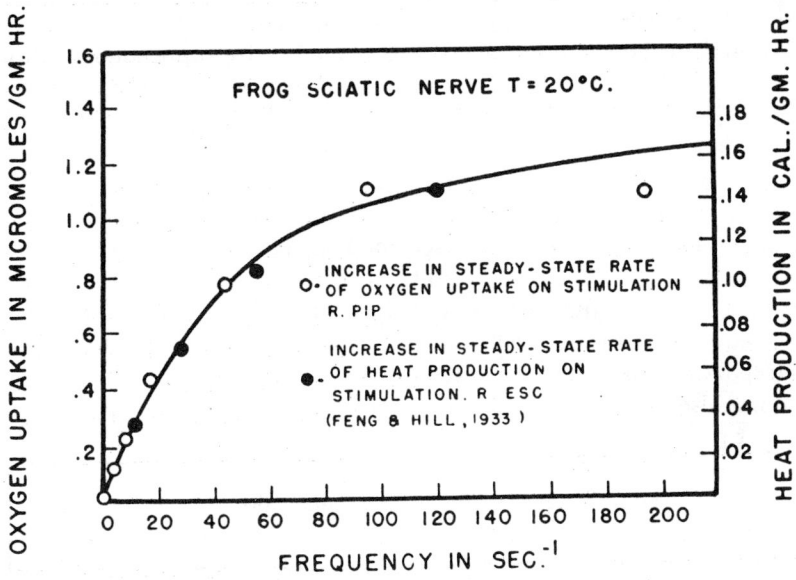

FIG. 7. Average increases in steady-state rates of oxygen uptake as a function of frequency of response. Only the α-fibers were active for the oxygen measurements. These points are average results from ten different sciatic nerves from *R. pipiens*. The solid line was calculated from the model discussed in the text. The heat data are superimposed for comparison by fitting one of the points to the curve for oxygen. An exact comparison of absolute values of these rates of heat production and of oxygen uptake is not desirable because different species of frog were used and the nerves were under different conditions for the two types of measurement.

mechanism imposes a limiting or maximum rate on the flux of material through the sequence of reactions that is started by the nerve impulses.

The theory of open steady-state systems seems appropriate for developing a model reaction system that has the same kinetic properties as the reaction system in the nerve. At present this method of analysis is employed merely as an aid to selecting further experiments from a seemingly endless list of experiments that can be done on the relation between activity and oxygen uptake.

The principal assumption relating to this model is that the impulses increase a specific rate constant k from zero to some finite value, thus opening a reaction pathway utilizing oxygen. The flux of matter through this reaction system then reaches a steady-state rate R that depends upon the value of k and other parameters of the reaction system, Fig. 8. The material being oxidized is represented as coming from an internal source S that is held essentially constant by the chemical reaction mechanism of the resting nerve. The product, x_2, of this new reaction pathway reacts with the oxidized form of an enzyme E that, in turn, is reoxidized by oxygen. Thus, in the steady state, the flux of substrate from S is measured by the uptake of oxygen. However, under the conditions of our experiments the kinetics of the reaction system are at all times independent of the oxygen concentration. This is represented in the schema by supposing the flux of substrate to be small enough so that the enzyme E is maintained entirely in the oxidized state. In other words, the enzyme is "saturated" with oxygen so that oxygen concentration cannot affect the rate appreciably. Under the conditions of oxygen saturation just described the theoretical rate of O_2 uptake for this model equals the rate of production of x_2 at all times.

Is is readily apparent how this scheme can represent the measured rate of oxygen uptake in the steady state at various frequencies of impulses. We suppose, as one possibility, that k is zero in a resting nerve and is some finite value for a brief period during each impulse. The reaction system then responds as though the average k was a linear function of the frequency of the impulses over the frequency range available for study in the nerve. (It is not possible to maintain all the α-fibers of a frog's sciatic nerve in a steady state of activity for sufficient time to reach a steady state of metabolism if the frequency is greater than about 100 impulses per second.) At low frequencies a steady-state rate R is attained after some time, which is greater the higher the average value of k. When k is very large compared to k_0 then R will not increase further but approaches a maximum rate given by $k_0 S$, which is the maximum flux possible from the source S. This is one way to represent in a reaction mechanism the experimental fact that the steady state rate of oxygen consumption does not increase linearly with the frequency of the impulses.

These statements can be represented in an equation by solving the differential equations for the processes indicated in the schema. These are

$$\frac{dS}{dt} = 0$$
$$\frac{dx_1}{dt} = k_0 (S - x_1) - k\, x_1$$
$$\frac{dx_2}{dt} = k\, x_1 - k'_2\, E\, x_2 \qquad (1)$$
$$\frac{dE}{dt} = 0$$

The last equation describes the supposition that all the enzyme is maintained at a constant concentration in the oxidized form. At steady state all rates of change are equal to zero. Thus x_1 can be eliminated and at the steady state

$$k'_2 E x_2 = \frac{k k_0 S}{k_0 + k_1} = \frac{k_0 S}{1 + \frac{k_0}{k}}$$

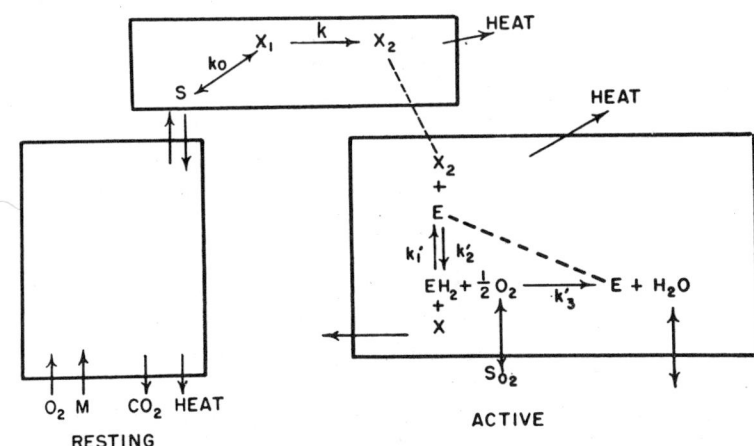

FIG. 8. A more detailed model of processes in axons analogous to the less explicit block diagram in Fig. 3. In the resting state the indicated flux of oxygen, CO_2, and heat represents the metabolic maintenance of the electrically defined structures. During activity the reaction system leading from substrate S to oxygen through the enzyme E represents a new reaction pathway set into operation when k becomes greater than zero during activity in the membrane.

According to our basic assumption

$$k = mf$$

Thus

$$k'_2 E x_2 = \frac{k_0 S}{1 + \frac{k_0}{mf}}$$

When the system is saturated with oxygen as described before, the rate of oxygen consumption R_{O_2} is at all times

$$R_{O_2} = k'_2 E x_2$$

By using this equation the steady-state relation becomes:

$$R_{O_2} = \frac{k_0 S}{1 + \frac{k_0}{mf}} = R_{ss}$$

When the frequency is very high, the rate approaches $k_0 S$ as a limit. This may be defined as $R_{\max}$.

Then

$$\frac{R_{ss}}{R_{max}} = \frac{1}{1 + (k_0/mf)} \tag{2}$$

This equation has been fitted to the steady-state data in Fig. 7. The solid line therein is calculated from:

$$R_{ss} = \frac{1.5}{1 + \dfrac{50}{f}}$$

Hence

$$k_0 S = 1.5 \text{ micromols}/\text{g}/\text{hr}$$
$$k_0/m = 50 \text{ impulses/sec}$$

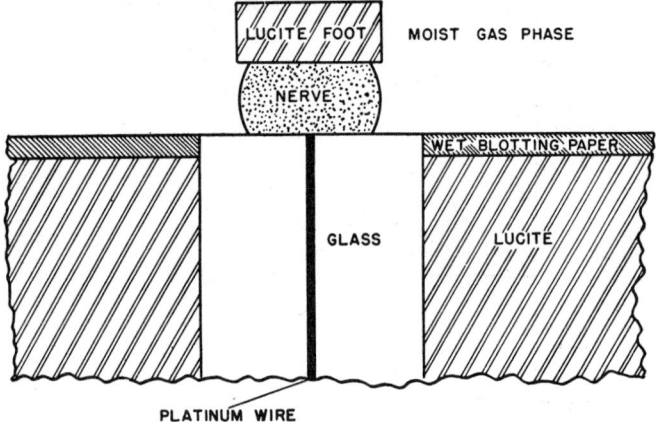

FIG. 9. A sciatic nerve pressed gently against an oxygen cathode by a flat piece of lucite is a partially closed reaction chamber. Such an arrangement forms a rapidly responding respirometer that can be used to measure the initial rate of increase in the rate of oxygen consumption produced by activity. The method is suitable for studying relative rates at different frequencies and can be calibrated for estimating absolute rates. (C. M. Connelly, *Rev. Sci. Instruments*, 1953).

The rate of oxygen consumption in a stimulated nerve does not change immediately to its steady-state value at each frequency. Indeed it requires about 25 minutes to achieve more than 95% of the total increase in the rate of oxygen consumption. The kinetic schema outlined above and fitted to the steady-state data can be further tested by measuring the time constants for this transition and the initial rate of change of R_{O_2}. The value of k_2E can be estimated independently. When the stimulated nerve is in a steady state and the activity is terminated, the rate constant for the return of R_{O_2} to the resting level is determined by the value of k_2E.

The measurement of the initial increase in rate requires a more rapidly responding respirometric method than the flow respirometer. Dr. C. M. Connelly (1948) has developed such a method. An oxygen cathode is placed against the nerve and the flat surface of a piece of lucite is brought gently against the upper surface of the nerve as in Fig. 9. The system is partially closed in that oxygen can diffuse into the region of nerve near the electrode only from the sides. After the system is in a steady state of diffusion of oxygen, any rapid change in oxygen utilization lowers the oxygen concentration at the electrode at a rate equal to the changed rate of utilization. After about 60 seconds or so the inward diffusion of oxygen from the gas

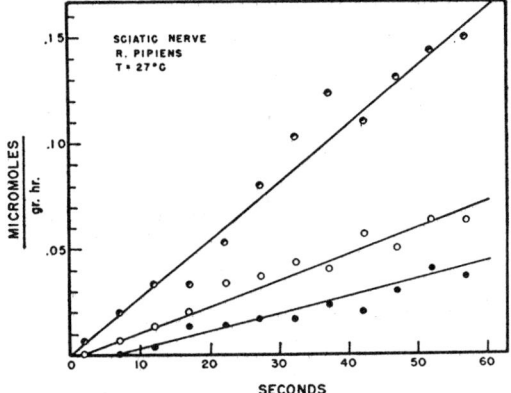

FIG. 10. Initial time course of the increase in rate of oxygen consumption during stimulation: ● 10 volleys per second; ○ 20 volleys per second; ◐ 50 volleys per second. The rate of increase of the rate of oxygen uptake is estimated from the slopes of these lines. Measurements made in moist chamber respirometer, Fig. 9.

phase begins to oppose this decrease appreciably. This method was used to measure the initial increase of the rate of O_2 uptake at various frequencies of response (Fig. 10). The preliminary studies indicate that $(dR/dt)_0$ changes with frequency of impulses, but the change is not linear as assumed.*

However, the model reaction system provides an approximate representation of the observed kinetics for frequencies up to 100 per second. By selective use of enzyme inhibitors it will be possible to introduce modifications so that the points of attack of such inhibitors can be indicated. This kind of study may permit physical identification of the proposed intermediate substances in the schema. Further experiments will suggest appropriate modifications of the model and thus extend the number of facts about nerves that are represented.*

---

* The study of this problem has been experimentally extended since this paper was written (cf. *Symposium on Quantitative Biology*. Cold Spring Harbor, 1952. In press.)

In this reaction schema the activity opened a new reaction pathway supplying oxidizable material to oxygen. To suppose an increase in the flux of matter through the oxidative chain of the resting nerve would seem a more obvious assumption. This possibility has not been excluded, but certain experiments make it seem unlikely. Thus, in some axons, such as the α-fibers of a frog's sciatic nerve, it is possible to prevent the increase of oxygen uptake during activity without interfering with the resting respiration or the activity (Bronk, 1949). This can be done with about 10^{-4} molar sodium azide dissolved in the Ringer's solution bathing the nerve. The capacity for activity even at one hundred volleys per second seems unimpaired. This observation was confirmed by Doty and Gerard, 1949, who added the further information that methylfluoroacetate can inhibit a large part of the resting respiration without affecting the magnitude of the increased oxygen uptake following conduction of impulses. We have recently observed that chloretone at 3 mM concentration acts somewhat like azide but, in addition, it reduces the resting respiration about 20% at the concentration that eliminates the extra respiration of activity. These facts suggest, but do not prove, that conduction of nerve impulses opens an oxidative reaction pathway that is distinguishable from that of the resting nerve since it involves enzymes that can be differentially inhibited.

There is another aspect of these experiments with azide and chloretone that requires discussion. When the increased oxygen uptake of active nerve was first discovered it seemed natural to suppose that the purpose of the extra chemical reactions was to supply energy to restore the active nerve to its previous metastable resting state. There is, however, nothing in these early experiments that requires such a conclusion. It now seems that the α-fibers of a frog's sciatic nerve can function indefinitely (up to 8 hours) without benefit of extra oxygen utilization. In addition our work with chloretone indicates that these nerves can conduct impulses at 100 per second for long periods with half of the resting respiration suppressed. This suggests that only a part of the resting respiration is indispensable for maintenance of the functional capacity of the nerve fibers. What then is the significance of an increased oxygen uptake normally observed in active axons? It would seem that proper perspective is preserved if we merely say that the data reveal a chemical accompaniment of repeated depolarization of axons which yields reduced products for reaction with oxygen through an enzymatically controlled reaction pathway. Further study of the kinetics and of the chemical components of this reaction chain is required to discover the significance of this chemical connection between the changes of the surface structure associated with activity and the oxidative reaction system of nerve.

In these metabolic studies we have concentrated upon one connecting link between the surface structure of an axon and the enzymatic reactions involving oxygen. During repetitive activity, the average electrical potential and resistance of the membrane change and the average rate of oxygen uptake adjusts to a higher value. It is probable that this increased rate of oxidation occurs through an increase of a rate constant associated with the physical changes in surface structure. This is but one of a multitude of connections implied in the block diagram of Fig. 3. Prevailing research efforts in this field can be classified in relation to a study of these connections and to the filling in of the structure of the several blocks in that diagram. Another example of such an experimental program will illustrate an attempt to give molecular content to the metastable structure and some connections of this system to the others.

METASTABLE STATES OF RESTING AXONS

One of the well-established facts about an axon is that it can be excited by an imposed electrical current. The outward flow of current across the membrane leads to local depolarization. For excitations to occur the imposed current must be of sufficient intensity and duration, that is to say, the axon structure has a barrier to excitation. It is as though the process of local depolarization must be brought to a certain degree within a certain time for the process to spread in a self-sustained manner. Whenever this activated state of the nerve membrane is of sufficient intensity a response occurs such that sufficient current from adjacent polarized areas of axon flows through the area with reduced or reversed polarization, and the active area grows until a propagated wave of depolarization travels along the axon. Such observations are the basis for the concept that the axon contains a metastable polarizable structure.

The physiological role of this communication line is dependent upon this threshold for excitation. The structure remains quiescent unless a signal is imposed at the proper terminations. Casual fluctuations in the environment of an axon while modifying its stability do not activate it. However, there are experimentally demonstrable limits to the range of thresholds compatible with this normal performance. On the one hand the threshold may be so high, or the structure so stable, that action currents may be insufficient to excite it. The line is blocked at such a point, and conduction is terminated. In addition, the action currents may be reduced in intensity by decreasing the membrane electromotive force so that propagation is impossible. On the other hand, if the threshold is lowered sufficiently, self-excitation occurs. Then normal function as a communication line is seriously impaired by unauthorized impulses, as described later. All these conditions can be realized experimentally and the limits of normal operation of an axon studied. All these conditions can be produced in animals and constitute pathological states of the nervous system. The importance of studying the limits of normal operation of axons is that significant features of the intra-cellular organization can be deduced from studies of the stable and self-excited states that bound the domain of metastable states. Only the latter states are compatible with the physiological functioning of the axon, as defined in the section, the "Physiological Function of the Axon."

A common observation in neurology, both clinical and experimental, is the variation of the excitability of nerve tissue with changes in the concentration of ionized calcium in the fluid bathing it. In presenting here some detailed evidence for the importance of calcium in stabilizing the reactive structure in axons we are contributing to a field of cell physiology whose broad outlines have been known for a long time. This aspect of nerve physiology seems to have arrived at a maturity where only quantitative data of sufficient precision for testing the various theories of excitation are useful. We do not wish to give the impression that calcium is regarded as the sole

TABLE 1
A-fibers Sciatic Nerve Frog

$10^3 \times K$	°C	pH	Buffer
1.6	26°	7.1	Carbonate
0.62	24°	7.1	Carbonate
1.9*	29°	7.1	Carbonate
0.89*	27°	7.1	Carbonate
1.3	25°	7.1	Phosphate
1.2	25°	7.1	Phosphate
1.1	26°	6.5	Phosphate
0.86	26°	6.5	Phosphate
0.57	25°	6.5	Phosphate
0.28	25°	6.5	Phosphate

$10^3 \times K$	Substance	Source
0.60	Calcium citrate	Hastings et al., J. Biol. Chem., **107**, 351 (1934).
0.40	Calcium proteinate	Drinker and Zinsser, J. Biol. Chem., **148**, 187 (1943).
1.27	Calcium cephalinate	Drinker and Zinsser, J. Biol. Chem., **148**, 187 (1943).

substance of importance in regulating excitability. On the contrary it seems evident that emphasis on any single substance must be replaced by experimental evaluation of the quantitative contribution of each of the known variables to the integrated cellular property, excitability.

It is a characteristic of contemporary writing on the functional properties of nerves that any mention of the protein-lipoid constitution of the nerve is usually avoided. This contrasts sharply with earlier discussions of excitation, response, and recovery in terms of colloidal protein structures. This diffidence probably arises from our greater knowledge of the constituents of cells and their multitudinous potentialities as functional aggregates. A further deterrent to speculation is the realization of the limitations of available data. From the changes with time of various measurable quantities a fairly satisfactory kinetic description of a system is possible with no knowledge of the physical nature of the variables. However, this avoidance of the problem does not solve it. Consequently, whenever a tentative hypothesis regarding the molecular basis of a cellular property can be formulated and tested, the result is important. It is in this spirit that the following section is presented.

It has been assumed that the property of excitability in a nerve cell is referable to a metastable molecular structure. One mechanism by which the excitability of such a structure could be modified by chemical agents would be a direct reaction of the agent with the molecules of the reactive structure. The action of calcium ions on the excitability of nerve will be examined from the point of view that dissociable salts of calcium are formed by reaction with such molecular components of the nerve.

Since the quasi-fluid structures of cells are primarily composed of protein and phospholipid building units, it is the calcium salts of such molecules that are of interest. The suggestion that these are involved in the excitation process is an old one. Recently, however, values for the dissociation constants of such organic salts of calcium have become available for purified substances and for mixtures. In general, it has been shown that calcium salts of proteins and certain phospholipids ionize in accordance with the mass action law in Eq. (3.)

$$K = \frac{(Ca^{++})\,(P)}{(CaP)} \tag{3}$$

In Table 1 are listed the values of K measured for the calcium salts of citrate, serum protein, and cephalin. These values, though different, are of the same order of magnitude. It is expected, therefore, that regions of the cell structure that contain substances like the above molecules in the proper spatial arrangements will bind ions of calcium, and the dissociation constant will be about 10^{-3} mol per liter. Therefore, when the solution bathing such structures contains calcium ions at a concentration of about 1 millimolar, the available binding sites of this kind will be half saturated. On the other hand, binding sites reacting with calcium ion as does oxalate or sulfate would be half saturated at much lower concentrations.

Within the framework of these ideas and these physical data we are able to examine experimentally the hypothesis that the irritable structures of nerve contain molecular groups that bind calcium, as do molecules of protein or cephalin.

It is known that the threshold intensity of a direct current (R) is a function of the calcium ion concentration in the fluid bathing the nerve. Thus

$$R = f\,(Ca^{++}, H^+, \ldots T, P)$$

As a first approximation, the other variables contributing to the threshold will be considered constant and additive with the contribution from calcium compounds. Thus

$$R = f_1(Ca^{++}) + f_2(H^+, \ldots T, P) \qquad (4)$$

In Fig. 11 is shown the average measured rheobase for the α-fibers as a function of the concentration of ($CaCl_2$) in the fluid bathing an excised sciatic nerve of a frog. The threshold becomes less dependent upon calcium as the concentration increases. One interpretation is that the nerve struct-

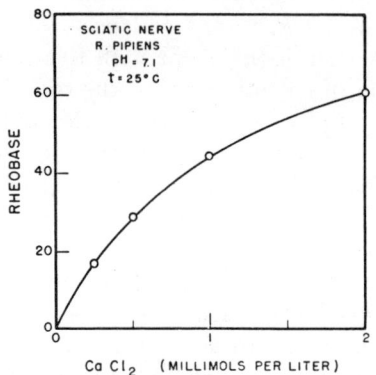

FIG. 11. Threshold strength of a constant current required to stimulate 50% of the α-fibers in a frog's sciatic nerve varies with the concentration of $CaCl_2$ in the solution bathing the nerve. Nerve excised. Modified Ringer's solution at pH 7.1. The solid line is calculated from K and R_m estimated from graph in Fig. 12.

ures determining the threshold intensity of current are gradually becoming saturated with calcium. This is equivalent to the assumption that the threshold is a measure of the concentration of the dissociable calcium salt within the excitable structure of the nerve.

By analogy with the mass law for interaction of Ca^{++} with isolated cellular constituents such as protein and cephalin the amount of bound calcium in the cell is given by

$$CaX = \frac{(Ca^{++})(X^=)}{K} \qquad (5)$$

Since X is a structural constituent of the cell the total concentration of it is assumed to be constant during an experiment. Hence,

$$(CaX) + (X^=) = (X_T)$$

Therefore,

$$\frac{(CaX)}{(X_T)} = \frac{1}{1 + \dfrac{K}{(Ca^{++})}}$$

or

$$\frac{(X_T)}{(CaX)} = 1 + \frac{K}{(Ca^{++})} \quad (6)$$

By comparison with experiment it is seen that this plausible hypothesis describes the calcium effect on the rheobase of nerve if it be assumed that:

$$\text{Rheobase} = a(CaX)$$

Under these assumptions,

$$\frac{R_m}{R} = 1 + \frac{K}{(Ca^{++})} \quad (7)$$

where R = corrected rheobase
R_m = maximum corrected rheobase.

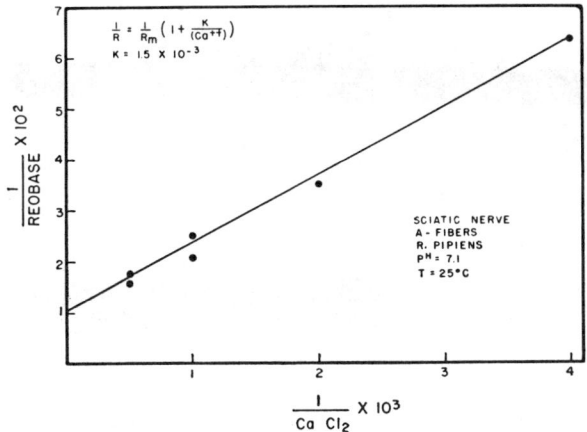

FIG. 12. Data from Fig. 11 plotted according to Eq. (5) of text in order to estimate K and R_m from slope and intercept.

$$\text{Ordinate}: \frac{1}{\text{Rheobase}} \times 10^2$$

$$\text{Abscissa}: \frac{1}{CaCl_2} \times 10^3$$

According to Eq. (4) the actual threshold must be corrected for the effects of other variables. The correction is an unspecified function of other variables assumed constant in an experiment where $CaCl_2$ in the bathing fluid is the independent variable. The nature of this correction is clear if the observed changes in rheobase are to be described by Eq. (7). This equation requires that R approach zero as the concentration of ionized calcium approaches zero. Therefore, in any one experiment the constant $f_2(x,y,z,$ etc.) is a value, expressed in the same units as the rheobase, that must be added to or subtracted from the measured rheobase in order to make the resulting curve approach the origin in a graph such as Fig. 12.

In making this correction it is assumed that the stability of the excitable structure is compounded by linear addition of the contributions from calcium salts and the other sources. This approximation can be avoided if other variables are adjusted so that the measured rheobases extrapolate to zero as the calcium concentration approaches zero.

However, the α-fibers in a sciatic nerve of a frog are usually spontaneously active at pH 7.1 if the calcium ion concentration is below 0.4 millimolar (Brink, Bronk, and Larrabee, 1946). If the pH is lower the threshold is higher. At pH 6.1 a nerve bathed in calcium-free solutions has a finite

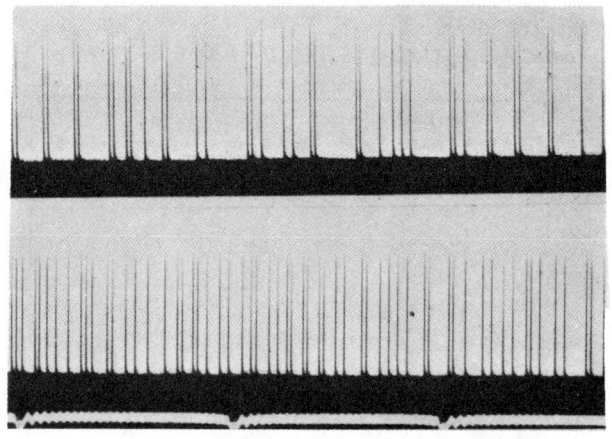

Fig. 13. Impulses recorded from a single axon were initiated by a process of self-excitation developed in a calcium deficient region of the axon. *Top.* Calcium removed by diffusion with modified Ringer's solution containing no added $CaCl_2$. *Bottom.* Increased frequency of impulses after treatment of same region with a solution containing some sodium citrate, a calcium binding agent. Time in 1/5 seconds.

threshold. In addition, the lower the pH the higher the Ca concentration can be before precipitation of components of the buffering system begins. Thus it is possible to extend the total range of concentration of calcium over which a finite rheobase can be measured. To estimate a dissociation constant for the supposed calcium compound under these conditions it is necessary to correct the observed rheobase. This is done, according to Eq. (4), by subtracting the estimated value of R at zero concentration of calcium ion, as described above.

The constant K and the value of R_m can be estimated as follows. The reciprocal of the rheobase (corrected if necessary) is plotted as a function of the reciprocal of the calcium ion concentration in the solution bathing the nerve. The theory requires the relation to be linear (Fig. 12). The intercept is the value of R_m and the slope divided by the intercept is the value of K in mols per liter. The calculated values of K are listed in Table 1.

Obviously, the relation in Fig. 11 can be interpreted as arising from the dissociation of a calcium compound in nerve which determines the threshold for electrical excitation by direct current. One cannot prove that this is a unique interpretation of the data, but it seems adequate and is in accord with demonstrated properties of protein and phospholipid molecules, Table 1. Therefore, in discussing the molecular nature of the metastable structure of axons it seems desirable to consider seriously the idea that a dissociable calcium salt of these structural molecules has a key position in determining the degree of stability.

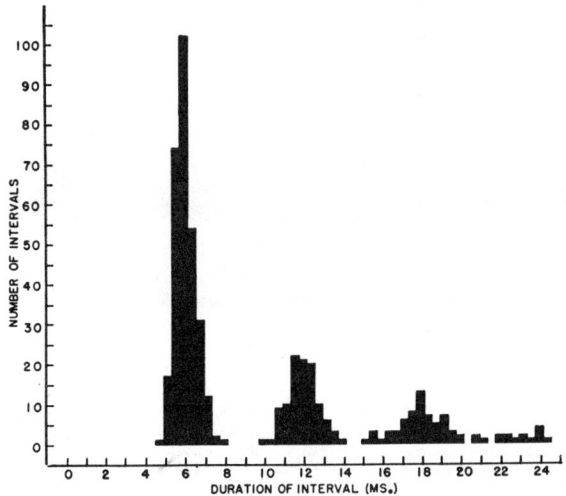

FIG. 14. The duration of the intervals between impulses during chemical excitation, as in records of Fig. 13, have a statistical regularity. Successive intervals between 500 impulses were measured during a steady state of activity when the average number of impulses per second was constant. Ordinates are number of impulses having a value in each 0.4-millisecond interval. Intervals greater than 24 milliseconds not shown. The least interval has a most probable value of 6 milliseconds and the other peaks are integral multiples of this value.

STEADY STATES OF SELF-EXCITATION

When the calcium ion concentration in the fluid bathing a nerve is lowered sufficiently, the axons begin to discharge impulses. This activity, in a single nerve fiber, is a series of discrete impulses separated by variable time intervals, Fig. 13. Statistical analysis of the duration of these intervals between impulses has revealed an inherent order. The longer intervals are multiples of a set of least intervals (Fig. 14). This is evidence for an oscillatory process of excitation in the chemically defective area that imposes the observed temporal pattern upon the impulses conducted along the axon (Brink, Bronk, and Larrabee, 1946). This capacity for rhythmic action is inherent in the surface structure of all parts of the axon. There is evidence in some nerves that the excitation develops from a rhythmic local electrical oscillation. In these nerves the frequency of the oscillations are apparently determined by the impedance properties of the membrane through which the current must flow. (cf. Brink, Bronk, Larrabee, 1946; Cole, 1942).

It has been observed that trains of impulses entering a calcium-deficient-area are propagated to recording electrodes on the far side. Thus it is possible to study the effect of such a chemically modified area upon a train of impulses of known frequency. For this purpose the average number of impulses arriving at the recording electrodes can be measured as a function of the frequency of conducted impulses sent along the axon by means of electrical excitation, Fig. 15.

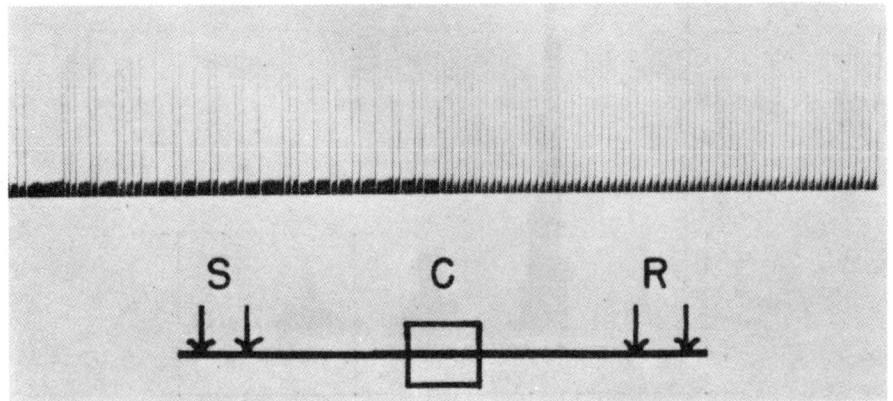

Fig. 15. Left side of record, typical temporal distribution of impulses recorded from single fiber at R when excitation develops by removal of calcium from fiber at C. When the axon is stimulated by equally spaced electrical shocks at S one impulse is recorded at R for each stimulus at S, right side of record. The axon conducts impulses from S to R through the chemically defective area C and the temporal pattern of the signals is not modified by the self-excitation process at C.

In Fig. 16 the results from two different fibers are shown. On the abscissas is the frequency of impulses initiated electrically and on the ordinate is the average frequency of impulses recorded. The initial frequency of chemically initiated impulses is given along the ordinate. When the electrical excitation is at high enough frequency, the points fall on the 45° line. This means that the frequency of impulses recorded equals the frequency of impulses started electrically. Then the axon is conducting the signals without distortion.

The experimental curves deviate from this line at low driving frequencies because of the self-excitation occurring in the treated area of the axon. No detailed analysis has been made, but it is clear that the higher the frequency of spontaneous impulses the higher the frequency of incoming signals required for a one-to-one relation between impulses started at A and impulses counted at B. Thus, in an axon with such a chemical defect, there is a range of frequencies within which impulses will be transmitted without frequency distortion. The range of signal frequency that is faithfully transmitted is, however, severely restricted.

Such self-excitation within the transmission line obviously interferes with the physiological role of the axons. Extra impulses will modify the temporal pattern and the average frequency of signals developed at normal points of

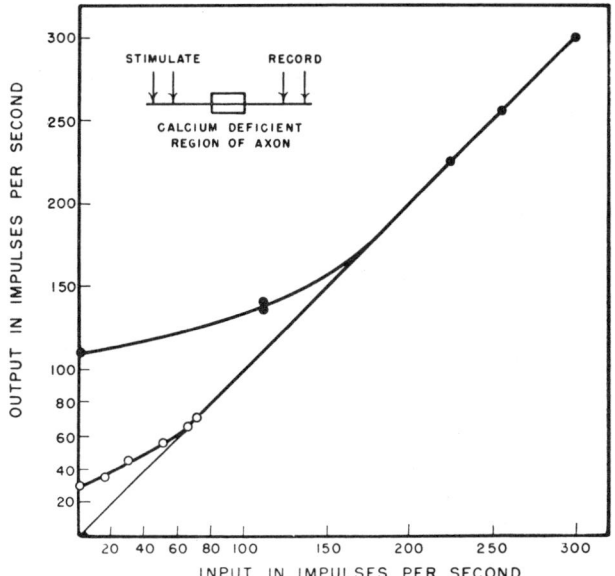

Fig. 16. Same experiment as in Fig. 15 is repeated at various frequencies of electrical stimulation, indicated on abscissae. ● Initial average frequency of chemically initiated impulses is about 110. ○ Another axon with a lower average frequency of chemically initiated impulses. The diagonal line represents one-to-one relation between electrical stimuli and recorded impulses. Deviations from this line represent distortion of temporal pattern of impulses started at S by the process of self-excitation in the chemically defective area.

excitation. Thus the degree of hyperexcitability that permits self-excitation places a limit on the range of normal operation of an axon. A metastable state with a finite threshold for activation is essential if trains of impulses are to be propagated without modification by minor environmental disturbances along the course of the axon.

Résumé

Physicists are well aware that the properties of a physical system can be discussed within the framework of more than one theory. The properties of axons measured electrically have been interpreted within the framework of electrically defined terms. The metastable structure that goes through a complete cycle of change with each impulse is even more loosely defined in terms suitable for discussing any metastable system. These are activated states, change of state, refractoriness, and reconstitution. In effect, the changes in stability are expressed in a mathematical theory of excitation without physical definition of the dependent variables (cf. Katz, 1939). The third prominent set of neuronal properties that change during activity are the rates of metabolic reactions. The measurements are made in the environment and indicate the flux of materials into and out of the axon. The measurements can be interpreted in terms of the theory of open steady-state reaction systems. Since only the flux from sources and into sinks are measured, the kinetic system is merely an equivalent mechanism that: (1) accounts for observed transient and steady-state flux of oxygen during activity, and (2) is consistent with the postulates of reaction rate theory.

When we wish to analyze an experimental correlation between quantities defined separately within any two of these characterizations of nerve properties, some assumptions are obviously required. The only alternative is the experimental development of a common set of operational definitions that would serve to discuss all three aspects of axonal activity in the same basic terms. This would be equivalent to a molecular description of the events. We are far from this desirable state of affairs.

Acknowledgments

The original data published here for the first time are from experiments made by various members of a group working with Dr. D. W. Bronk on problems in neurophysiology. The men whose work has been discussed are Dr. Bronk, Dr. C. M. Connelly, Dr. F. D. Carlson, and myself. This experimental program is generously supported by grants to Dr. Bronk from the Supreme Council, Scottish Rite Masons, and from the American Philosophical Society.

REFERENCES

1. Adrian, E. D., 1928, The Basis of Sensation. Christophers, London.
2. Adrian, E. D., and Bronk, D. W., 1928. The Discharge of Impulses in Motor Nerve Fibers. Pt. I. *J. Physiol. (London)* **66**, 81.
3. Bernstein, J., 1866. Die Fortpflanzungsgeschwindigkeit der negativen Schwankung im Nerven Zentrolblatt fur die Medizinischen Wissenschaften. Page 597.
4. Brink, F., Bronk, D. W., and Larrabee, M. G., 1946. Chemical Excitation of Nerve. *Ann. N. Y. Acad. Sci.* **47**, 375.
5. Bronk, D. W., and Stella, G., 1932. Afferent Impulses in the Carotid Sinus Nerve. I. The relation of the Discharge from Single End Organs to Arterial Blood Pressure. *J. Cellular Comp. Physiol.* **1**, 113.
6. Bronk, D. W., 1949. Croonian Lecture. To be published in *Proc. Roy. Soc. B*.
7. Carlson, F. D., Brink, F., and Bronk, D. W., 1950. A Continuous Flow Respirometer Utilizing the Oxygen Cathode. *Rev. Sci. Instruments* **21**, 923.
8. Cole, K. S., and Curtis, H. J., 1939. Electrical Impedance of the Squid Giant Axon During Activity. *J. Gen. Physiol.* **22**, 649.

9. Cole, K. S., 1942. Rectification and Inductance in the Squid Giant Axon. *J. Gen. Physiol.* **25,** 29.
10. Connelly, C. M., 1948. Unpublished work. Cf. *Cold Spring Harbor Symposia Quant. Biol.* **17** (1952).
11. Curtis, H. J., and Cole, K. S., 1942. Membrane Resting and Action Potentials from the Squid Giant Axon. *J. Cellular Comp. Physiol.* **19,** 135.
12. Davies, P. W., and Brink, F., 1942. Microelectrodes for Measuring Local Oxygen Tension in Animal Tissues. *Rev. Sci. Instruments* **13,** 1.
13. Doty, R. W., and Gerard, R. W., 1949. Separate Inhibition of Resting and Active Oxygen Consumption of Functioning Nerves. *Federation Proc.* **8,** 35.
14. Erlanger, J., and Gasser, H. S., 1937. Electrical Signs of Nervous Action. University of Pennsylvania Press, Philadelphia.
15. Feng, T. P., and Hill, A. V., 1933. The Steady State Heat Production of Nerve. *Proc. Roy. Soc. (London)* **B113,** 356.
16. Fenn, W. O., 1927. The Oxygen Consumption of Frog Nerve During Stimulation. *J. Gen. Physiol.* **10,** 767.
17. Gerard, R. W., 1927. Studies on Nerve Metabolism. II. Respiration in Oxygen and Nitrogen. *Am. J. Physiol.* **82,** 381.
18. Gerard, R. W., 1932. Nerve Metabolism. *Physiol. Revs.* **12,** 469.
19. Hartline, H. K., and Graham, C. H., 1932. Nerve Impulses from Single Receptors in the Eye. *J. Cellular Comp. Physiol.* **1,** 277.
20. Hill, D. K., and Keynes, R. D., 1949. Opacity Changes in Stimulated Nerve. *J. Physiol. (London)* **108,** 278.
21. Hodgkin, A. L., 1937. Evidence for Electrical Transmission in Nerve. Pt. I. *J. Physiol. (London)* **90,** 183; Pt. 11. *Ibid.* **90,** 211.
22. Hodgkin, A. L., and Huxley, A. F., 1945. Resting and Action Potentials in Single Nerve Fibers. *J. Physiol. (London)* **104,** 176.
23. Katz, B., 1939. Electric Excitation of Nerve. Oxford Press, London.
24. Lorente de Nó, R., 1947. A Study of Nerve Physiology. *Studies Rockefeller Inst. Med. Research* (a) **131,** (b) **132**.
25. Matthews, B. H. C., 1933. Nerve Endings in Mammalian Muscle. *J. Physiol. (London)* **78,** 1.
26. Pumphrey, R. J., 1935. Nerve Impulses from Receptors in the Mouth of the Frog. *J. Cellular Comp. Physiol.* **6,** 445.
27. Schmitt, F. O., 1936. The Oxygen Consumption of Stimulated Nerve. *Cold Spring Harbor Symposia Quant. Biol.* **4,** 188.
28. Schmitt, F. O., and Schmitt, O. H., 1940. Partial excitation and Variable Conduction in the Squid Giant Axon. *J. Physiol. (London)* **98,** 26.
29. Shanes, A., 1949. Electrical Phenomena in Nerve. I. Squid Giant Axon. *J. Gen. Physiol.* **33,** 59.

BIOLUMINESCENCE AND THE THEORY OF REACTION RATE CONTROL IN LIVING SYSTEMS

Frank H. Johnson

Princeton University, Princeton, New Jersey

Luminescence, as manifested in the production of "cold light" by living organisms, is one of the earliest problems in biophysics. The first noteworthy scientific investigation of bioluminescence was, in fact, undertaken by the classical physicist, Robert Boyle, who in the seventeenth century conducted numerous experiments on "shining fish" and "shining meat" in his home. After nearly 300 years, the full nature of the reaction remains to be established, e.g., the chemical identity and structure of the reactants, but a great deal has been learned about the light-emitting system and in the past several years it has provided an efficient research tool that has led to certain advances in the theory of biological reactions in general. The purpose of this paper is to discuss briefly some of these advances and the contributions of bioluminescence thereto. The more detailed aspects of the luminescent system need not be considered here; they may be found in recent reviews (Johnson, 1947; Johnson and Eyring, 1948) and in monographs by Harvey (1940, 1952), and by Johnson, Eyring, and Polissar (1954).

In the 1660's, when Boyle studied the luminescence of dead fish and meat, he was not aware of its living source, which we now know must have been in luminous bacteria. Indeed, this was a few years prior to the first observation of any bacteria, by van Leeuwenhoek in 1683. Yet Boyle made some observations of fundamental significance. In his report to the newly established Royal Society, he wrote: "But notwithstanding the vividness of this light, I could not, by the touch, discern the least degree of heat . . . neither I, nor any of those about me, could perceive, by the smell, the least degree of stink . . . pouring on it a little pure spirit of wine, . . . I found that the light was vanished. But water would not so easily quench our seeming fires; . . . " (Boyle, 1672). In other experiments, Boyle showed that "air" (oxygen) was necessary for luminescence, and that, in contrast to the incandescence of a glowing ember, when extinguished by exhausting the air with a vacuum pump, the light of "shining fish" shone forth again as soon as air was readmitted. Boyle evidently sensed that he was dealing with something beyond the ordinary in physical phenomena, and it is still true that the occurrence of a process in a living cell places it in a special category, of a higher order of complexity.

In many instances, biological processes, e.g., growth, reproduction, differentiation of tissues, and intellectual activity, appear so complex that it would not seem possible to analyze them, especially from a kinetic point of view, in terms of the same laws that apply to simpler reactions in a test tube. The observed biological result is achieved through a subtle interplay of a multitude of reactions in the highly organized, polyphasic system of protoplasm. A complete analysis requires both the identification of all the reactants concerned and the mechanisms which control their reaction rates. Although in recent years biochemistry has made enormous strides in working

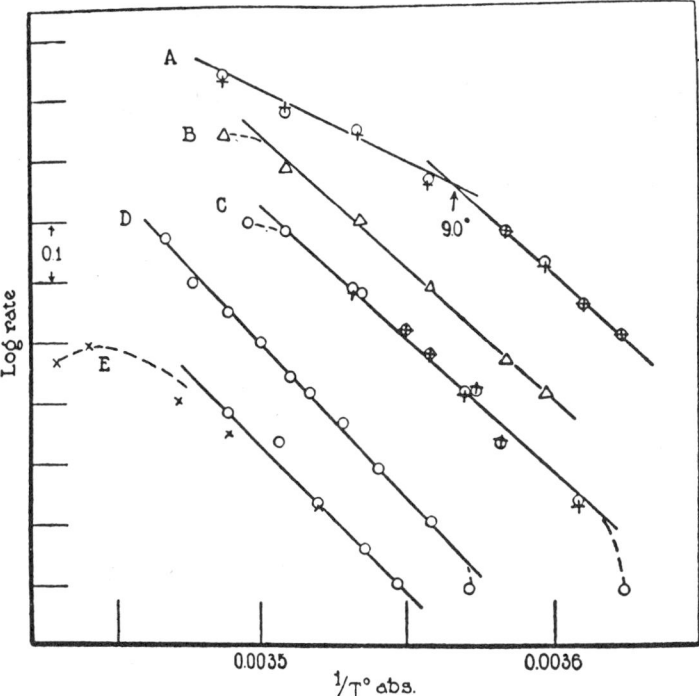

Fig. 1. Influence of temperature on the rate of early development of teleostean fishes, plotted in accordance with the Arrhenius equation (Needham, 1931, Fig. 82, p. 520). The data are from several investigators, and represent four different genera of fishes.

out the intermediary pathways of metabolism, there is still a very long way to go. Fortunately, a great deal can be learned about the kinetic mechanisms without knowing the chemical identity of the reactants, and progress in the interpretation of kinetics advances the understanding of the nature of the reactants.

In spite of the complexity that characterizes the processes taking place in living organisms, it has been known for a long time that to a certain extent they behave, as a whole, in the manner of a simple chemical reaction. For example, the influence of temperature over a limited range, from relatively low temperatures up to nearly the maximum for the process in question, is generally in quantitative accord with the Arrhenius' relation between temperature and the velocity of a single reaction:

$$v = Ae^{-\mu/RT} \tag{1}$$

Growth is a highly complicated phenomenon, yet the early development of teleostian fishes (Fig. 1) and many other organisms has been shown to be accelerated by a rise in temperature in a manner such that the logarithm of the rate is a linear function of the reciprocal of the absolute temperature (Needham, 1931), as one would expect for a simple reaction in accordance

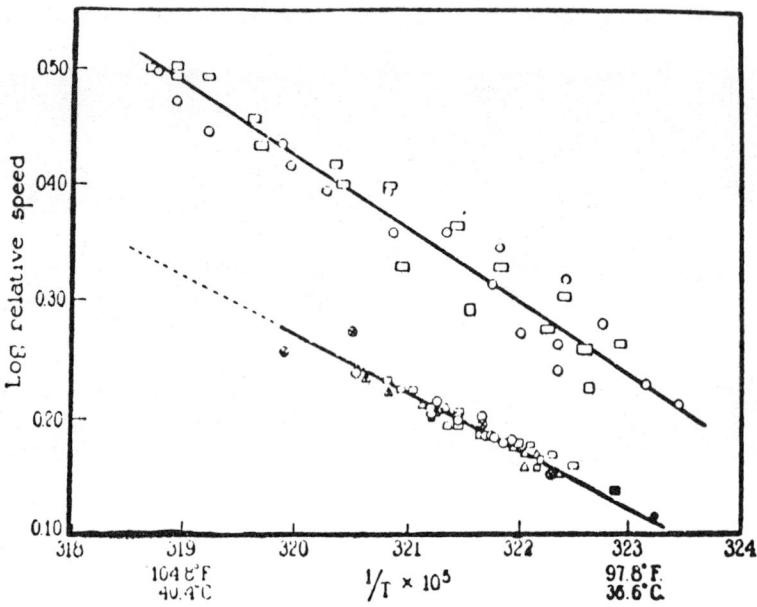

FIG. 2. Relation between body temperature and rate of heart beat (upper line) and rate of tapping of intervals estimated as one second (lower line), in human subjects, plotted in accordance with the Arrhenius equation. (Hoagland, 1936, Fig. 15, p. 274.)

with equation (1). A similar relation obtains for mental activity in humans, to the extent that it can be measured by recording the rate at which agreeable individuals tap what they estimate as one-second intervals, while their body temperatures are raised by diathermy, or vary naturally with fever (Fig. 2, Hoagland, 1936). Thus, one's sense of time has a temperature coefficient, and time seems to drag when one has a fever. Numerous and apparently unrelated other processes have been shown especially by Crozier and others since 1924, to conform to the Arrhenius relation, sometimes with similar and sometimes with widely different slopes of the line, or μ values, e.g., the frequencies of cricket chirping, of firefly flashing, and of heart beats. The simplest interpretation has been that, although many reactions are involved, one acts as a pace-setter, limiting the velocity of the whole process and making it behave with respect to temperature in the manner of a single reaction. Fundamentally, there seems to be no objection to this interpretation, but it is not sufficient to account for the fact that, in nearly every instance, the apparent activation energy, or μ value, progressively decreases at the higher temperatures, until the rate is no longer accelerated with rise in temperature but reaches a maximum, beyond which it rapidly decreases if the temperature is raised still further.

One of the problems which has been partially answered by recent work is the mechanism of the biological temperature-activity curve. In reaching a clearer understanding of the temperature relationships, bioluminescence

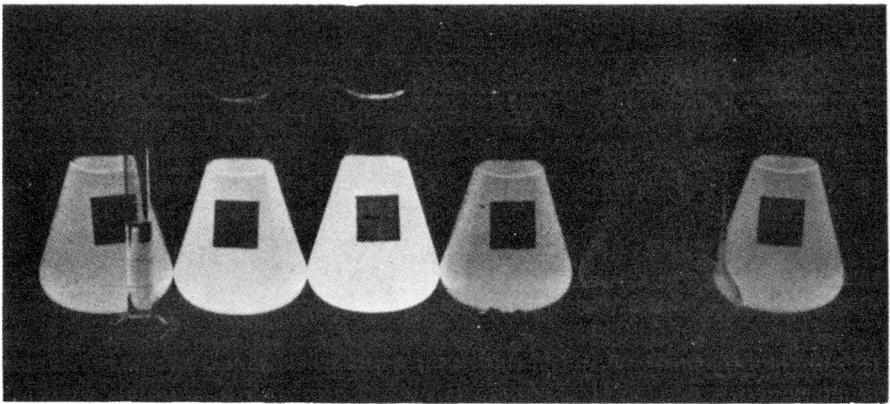

FIG. 3. The effect of various drugs on the brightness of bacterial luminescence. From left to right, small concentrations of alcohol, aspirin, salt solution (control), procaine, ether, and sulfanilamide, respectively, were added to the flasks before photographing them by the light of the bacteria themselves. (Johnson, 1948, Fig. 9, p. 232.)

has proved to be a most useful tool. A second tool, of more fundamental importance, has been the theory of absolute reaction rates (Eyring, 1935; Glasstone, Laidler, and Eyring, 1941). The reasons are as follows.

First, with respect to luminescence, extensive investigations, principally by Harvey and his associates, indicate that light emission results from an oxidative enzyme reaction, in which the enzyme *luciferase* catalyzes the oxidation of a substrate, *luciferin*, in the presence of water and molecular oxygen. Part of the energy liberated in this reaction is dissipated in the form of radiant energy of visible wavelengths, generally in the blue end of the spectrum. The important feature from the point of view of kinetic studies is that, under given conditions, the brightness of the emitted light is proportional to the reaction velocity between luciferin and luciferase and this, as shown by the flow method studies of Chance *et al.* (1940) is the slowest step in a series that leads to radiation. Thus, virtually unique in nature, here is a biological reaction with a natural indicator of its own velocity, and one which can be easily and accurately measured by any of several modern devices, such as the photomultiplier.

From a few of the multitude of luminescent organisms, most of which are marine in origin, the luciferin and luciferase can be separated and the reaction studied *in vitro*. In others, e.g., the luminous bacteria, efforts to extract the light-emitting system have thus far failed. But bacterial luminescence in the intact living cells has certain advantages for kinetic studies that are lacking in the isolated system. The latter proceeds as a first order reaction, and the unstable substrate is rapidly exhausted. In bacteria, the luminescent reaction is in a steady state, such that under favorable physiological conditions the brightness of the light remains uniform over considerable periods of time, the luciferin being reduced or regenerated in some manner as fast as it is oxidized with light emission. Each recording of light intensity is equivalent to measuring the velocity of a first order reaction, and changes in reaction velocity under the influence of various environmental factors, such as temperature, *p*H, and salts, can be immediately followed, even when these changes are very rapid. Moreover, the luminescent oxidation appears to be relatively isolated from the main respiratory pathway leading to oxygen consumption, thus eliminating some of the complications arising from a long series of consecutive reactions responsible for the measured result.

Finally, the luminescent oxidation shows the general temperature relationships characteristic of so many other enzyme reactions and biological processes and it is susceptible to reversible inhibition by drugs and other agents (Fig. 3). Obviously, it provides a much simpler and more efficient tool for investigating the fundamental mechanisms controlling biological reactions than processes so complicated as, for example, growth.

With regard to the second tool, the theory of absolute reaction rates provides at the same time the modern and the only completely rational theory of chemical reaction rates. In essence it states that every chemical reaction is in fact an unstable equilibrium, designated $K^{\ddagger}$, between the reactants and an activated complex. Once formed, the activated complex decomposes at a universal frequency, given by the expression

$$\frac{kT}{h} \qquad (2)$$

in which k is the Boltzmann constant, T the absolute temperature, and h is Planck's constant. A transmission coefficient, κ, usually equal to unity, is introduced to take into account the probability that the formation of the activated complex will actually lead to reaction, rather than to a reconstitution of the reactants. Thus, the specific reaction rate, k', may be written:

$$k' = \kappa \frac{kT}{h} K^{\ddagger} \qquad (3)$$

The equilibrium constant of activation, $K^{\ddagger}$, for normal to activated states, has mathematically the same form as the equilibrium constant, K, for initial to final states known in thermodynamics, i.e.,

$$K^{\ddagger} = e^{-\Delta F^{\ddagger}/RT} = e^{-\Delta H^{\ddagger}/RT} e^{\Delta S^{\ddagger}/R} = e^{-\Delta E^{\ddagger}/RT} e^{-p\Delta V^{\ddagger}/RT} e^{\Delta S^{\ddagger}/R} \quad (4)$$

in which R is the gas constant, T the absolute temperature, p the pressure, and $\Delta F^{\ddagger}$, $\Delta H^{\ddagger}$, $\Delta E^{\ddagger}$, $\Delta V^{\ddagger}$, and $\Delta S^{\ddagger}$ represent the free energy, the heat, the energy, the volume change, and the entropy of activation, respectively.

From Eq. (4) it is apparent that the rate of any reaction which involves a considerable volume change will be profoundly influenced by hydrostatic pressure. Ordinarily, in simple reactions, the volume change of activation is small, of the order of not more than a few cubic centimeters per mol. With biological processes, however, the volume change may be quite large, amounting to as much as 100 cc per mol, for they depend ultimately on very large molecules, such as proteins. Thus, to understand the mechanism of rate control, it is essential to determine the influence not only of temperature but also of hydrostatic pressure. Furthermore, since the rate may be altered through equilibrium combinations between the enzyme catalysts and chemical agents which combine reversibly with them, the influence of temperature and pressure, as well as concentration of such agents, requires analyzing.

Prior to the theory of absolute reaction rates, an extensive literature, embodying a vast amount of data concerning the quantitative effects of hydrostatic pressure on biological processes had already accumulated, but no rational interpretation of the chemical mechanism had been successfully applied. Likewise, it was not known that the action of certain drugs, such as urethane or alcohol, could be profoundly modified by an increase in hydrostatic pressure of several hundred atmospheres, nor was the influence of temperature on the potency of their effects clearly interpreted in most instances, nor the significance of reversible protein denaturation in the temperature-activity curve itself.

It remained for the study of bioluminescence to furnish a key to the interrelationships of the temperature, pressure, and drug or enzyme-inhibitor parameters, and at the same time to demonstrate the unanticipated reversal, as well as retardation, of thermal denaturation of proteins. These observations have led to some new notions concerning protein reactivity and biological rate control. Although the interpretations were based at first on purely kinetic data pertaining for the most part to luminescence in living bacterial cells, they have been supported in subsequent studies by finding similar effects through biochemical and other lines of approach. The important points, in brief, are as follows.

Like other biological processes, the rate of the luminescent oxidation and hence the brightness of the light, increases as the temperature of a bacterial suspension is raised. Over a considerable range in temperatures below that of the maximum, or "optimum," the logarithm of the light intensity is a nearly linear function of the reciprocal of the absolute temperature. The slope of the line indicates a μ value of *ca.* 15,000 cal in the Arrhenius equation, within the range typical of many enzyme reactions and more complex biological phenomena. At the low temperatures, the application of hydrostatic pressure immediately and reversibly diminishes the brightness, showing that the reaction is markedly retarded under pressure and, there-

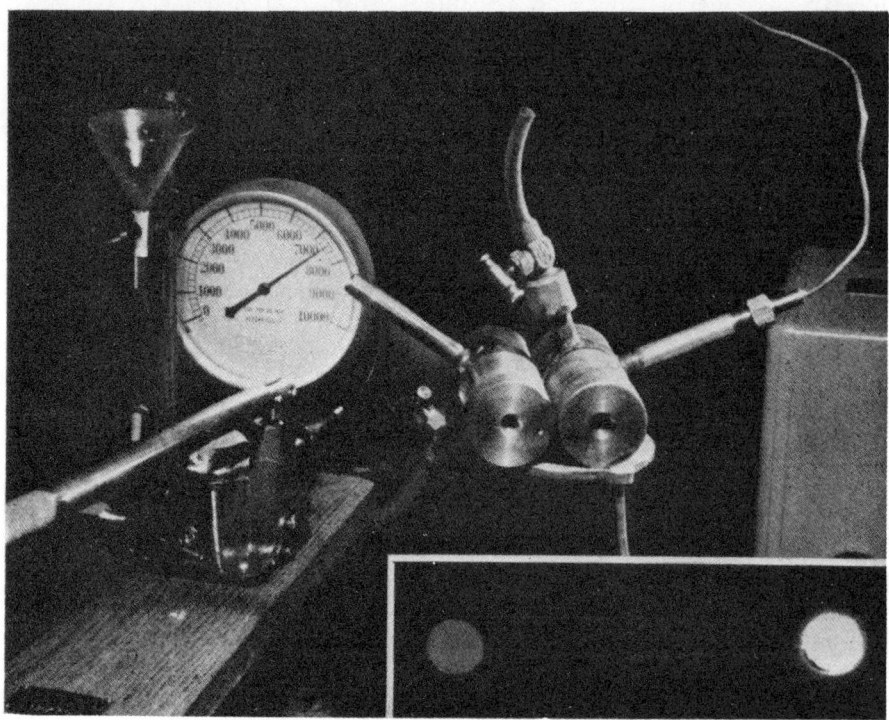

Fig. 4. Apparatus and (inset) effect of hydrostatic pressure on the brightness of bacterial luminescence at a temperature several degrees above the normal "optimum." The small pressure chamber on the left is at atmospheric pressure while that on the right is maintained at 7400 psi by the hydraulic pump. The inset shows a close up of the windows of the pressure chambers, photographed by the light of the bacterial suspensions inside. (Johnson, 1948, Fig. 7, p. 231.)

fore, that an essential reaction is characterized by a large volume increase of activation. The simplest interpretation is that this reaction is that of the enzyme itself, which limits the overall velocity of the oxidation. At temperatures higher than the optimum, the light may be reduced to only a few per cent of its maximum intensity, but if the suspension of bacteria is held only momentarily at the high temperatures and then is quickly cooled, the intensity returns at once to its former value, showing that the thermal diminution could scarcely result from an irreversible destruction of heat labile constituents. The characteristics of the decreasing intensity with rise in temperature above the optimum leave little doubt that very large molecules are concerned, and the ready reversibility of the change indicates that the reaction is a mobile equilibrium. The simplest interpretation again is that the enzyme itself is directly affected, and undergoes a reversible denaturation of its protein. Moreover, at the high temperatures, the application of pressure immediately and reversibly counteracts the diminution in enzyme activity, showing that the equilibrium change to an inactive form is accompanied by a large volume increase of reaction. A photographic illustration of the type of apparatus used in pressure experiments, together with the change in intensity of luminescence, photographed by its own light, is shown in Fig. 4.

On the basis of these observations, and the assumption of only the two limiting reactions, i.e., the oxidative enzyme reaction and a reversible protein denaturation of the catalyst, both involving large molecular volume increases, an equation for the intensity (I) of luminescence over the whole range of temperatures from well below to well above the optimum, follows directly (Johnson, Brown, and Marsland, 1942; Brown, Johnson and Marsland, 1942; Eyring and Magee, 1942; Johnson, Eyring and Williams, 1942; cf. also, Johnson, Eyring, and Polissar, 1954):

$$I = \frac{b\,k'\,(LH_2)\,(A_t)}{1 + K_1} \tag{5}$$

In Eq. (5), b represents a proportionality constant; LH_2 the substrate luciferin (in effect constant); A_t the total amount of luciferase, active plus inactive forms (also constant unless subjected to prolonged heating or excessively high temperatures); k' the specific reaction rate constant; and K_1 the equilibrium constant for the reaction of native, active to denatured, inactive forms of the enzyme. Substituting the meaning of k', from the modern rate theory, and of K_1 from thermodynamics, Eq. (5) becomes:

$$I = \kappa\,\frac{kT}{h}\,\frac{b\,(LH_2)\,(A_t)\,e^{-\Delta E^{\ddagger}/RT}\,e^{-p\Delta V^{\ddagger}/RT}\,e^{\Delta S^{\ddagger}/R}}{1 + e^{-\Delta E_1/RT}\,e^{-p\Delta V_1/RT}\,e^{\Delta S_1/R}} \tag{6}$$

This involves some unknown constants which are difficult to determine for reactions in living cells, i.e., the absolute concentration of luciferin and luciferase, and the entropy of activation, but they may be put together, along with κ, k, and h, in one unknown constant, c:

$$I = \frac{cT e^{-\Delta E^{\ddagger}/RT}\,e^{-p\Delta V^{\ddagger}/RT}}{1 + e^{-\Delta E_1/RT}\,e^{-p\Delta V_1/RT}\,e^{\Delta S_1/R}} \tag{7}$$

Numerical values of the constants for energy, volume change, etc., of activation and of reaction may be obtained from a few measurements of the variation in light intensity with temperature and pressure, and this equation is then found to account with considerable accuracy for the temperature-activity curves at both normal and increased pressures. Figure 5 shows the curves calculated by Eyring and Magee, along with the data obtained in the experiments of Brown, Johnson, and Marsland. The calculated curves for luminescence at increased pressures take into account a temperature dependence of $\Delta V^{\ddagger}$ and of ΔV_1, which results in a closer fit between the theoretical and experimental curves, but indicates that the theory is somewhat oversimplified, i.e., additional reactions are involved which are not included in Eq. (7).

The conceptual advances that have arisen from the foregoing observations consist of (1) the clear recognition and demonstration that reversible protein denaturation, at ordinary temperatures, may be a controlling factor in the physiological response to temperature variations, (2) the notion that the

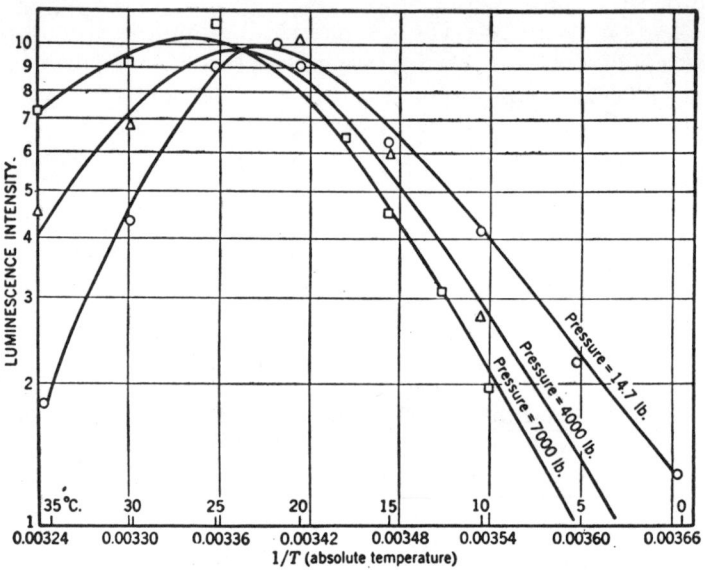

FIG. 5. The temperature-activity curves of luminescence in *Photobacterium phosphoreum* at normal and increased pressures. The lines represent mathematical curves, calculated by Eyring and Magee (1942) in accordance with Eq. (7) of the text, and the points represent data from the experiments of Brown, Johnson, and Marsland (1942). (Johnson, 1947, Fig. 7, p. 236.)

catalytic activity of enzymes may take place with a large molecular volume increase of activation, (3) that the thermal denaturation of proteins may take place with an even larger volume increase, (4) that the two reactions, involving the same component of the system, i.e., the enzyme, are sufficient to account in large part for the temperature-activity curve of the process. The reversible denaturation of single proteins and purified (crystallized) enzymes had been known for some time, from the work chiefly of Northrop and Kunitz (Northrop, 1939) and of Anson and Mirsky (1934), but its possible role in living cells had not been specifically illustrated. Subsequent studies by Chase (1946) have demonstrated a quantitatively reversible thermal inactivation of the partially purified luminescent system extracted from the marine crustacean, Cypridina, but the influence of pressure on this system *in vitro* remains to be thoroughly investigated.

From Eq. (7) it is evident that any chemical agents which combine reversibly with the enzyme in a manner that reduces its catalytic activity may at the same time influence the temperature-activity curve. The manner in which they do so will depend upon the manner in which they combine, and the particular thermodynamic constants of heat, volume change, and entropy of reaction in the equilibrium established. On a priori considera-

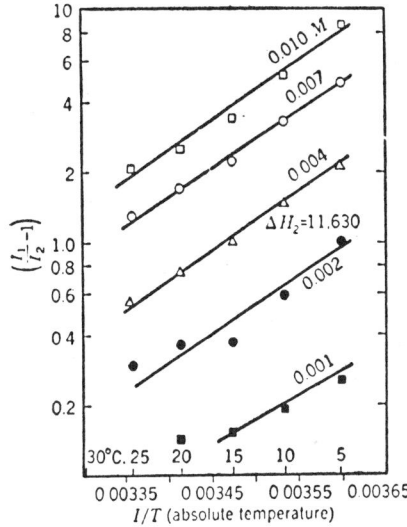

FIG. 6. The sulfanilamide inhibition of bacterial luminescence in relation to temperature, analyzed for the heat of reaction at different concentrations of sulfanilamide, from 0.001 to 0.010 M as indicated on the graph, in accordance with Eq. (8) of the text. (From Johnson et al., 1945; Johnson, 1947, Fig. 10, p. 239.)

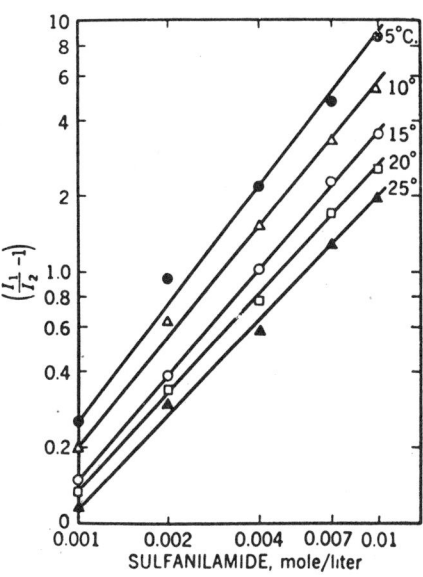

FIG. 7. Relation between concentration of sulfanilamide and inhibition of bacterial luminescence at different temperatures, plotted in accordance with Eq. (8) of the text. (From Johnson et al., 1945; Johnson, 1947, Fig. 9, p. 239.)

tions, it would seem likely that, among different drugs, some would combine in a manner independently of the ordinary temperature inactivation, as by combining with a prosthetic group, while others would act in a manner closely related to this reaction, as by catalyzing the reversible thermal inactivation. Actually, because of the enormous complexity of the highly ordered, large enzyme molecule, it is capable of entering into a great variety of reactions which affect its catalytic activity, and many relationships between temperature or pressure and the amount of inhibition caused by a given concentration of a drug are possible. In the simplest cases, however, two distinct types of action may be recognized, and formulations for their analysis follow directly from the mechanisms postulated (cf., Johnson,

Brown, and Marsland, 1942; Johnson, Eyring, and Williams, 1942; Johnson, Eyring, Steblay, Chaplin, Huber, and Gherardi, 1945; Johnson, Eyring, and Polissar, 1954). The formulations, whose derivations may be found in the original papers, are given below (Eqs. 8 and 9).

The first type of inhibition is illustrated by the action of sulfanilamide on bacterial luminescence. The effect is reversible on dilution of the medium

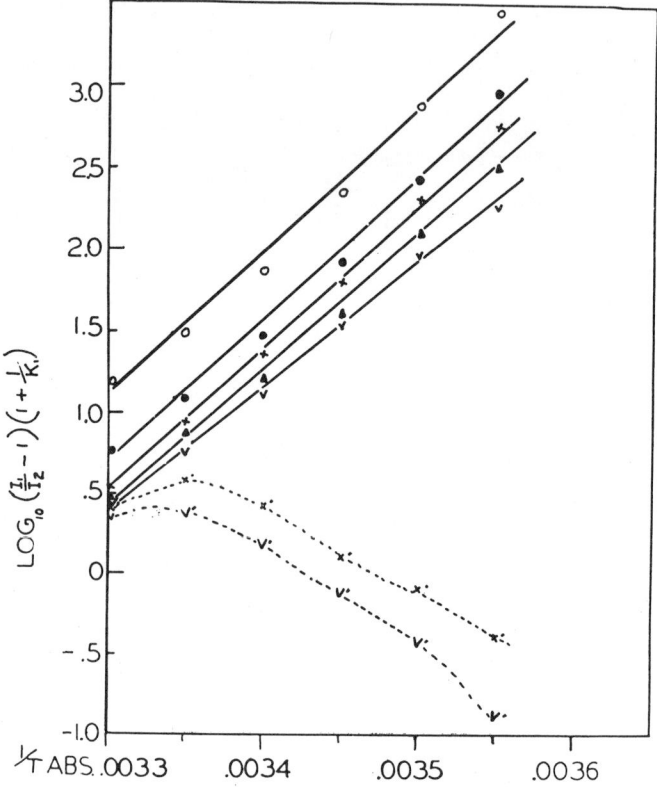

FIG. 8. The relation between temperature and the inhibition of bacterial luminescence by different concentrations of alcohol, ranging (bottom line to top line) from 0.4 to 0.8 M. The solid lines represent the data plotted in accordance with the appropriate Eq. (9) of the text; the broken lines show the results of plotting some of the same data according to Eq. (8). (Johnson et al., 1945, Fig. 36, p. 518.)

containing the bacteria or on centrifuging the cells and resuspending them in a drug-free medium (buffered salt solution). An equilibrium combination is evidently concerned, apparently with the light-emitting system directly, inasmuch as sulfanilamide also inhibits, at similar concentrations, the luminescence of the extracted and partially purified system of *Cypridina* (Johnson and Chase, 1942). The amount of the inhibition under given conditions depends upon the temperature as well as the concentration of the drug added. With a given concentration in the medium, the amount of the inhibition decreases as the temperature is raised toward that of the optimum. Pressure, on the other hand, has little effect. Assuming that in this case the drug combines according to equilibrium constant K_2 with the enzyme in a manner that stops its catalytic activity, combining perhaps with a prosthetic group not involved in the thermal denaturation equilibrium, the following formulation applies:

$$\left(\frac{I_1}{I_2} - 1\right) = K_2 (X)^r = (X)^r e^{-\Delta H_2/RT} e^{\Delta S_2/R} \tag{8}$$

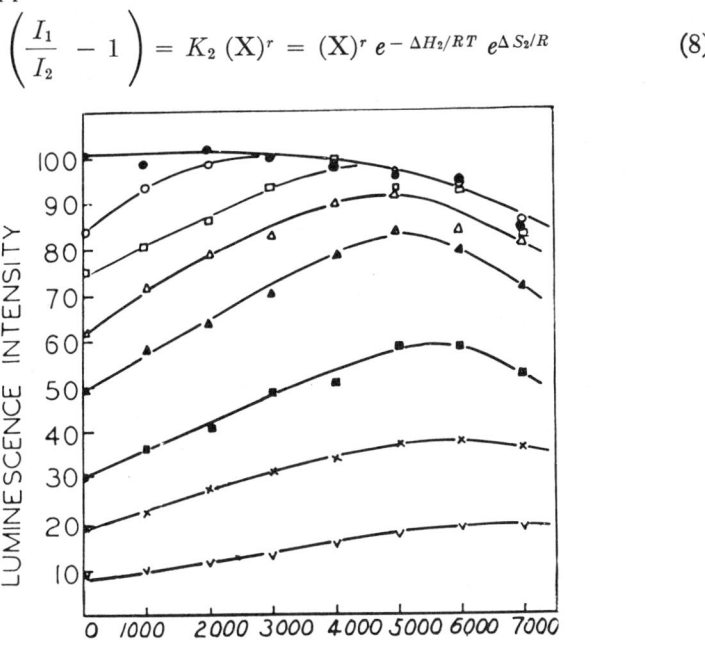

FIG. 9. Influence of pressure on the brightness of bacterial luminescence in suspensions of cells to which alcohol has been added in concentrations (top line to bottom line) of 0.0, 0.2, 0.4, 0.5, 0.6, 0.8, 1.0, and 1.5 M, respectively, at near the normal optimum temperature. (Johnson et al., 1945, Fig. 37, p. 519.)

in which I_1 refers to the luminescent intensity of a drug-free control, I_2 to that of a corresponding portion of the cell suspension containing a given molar concentration of sulfanilamide (X), and r represents the ratio of the number of molecules of sulfanilamide combining per enzyme molecule. Thus, by plotting the logarithm of the expression on the left successively against the reciprocal of the absolute temperature and the log molar concentration of drug, the constants of the equilibrium may be obtained (Figs. 6 and 7). The straightness of the lines attest the essential correctness of the mechanism postulated. The heat of reaction, according to the slopes of the lines in Fig. 6 is about 11,000 cal, and the ratio of molecules in the combination, according to the data of Fig. 7, is close to 1.

The second type of inhibition is illustrated by urethane or ethyl alcohol. In both cases, under given conditions, pressure causes a decrease in the amount of inhibition, while a rise in temperature tends to increase the inhibition. Here, the inhibitor combines with equilibrium constant K_3, in a manner that promotes the thermal denaturation which ordinarily proceeds with a large volume increase. The change is susceptible of reversal by pressure. Moreover, although a rise in temperature again dissociates the enzyme-inhibitor complex, it is more effective in making available the bonds at which the inhibitor combine, the net result depending upon the thermodynamic constants of the two equilibria. The following formulation applies:

$$\left[\left(\frac{I_1}{I_2} - 1\right)\left(1 + \frac{1}{K_1}\right)\right] \begin{array}{l} = K_3 \, (U)^s \\ = (U)^s \, e^{-\Delta H_3/RT} \, e^{\Delta S_3/R} \\ = (U)^s \, e^{-\Delta E_3/RT} \, e^{-p\Delta V_3/RT} \, e^{\Delta S_3 R} \end{array} \quad (9)$$

in which the expression $[(I_1/I_2) - 1]$ has the same meaning as previously (Eq. 8) with reference to the effect of alcohol or of urethane in molar concentration (U), while s represents the ratio of inhibitor to enzyme molecules in the combination formed. Numerical values for the constants in Eq. (9) may be obtained by plotting the logarithm of the expression on the left successively against the reciprocal of the absolute temperature, hydrostatic pressure, and log molar concentration of the drug. Examples of the data and analytical treatment, for alcohol, are illustrated in Figs. 8, 9, and 10. With alcohol the heat of reaction is much higher than with sulfanilamide amounting to $-37,000$ cal, and several alcohol molecules evidently combine with one enzyme molecule. The average ratio of combining molecules varies according to the hydrostatic pressure (Fig. 10), ranging from 2.8 at atmospheric pressure to about 4.0 under 7000 psi. Other data indicate that the value of ΔV_3, computed from the quantitative effects of high concentrations of alcohol that cause great inhibitions, also varies with concentration of the drug, other things remaining constant. Such variations indicate that these inhibitions involve more than one equilibrium reaction. Evidence for a similar phenomenon has been found in connection with the action of urethane in catalyzing the thermal denaturation of tobacco mosaic virus, to be discussed presently.

Equation (9) provides a precise interpretation of the action of inhibitors which act by the assumed mechanism and it is of particular interest that it includes the equilibrium constant, K_1, for the reversible thermal denaturation itself. No complete understanding of this type of action would be possible without taking into account the significance of this equilibrium. Yet the equation is oversimplified in including only the one equilibrium combination between drug and enzyme. The additional reactions, however, are not prominent except at relatively high drug concentrations, high temperatures, or to some extent at different pressures. At high concentrations or temperatures, another complicating factor becomes evident, in that the same agents which promote a reversible denaturation of the protein enzyme may also catalyze an irreversible denaturation, with the result that a progressive destruction of the system takes place.

The importance of temperature and pressure data in distinguishing, as well as understanding, the two mechanisms of inhibition is apparent from

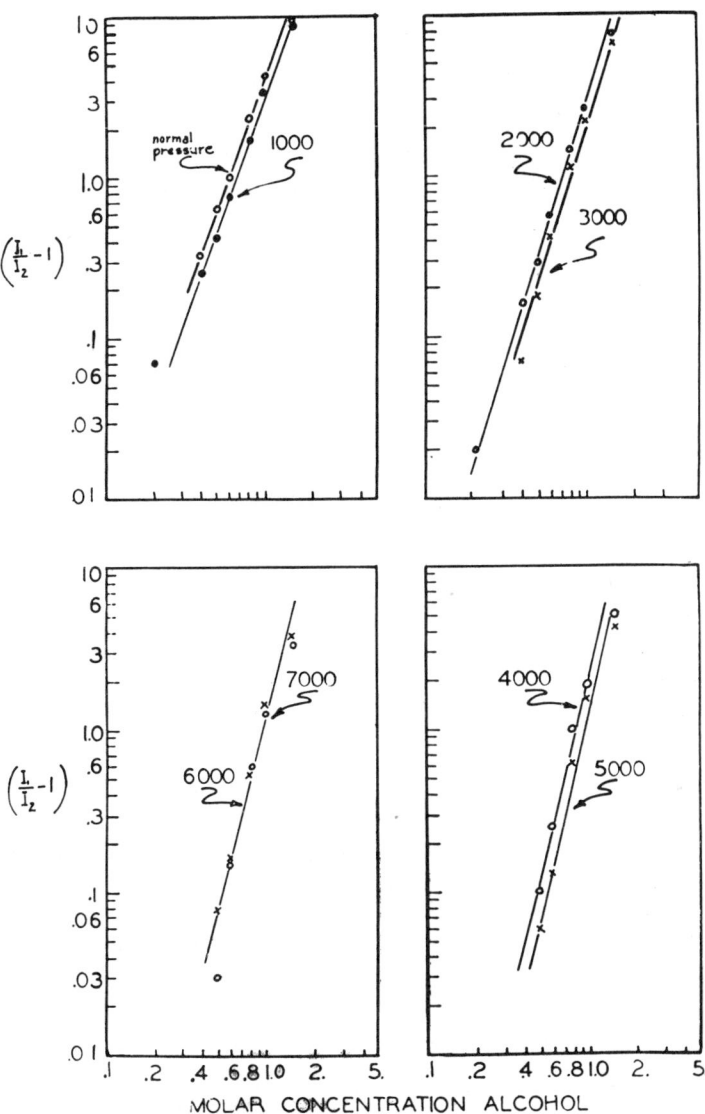

Fig. 10. Analysis of the data shown in Fig. 9, in accordance with Eq. (9) of the text, for the relation between concentration of alcohol and inhibition of luminescence at different hydrostatic pressures. (Johnson et al., 1945, Fig. 38, p. 521.)

Eqs. (8) and (9). In either case, a straight line relationship results when the log of the amount of inhibition is plotted against the log molar concentration of inhibitor at a given temperature and pressure, the slope of the line indicating the average value of the exponent r or s, respectively, i.e., the ratio of drug-enzyme molecules in the combination formed. The expression U/I, or the "per cent uninhibited" divided by the "per cent inhibited," derived by Fisher and Öhnell (1940) leads to the same information (when plotted as the logarithm against log molar concentration of inhibitor), since their expression is the reciprocal of the expression $[(I_1/I_2) - 1]$.

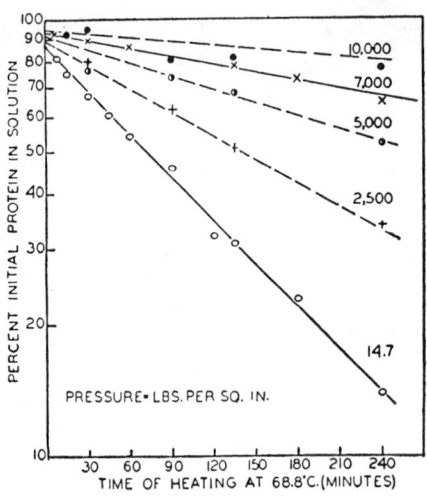

FIG. 11. The rate of precipitation of tobacco mosaic virus at 68.8° C, at normal and increased pressures. All points represent data from experiments. The solid lines were determined experimentally, and the broken lines were calculated on the basis of these data for pressures of 14.7 and 7000 psi, through the theory of absolute reaction rates. (Johnson, Baylor, and Fraser, 1948, Fig. 2, p. 241.)

The foregoing theory and conclusions based on kinetic evidence obtained largely with respect to bioluminescence, have now been applied to other processes, both simple and complex. The following representative examples illustrate some conclusions of general interest, viz., that (a) hydrostatic pressure may reverse or prevent the thermal denaturation of proteins and enzymes, (b) reversible, as well as irreversible protein denaturation may be a significant factor controlling biological reactions under physiological conditions, and (c) certain drugs, such as alcohol or urethane act by the mechanism of catalyzing the denaturation through one or more reactions with the same component of the system affected.

With regard to (a), an interesting example is that of tobacco mosaic virus denaturation (Johnson, Baylor, and Fraser, 1948). In a solution of the pure virus protein, buffered at neutrality, precipitation occurs at high temperatures as a first order reaction (Lauffer and Price, 1940). The rate of precipitation, determined by micro-Kjeldahl analyses of the amount of protein remaining in solution after successive intervals of time at 68.8° C, is much slower at 7000 psi (Fig. 11). From the difference in rates at normal and 7000 psi pressure, the volume change of activation in the limiting reaction can be calculated, by the theory of absolute reaction rates, and turns out to be very close to +100 cc per mol. Using this value, the rates at different pressures at the same temperature were then predicted with considerable

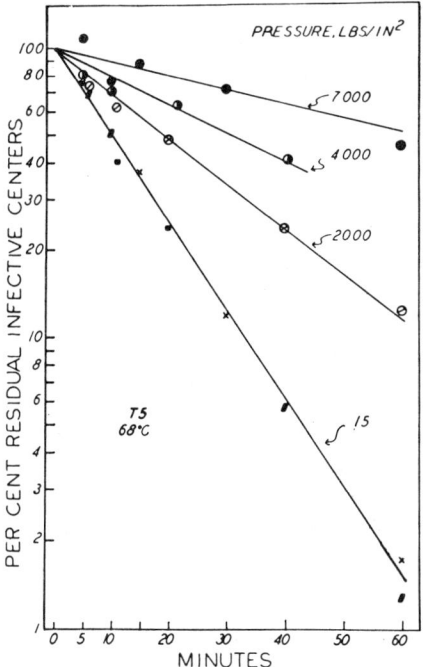

FIG. 12. Influence of hydrostatic pressure on the rate of thermal inactivation of colibacteriophage T5. (Foster, Johnson, and Miller, 1949, Fig. 3, p. 6.)

accuracy (broken lines in Fig. 11). Furthermore, from a knowledge of the activation energy (Lauffer and Price, 1940), it was possible to calculate with equally satisfactory accuracy the rates at a higher temperature and still other pressures. With such a large molecule as the tobacco mosaic virus it is not surprising to find a large volume change of activation in the denaturation reaction, although in terms of the per cent change in size of an individual molecule it is very small indeed, amounting to not more than a few hundredths of a per cent. The fact that the volume change is not even larger may be interpreted to mean that relatively few of many possible bonds are broken in the pressure-sensitive reaction that limits the overall rate of denaturation. Other factors possibly also are involved (cf. Johnson, Eyring, and Polissar, 1954).

A slightly larger volume increase of activation occurs in the thermal destruction of coli-bacteriophage T5 at 68° C (Fig. 12, Foster, Johnson, and Miller, 1949). The value of $\Delta V^{\ddagger}$ in the limiting reaction, calculated from data illustrated in Fig. 12, amounts to $+113$ cc per mol, and is the same at different pressures up to 7000 psi. A different phage, T7, for the same bacterial organism, does not show the same phenomenon. Under similar conditions, T7 is more sensitive to destruction by heat, and pressure slightly accelerates, rather than retards the rate. In both cases, however, small concentrations of urethane accelerate the rate of destruction.

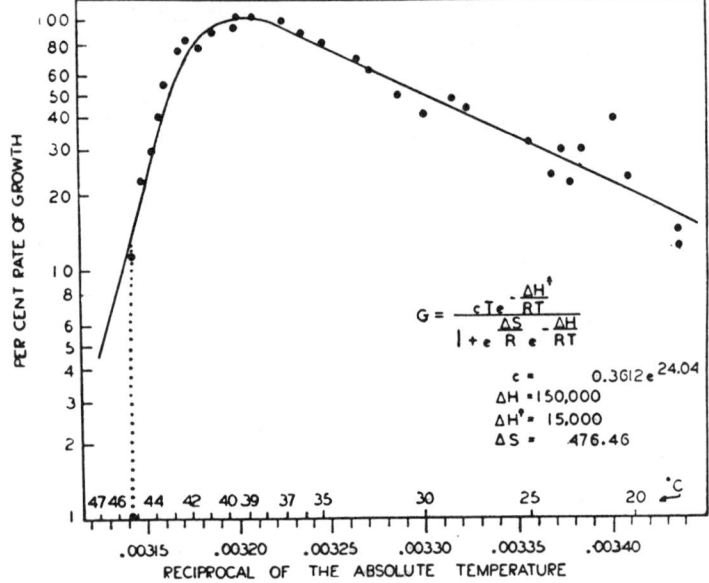

FIG. 13. Relation between temperature and rate of growth of *Escherichia coli* in a "synthetic" medium. The smooth curve was calculated according to the equation and constants shown in the figure (Eq. (7) of the text), while the points represent the experimentally observed data. (Johnson and Lewin, 1946, Fig. 11, p. 61.)

With regard to the second generality, that the reversible protein denaturation of an essential system may become a predominantly limiting reaction controlling the total rate of a complex process under physiological conditions, quantitative studies of bacterial growth rates support the conclusions reached in the studies of luminescence. Thus, during the early "logarithmic phase" of growth, the rate of cell division of *Escherichia coli* is constant under given conditions and increases with rise in temperature in the manner of a simple chemical reaction, as if limited by the activity of a single system (Fig. 13). As the temperature is raised, the rate goes through a maximum at 37 to 39° C, and then rapidly decreases with further rise in temperature, virtually stopping at about 45° C. This cessation in reproduction at 45° C is immediately reversible, however, if the culture is maintained at the bacteriostatic temperature for not more than a few minutes; reproduction is resumed at once when it is transferred back to 37° C. The same Eq. (7) used for the brightness of luminescence in relation to temperature is adequate to account for the change in rate of growth as a function of temperature throughout most of the range from low temperatures to those high enough to stop growth. There are some similar effects of pressure, also, but the phenomenon of growth is more complicated than luminescence, and the influence of pressure is not so readily accounted for by the simple

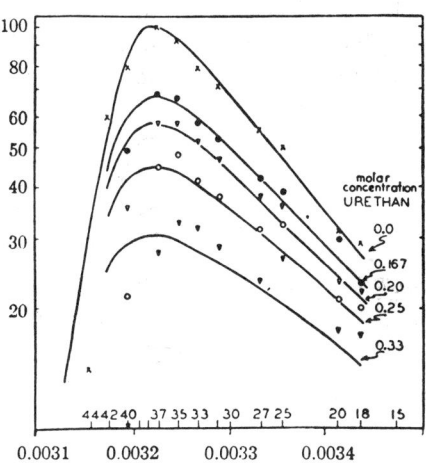

Fig. 14. Relation between temperature, urethane concentration, and rate of oxygen consumption by *Rhizobium*. The smooth curves were calculated in accordance with Eq. (9) of the text; the points represent data from the experiments. (Koffler, Johnson, and Wilson, 1947, Fig. 1, p. 1115.)

equation. Additional reactions, not included in the equation, evidently become limiting at increased pressures and the analysis is correspondingly more complicated. While the nature of these reactions remains to be determined, it is a fair guess that they include processes of synthesis and also sol-gel changes such as are known to occur in cell division of sea urchin's eggs (Marsland, 1948).

Finally, with reference to the fundamentally protein denaturing mechanism of action of certain drugs and the fact that multiple reactions with a given component of the system may be involved, both in living cells and extracted proteins, the following examples concerning the action of urethane provide pertinent illustrations.

The influence of urethane on either the aerobic or anaerobic respiration of intact cells of *Rhizobium* depends not only on the concentration of the drug added but also on the temperature. The rate of aerobic glucose oxidation, at different temperatures and concentrations of urethane, is illustrated by the data of Koffler, Johnson, and Wilson (1947) in Fig. 14. The points are the experimentally observed data. The smooth curves were calculated, using Eq. (9) that was derived for luminescence. The agreement between observed and calculated curves is for the most part satisfactory, but the theoretical curves predict rates that are too high in the regions of high temperatures, or high concentrations of urethane. The deviations are in the direction that would be expected if, in addition to promoting a

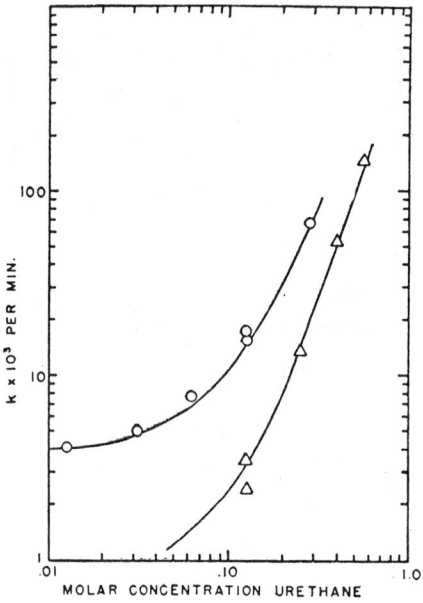

Fig. 15. The acceleration of tobacco mosaic virus denaturation at 68.8° C by urethane at normal pressure (circles) and at 7000 psi (triangles). The lines were calculated, while the points represent data from the experiments. (Fraser, Johnson, and Baker, 1949, Fig. 2, p. 317.)

reversible denaturation of the limiting enzyme, urethane also catalyzes an irreversible destruction of the system, which becomes more prominent at the higher temperatures or higher concentrations of the drug. Likewise, more than one reaction between the drug and the enzyme would be expected to contribute to such deviations. The fact that the ratio of combining molecules, according to the analysis of the data, appears to be 2.25, rather than an integer, is evidence of more than one reaction.

Evidence for multiple reactions between urethane and a given protein has been more clearly analyzed with reference to the catalysis of tobacco mosaic virus denaturation by urethane at increased temperatures (Fraser, Johnson, and Baker, 1949). The rate of precipitation of TMV protein at

68.8° C is accelerated by urethane and retarded by hydrostatic pressure, but the quantitative influence of pressure varies with the concentration of urethane, other factors remaining the same. The data are illustrated in Fig. 15, where the points represent the experimentally observed values, and the smooth curves were calculated on the assumption that three distinct reactions take place between urethane and the protein, the total rate of

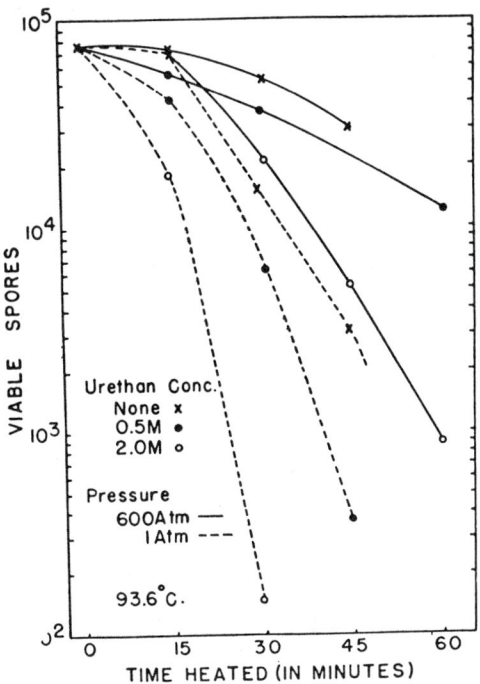

FIG. 16. The influence of urethane and hydrostatic pressure on the rate of disinfection of bacterial spores (*Bacillus subtilis*) at 93.6° C. (Johnson and ZoBell, 1949, Fig. 1, p. 361.)

precipitation representing the sum of the thermal denaturation rate itself plus the reactions catalyzed by urethane. The latter involve 1, 2, and 3, respectively, more molecules of urethane combined with the activated than the normal state of the molecule whose destruction results in precipitation, and volume increases of 100, 100 and 50 cc per mol, respectively. An approximate, less precise fit between the observed and calculated data may be obtained on the assumption of only two urethane catalyzed reactions, but in either case it is clear that multiple reactions are concerned.

A somewhat more complicated situation, in which the results are in general consistent with those just discussed, has been found in the disinfection of bacterial spores (Johnson and ZoBell, 1949). Although the kinetics of the process are not readily interpreted, it is evident (Fig. 16) that the rate of killing of the individuals in a population of spores at 93.6° C is accelerated by small concentrations of urethane and is retarded by hydrostatic pressure. At ordinary temperatures, even very high concentrations of urethane have no effect on the number of viable spores. The relations of temperature, pressure, and urethane suggest that protein denaturation is an important mechanism in the loss of viability.

The importance of temperature, as well as the concentration of chemical agents, in controlling biological reactions has long been recognized, and the foregoing discussions indicate that substantial progress is being made toward the rational analysis and understanding of the mechanisms concerned. The theoretical significance of hydrostatic pressure as another variable is apparent from the absolute reaction rate theory, but otherwise is not so obvious. The natural variations in hydrostatic pressure at the earth's surface are usually much too small to affect directly any ordinary equilibrium or reaction rate, although it is possible that they may be concerned in certain special cases, such as stimulation of nerves via tactile corpuscles. In the sea, the importance of pressure is more apparent, where it increases with depth by about a ton a mile. The activities of organisms living in the depths of the oceans, as well as the rates of chemical reactions taking place in solution there, are undoubtedly modified by the enormous pressures; likewise the growth and metabolism of bacteria which have been found to occur in deep oil well brines. The real significance of pressure in these instances remains largely to be worked out. What then is the usefulness of investigating the influence of pressure on biological reactions in general, beyond that of making the physical-chemical analysis more nearly complete?

A fully definitive answer to this question can scarcely be given at the present time. It depends on the precise interpretation of the effects that have been observed and of future studies that need to be carried out. It seems evident, however, that, in the first place, large molecular volume changes accompanying biological equilibria and rate processes are very widespread, and in the second place, the study of pressure effects is a useful, in some cases unique, tool for learning their significance.

Perhaps some further conclusions in regard to the meaning of pressure effects may be drawn now. The magnitude of the volume changes accompanying the activity of certain enzymes, e.g., in bacterial luminescence, and in reversible protein denaturation, is such as to indicate that drastic changes in configuration of the protein molecule take place, probably by unfolding from a globular to a more fibrous form. When this occurs in the formation of the activated complex in enzyme activity, it would appear that combination of the substrate takes place with groups that are normally folded up within the protein. With some enzymes, such as invertase (Eyring, Johnson, and Gensler, 1946), whose activity is insensitive to, or only very slightly affected by, pressure under ordinary conditions, the combining sites for the substrate are presumably at the surface of the protein, making any considerable unfolding of the protein unnecessary for the process of catalysis. With both types, however, the thermal denaturation, reversible or irreversible, is opposed by hydrostatic pressure, and it is reasonable to conclude that unfolding of the molecule is involved, accompanied by the exposure of hydrophobic groups from within. The exposure of the hydrophobic groups results in a decrease in water solubility and an effectively larger volume of the molecule to which water molecules are less strongly attracted. It is presumably by combination with these hydrophobic groups that the denaturation process is catalyzed by lipid soluble chemical agents such as alcohol, acetone, chloroform, urethane, and others.

In still another connection the reversible unfolding of large molecules, which might in general be expected to be accompanied by a considerable volume change and, therefore, to be sensitive to variations in hydrostatic pressure, would seem significant. For, there seems to be no other mechanism than that of a templet for the asymmetric synthesis of the optically active, biologically specific molecules of proteins, genes, viruses, and others. This mechanism requires that one molecule act as a templet, or pattern, for the construction of a second one. In order to act as a templet, it would be necessary for any three-dimensional, globular molecule to unfold to a one- or two-dimensional form, probably accompanied by a volume increase. Thus, it would be anticipated that the reproduction of genes and viruses, as well as whole cells, would be readily retarded by increased pressures. With cells in general this has been found to be the case, and recent work (Foster and Johnson, 1951), indicates that the multiplication of bacteriophage in susceptible host cells is much reduced at moderately increased pressures. In these phenomena other mechanisms than through influencing synthetic reactions involving the templet mechanism are also no doubt concerned in the observed effects of pressure. Moreover, the nature of the adaptation by which deep sea bacteria are enabled to grow at cold temperatures and under high pressures remains to be elucidated, but it seems a safe guess that in all cases the net result depends upon a complex interplay of many reactions which enter into the control of the overall rate and are subject to the action of hydrostatic pressure as well as temperature.

Three centuries ago, the study of bioluminescence began out of theoretical, almost purely philosophical interest. The problem is not yet fully solved, and its interest is still largely theoretical, but it has proved to be a useful tool. To an unusual extent among processes that take place in living cells, its kinetics can be analyzed in the manner of a simple chemical reaction, and the results of these analyses have anticipated phenomena with other, apparently unrelated systems of greater as well as less complexity. The ultimate solution of bioluminescence may be expected to contribute to other fundamental problems concerning the mechanisms of energy exchange in metabolic reactions.

The most important branches of physics today are nuclear physics and biophysics. The former, in recent years, has made staggering progress. The latter is still in its infancy and only one of the many aspects of its currently developing potentialities could be enlarged upon in this discussion. It is growing rapidly, and one may look forward with confidence that biophysics will bring results equally spectacular and of no less consequence to mankind.

REFERENCES

1. Anson, M. L., and Mirsky, A. E., 1934. The Equilibrium between Active Native Trypsin and Inactive Denatured Trypsin. *J. Gen. Physiol.* **17,** 393.
2. Anson, M. L., and Mirsky, A. E., 1934. The Equilibria between Native and Denatured Hemoglobin in Salicylate Solutions and the Theoretical Consequences of the Equilibrium between Native and Denatured Protein. *J. Gen. Physiol.* **17,** 399.
3. Boyle, Robert, 1672. Some Observations about Shining Flesh, Both of Veal and of Pullet, and That without Any Sensible Putrefaction in Those Bodies. *Phil. Trans. Roy. Soc. (London)* No. 89, 5108. Dec. 16, 1672.
4. Brown, D. E., Johnson, F. H., and Marsland, D. A., 1942. The Pressure-Temperature Relations of Bacterial Luminescence. *J. Cellular Comp. Physiol.* **20,** 151.
5. Chance, B., Harvey, E. N., Johnson, F. H., and Millikan, G., 1940. The Kinetics of Bioluminescent Flashes. A Study in Consecutive Reactions. *J. Cellular Comp. Physiol.* **15,** 195.
6. Chase, A. M., 1946. Reversible Heat Inactivation of Cypridina Luciferase. *J. Cellular Comp. Physiol.* **27,** 121.
7. Eyring, H., 1935. The Activated Complex in Chemical Reactions. *J. Chem. Phys.* **3,** 107.
8. Eyring, H., and Magee, J., 1942. Application of the Theory of Absolute Reaction Rates to Bacterial Luminescence. *J. Cellular Comp. Physiol.* **20,** 169.
9. Eyring, H., Johnson, F. H., and Gensler, R. L., 1946. Pressure and Reactivity of Proteins, with Special Reference to Invertase. *J. Phys. Chem.* **50,** 453.
10. Fisher, K. C., and Öhnell, R., 1940. The Steady State Frequency of the Embryonic Fish Heart at Different Concentrations of Cyanide. *J. Cellular Comp. Physiol.* **16,** 1.
11. Foster, R. A. C., and Johnson, F. H., 1951. Influence of Urethane and of Hydrostatic Pressure on the Growth of Bacteriophages T2, T5, T6, and T7. *J. Gen Physiol.* **34,** 529.
12. Foster, R. A. C., Johnson, F. H., and Miller, V. K., 1949. The Influence of Hydrostatic Pressure and Urethane on the Thermal Inactivation of Bacteriophage. *J. Gen. Physiol.* **33,** 1.
13. Fraser, D., Johnson, F. H., and Baker, R. S., 1949. The Acceleration of the Thermal Denaturation of Tobacco Mosaic Virus at Normal and Increased Pressure. *Arch. Biochem.* **24,** 314.
14. Glasstone, S., Laidler, K. J., and Eyring, H., 1941. *The Theory of Rate Processes.* McGraw-Hill Book Co., New York.
15. Harvey, E. N., 1940. *Living Light.* The Princeton University Press, Princeton, N. J.
16. Harvey, E. N., 1952. *Bioluminescence.* Academic Press, New York.
17. Hoagland, H., 1936. Some pacemaker aspects of rhythmic activity in the nervous system. *Cold Spring Harbor Symposia Quant. Biol.* **4,** 267.
18. Johnson, F. H., 1947. Bacterial Luminescence. *Advances in Enzymol.* **7,** 215.
19. Johnson, F. H., 1948. Bioluminescence: a Reaction Rate Tool. *Sci. Monthly* **67,** 225.
20. Johnson, F. H., Baylor, M. B., and Fraser, D., 1948. The Thermal Denaturation of Tobacco Mosaic Virus in Relation to Hydrostatic Pressure. *Arch. Biochem.* **19,** 237.

21. Johnson, F. H., Brown, D. E., and Marsland, D. A., 1942. A Basic Mechanism in the Biological Effects of Temperature, Pressure and Narcotics. *Science* **95**, 200.
22. Johnson, F. H., Brown, D. E., and Marsland, D. A., 1942. Pressure Reversal of the Action of Certain Narcotics. *J. Cellular Comp. Physiol.* **20**, 269.
23. Johnson, F. H., and Chase, A. M., 1942. The Sulfonamide and Urethane Inhibition of Cypridina Luminescence *in vitro*. *J. Cellular Comp. Physiol.* **19**, 151.
24. Johnson, F. H., and Eyring, H., 1948. The Fundamental Action of Pressure, Temperature, and Drugs on Enzymes, as Revealed by Bacterial Luminescence. *Ann. N. Y. Acad. Sci.* **49**, 376.
25. Johnson, F. H., Eyring, H., and Polissar, M. P., 1954. *The Kinetic Basis of Molecular Biology*. John Wiley and Sons, New York.
26. Johnson, F. H., Eyring, H., Steblay, R., Chaplin, H., Huber, C., and Gherardi, G., 1945. The Nature and Control of Reactions in Bioluminescence. With Special Reference to the Mechanism of Reversible and Irreversible Inhibitions by Hydrogen and Hydroxyl Ions, Temperature, Pressure, Alcohol, Urethane and Sulfanilamide in Bacteria. *J. Gen. Physiol.* **28**, 462.
27. Johnson, F. H., Eyring, H., and Williams, R. W., 1942. The Nature of Enzyme Inhibitions in Bacterial Luminescence. *J. Cellular Comp. Physiol.* **20**, 247.
28. Johnson, F. H., and Lewin, I. 1946. The Growth Rate of *E. coli* in Relation to Temperature, Quinine and Coenzyme. *J. Cellular Comp. Physiol.* **28**, 47.
29. Johnson, F. H., and ZoBell, C. E., 1949. The acceleration of spore disinfection by urethane, and its retardation by hydrostatic pressure. *J. Bact.* **57**, 359.
30. Koffler, H., Johnson, F. H., and Wilson, P. W., 1947. The Combined Influence of Temperature and Urethane on the Respiration of Rhizobium. *J. Am. Chem. Soc.* **69**, 1113.
31. Lauffer, M. A., and Price, W. C., 1940. Thermal Denaturation of Tobacco Mosaic Virus. *J. Biol. Chem.* **133**, 1.
32. Marsland, D. A., 1948. Protoplasmic Contractility. Pressure Experiments on the Motility of Living Eggs. *Sci. Monthly* **67**, 193.
33. Needham, J., 1931. *Chemical Embryology*. Vol. 1, p. 520. Cambridge University Press, Cambridge, England.
34. Northrop, J. H., 1939. *Crystalline Enzymes*. Columbia University Press, New York.

INDEX

Absorption at long wavelengths, 73, 113, 115, 117
Absorptivity, a function of wavelength, 73, 115, 118
Acceptors, 78, 104
Activation energy, 61
  in insulators and semiconductors, 76
Alkali halides, 71
Amino acids, 192
Axons, 213
  metastable states, 227
  physiological function, 214
  signalling, 217

Balloon altitudes in cosmic ray experiments, 11
Barium titanate, ceramics, 153
  single crystals, 157
Barriers, 59, 66, 72, 110
Bioluminescence, 238, 241, 247; see also *Luminescence*
Bohr magneton, 3
Bombardment of semiconductors, 106, 109

Carriers, density, 80, 86, 92, 96, 108
  lifetime, 71
Cloud chamber experiments, 11, 32
  counter-controlled, 13
Compounds, semiconducting, 64, 65, 81
Conduction in solids, 129
Conductivity, 57, 61, 66, 71, 75, 78, 86, 97, 105, 108
  intrinsic, 129
Conductivity and Hall effect, 86

Conductors, variable, 57
Contact rectification, 66
Crystal counters, 111
Curie-Weiss law, 151, 154
Cytochromes, 199

Detectors, 66, 67
Deuteron irradiation, 107
Dielectric constant, barium titanate, 153
  proteins, 196
  semiconductors, 73, 88, 112, 118, 196
Diffusion, 58, 64
  coefficients, 133
Dissociation energy, 61
Dissociation equilibrium, 78
Dodecylamine hydrochloride, turbidity, 176

Electron bombardment of semiconductors, 79, 110
Electron current, "multiplication," 72
Electron showers, 14, 29, 30
Energy band, 76, 109
Energy gap, 78, 92, 99, 113, 118
Enzymes, 196
Exciton theory, 207

F centers, 71
Fermi level, 79, 108
Ferroelectrics, 150
Films, germanium, 104

Index

g value, 3
Gas discharges, 66
Germanium, 76, 81, 86, 128
  films, 104
  lens, 118
  optical properties, 112
  rectifier, 68
  thermoelectric power, 97
Grain boundaries, 88, 102

Hall coefficient, 91, 95
Hall effect, 58, 62, 63, 86, 95, 96
Hall reaction, 204
Heat treatment, 102
High-frequency measurements, 75
Holes, in semiconductors, 133, 136

Insulators, 76
Interference, 180
Interstitial atom, 74, 81, 106
Irradiation curve, 108

Kinetics, chemical of biological systems, 189

Lattice defects, 71, 76, 78, 81, 102, 106
Lattice vacancy, 74, 81, 106
Lead compounds, conductivity, 69
Light scattering, 169
Lorentz force, 62
Luciferin, 195, 245
Luminescence, 195, 238

Magnetic moment, 3
Mass, effective, 59, 115
Mean free path, 59, 61, 88

Melting point, conductivity at, 59
μ-Meson, 36
π-Meson, 41
Metabolism, 219
Metabolites, 203
Mobility, 58, 59, 63, 70, 87, 91, 94, 115

N type conduction, 62, 65, 75, 107-109, 110, 118
Nerve impulses, 213
Neural phenomena, 199
Neutron irradiation, 108
Nuclear reactions, 46
Nuclei, heavy, 23

Optical absorption and free carriers, 73, 103, 112, 115
Optical properties, 73

P type conduction, 62, 75
Peptides in chemical kinetics, 192
Perovskite, 152
Photoconductivity, 68, 110
Photoelectrons, 69
Photographic emulsions, 11, 33
Photosynthesis, 204
Photovoltaic cells, 72, 110
Planck's constant, 3, 70
Polarizability, 170
Polymers, 169
Polystyrenes, molecular weight, 182
  turbidity, 172
Pressure influence on bioluminescence, 244, 249
Protein structure, 191

Radiation, cosmic, 28, 52
  effects on solids, 106
Radioactive impurities, 81, 86

Rectification, contact, 66
Rectifier, copper-cuprous oxide, 67
　dry disk, 72
　germanium, 68
　P-N junction type, 110
Refractive indices, semiconductors, 73, 118
Resistivity, semiconductors, 61, 86, 88, 98
Resonance-fluorescence theory, 207
Respirometer, 219
Rutherford scattering, 88

Scattering, 87, 88, 93, 95, 102
"Schubweg," 71
Selenium, 72; see also *Semiconductors*
Semiconductors, amphoteric, 75
　defects, 67, 71, 102, 105
　degeneracy, 94
　elementary, 65, 76, 81
　general survey, 74
　germanium, 76
　ideal, 81
　impurity, 78, 97
　interaction of imperfections, 61, 119
　materials for, 74-76
　reflectivity, 73, 112
　refractive indices, 73, 118
　single crystal, 81, 86
Silicon, 76, 82, 129
　dissociation energy, 61
　resistivity as function of temperature, 59
　rectifier, 68

Silver sulfide, conductivity, 57-58
Spectrum of atom, 4
Spin, 3, 6
Sulfanilamide and bioluminescence, 247
Sulfides, conductivity, 57-58
　photoconductivity, 69
　thermoelectric properties, 65, 66

Tellurium, 80, 81, 96, 105
Temperature-activity curves in bioluminescence, 241, 247
Thermoelectric effects in semiconductors, 65
Thermoelectric power, 64, 65, 97, 102
Transmissivity of silicon and germanium in the infrared, 112-113
Transistors, 128, 143
Trapping of carriers, 63, 71
Turbidity, 172, 176

Wavelength and particle size, 180
Wigner effect, 106
Work function in contact rectification, 68

Zeeman effect, 3

QC1 .A48 1970
American Ass / The present state of physics; a sym